AQA Science

Exclusively endorsed and approved by AQA

Jim Breithaupt • Ann Fullick • Patrick Fullick

Series Editor: Lawrie Ryan

GCSE Science

Nelson Thornes

GCSE Science — Contents

GCSE Science

Welcome to AQA Science!

How to use this book

This textbook will help you throughout your GCSE course and to prepare for AQA's exams. It is packed full of features to help you to achieve the best result you can. Each feature is designed to help you remember and apply your knowledge in a different way. All features are worth a look, but some might particularly suit the way you learn. As they are consistently colour coded throughout the book, it will be easy to find these on each page. This will help you to use this book in the best way for you.

a) What are the yellow boxes?

To check you understand the science you are learning, questions are integrated into the main text. The information to answer these is on the same page, so you don't waste your time flicking through the entire book.

LEARNING OBJECTIVES

By the end of the lesson you should be able to answer the questions posed in the learning objectives; if you can't, review the content until it's clear.

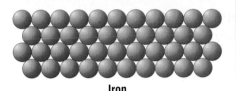

Iron

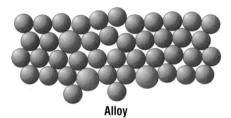

Alloy

Figure 1 Key diagrams are as important as the text. Make sure you use them in your learning and revision.

Key words

Important scientific terms are shown like this:

observation or **anomalous**

You can look up the words shown like this – **bias** – in the glossary.

Avoid common mistakes and gain marks by sticking to this advice.

KEY POINTS

If you remember nothing else, remember these! Learning the key points in each lesson is a good start. They can be used in your revision and help you summarise your knowledge.

PRACTICAL

Become familiar with key practicals. A simple diagram or photo and questions make this feature a short introduction, reminder or basis for practicals in the classroom.

E.g.

Geiger tube

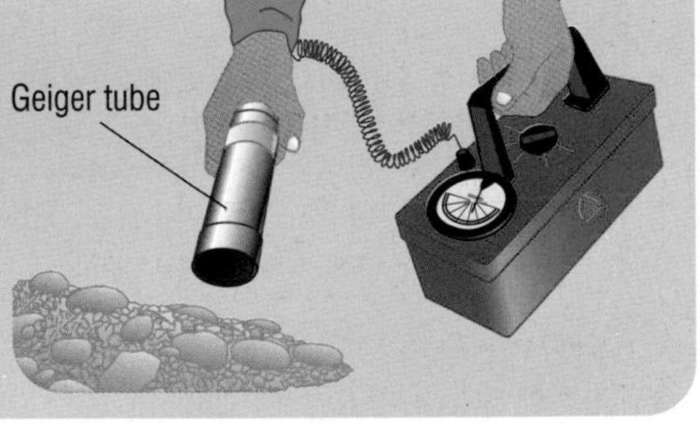

Figure 2 Using a Geiger counter

DID YOU KNOW?

Curious examples of scientific points that are out of the ordinary, but true…

FOUL FACTS

Some science is just too gruesome to ignore. Delve into the horrible yet relevant world of Foul Facts.

At the start of each unit you will find a double-page introduction. This reminds you of ideas from previous work that you will need. The recap questions and activity will help find out if you need some revision before starting.

SCIENCE @ WORK

When will you ever use science in 'real life'? Check this feature to find out.

NEXT TIME YOU…

…cook an egg, fall off your skateboard or listen to your mp3 player; this feature links the science you are learning to your everyday life. This will make it easier to remember and apply concepts you could be examined on.

SUMMARY QUESTIONS

Did you understand everything? Get these questions right, and you can be sure you did. Get them wrong, and you might want to take another look.

This first chapter looks at 'How Science Works'. It is an important part of your GCSE because the ideas introduced here will crop up throughout your course. You will be expected to collect scientific evidence and to understand how we use evidence. These concepts will be assessed as the major part of your internal school assessment. You will take one or more 45-minute tests on data you have collected previously plus data supplied for you in the test. These are called Investigative Skills Assignments. The ideas in 'How Science Works' will also be assessed in your examinations.

What you already know

Here is a quick reminder of previous work with investigations that you will find useful in this chapter:

- You will have done some practical work and know how important it is to keep yourself and others safe.
- Before you start investigating you usually make a prediction, which you can test.
- Your prediction and plan will tell you what you are going to change and what you are going to measure.
- You will have thought about controls.
- You will have thought about repeating your readings.
- During your practical work you will have written down your results, often in a table.
- You will have plotted graphs of your results.
- You will have made conclusions to explain your results.
- You will have thought about how you could improve your results, if you did the work again.

RECAP QUESTIONS

Helen wrote this about a practical she did:

I wanted to find out how strong an electromagnet could be with different lengths of wire wrapped around an iron bar. I thought that the more wire there was, the more iron filings the magnet would pick up.

I had to wear safety glasses to stop any iron filings getting in my eyes. I took an iron rod and wrapped a coil of wire around it. I connected the ends of the wire to a battery and then picked up some iron filings. I put the rod on an empty tray and switched the battery off. Then the iron filings dropped into the tray and I weighed them.

I then repeated this with 12 coils when I got 2.1 grams; 15 coils when I got 3.0 grams and 8 coils when I got 1.5 grams. With 6 coils I gathered 1.2 grams. I used the same battery throughout.

It was difficult because I couldn't scrape off all of the iron filings.

1 What was Helen's prediction?

2 What was the variable she chose to change? (We call this the **independent** variable.)

3 What was the variable she measured to judge the effect of varying the independent variable? (We call this the **dependent** variable. Its value *depends* on the value chosen for the independent variable.)

4 Write down a variable that Helen controlled.

5 Write down a variable Helen did not say she had controlled.

6 Make a table of her results.

7 Draw a graph of her results.

8 Write a conclusion for Helen.

9 How do you think Helen could have improved her results?

How science works for us

Science works for us all day, every day. You do not need to know how a mobile phone works to enjoy sending text messages. But, think about how you started to use your mobile phone or your television remote control. Did you work through pages of instructions? Probably not!

You knew that pressing the buttons would change something on the screen (**knowledge**). You played around with the buttons, to see what would happen (**observation**). You had a guess at what you thought might be happening (**prediction**) and then tested your idea (**experiment**).

If your prediction was correct you remembered that as a *fact*. If you could repeat the operation and get the same result again then you were very pleased with yourself. You had shown that your results were **reliable**.

Working as a scientist you will have knowledge of the world around you and particularly about the subject you are working with. You will observe the world around you. An enquiring mind will then lead you to start asking questions about what you have observed.

Science moves forward by slow steady steps. When a genius such as Einstein comes along then it takes a giant leap. Those small steps build on knowledge and experience that we already have.

Each small step is important in its own way. It builds on the body of knowledge that we have. In 1675 a German chemist tried to extract gold from urine. He must have thought that there was a connection between the two colours. He was wrong, but after a long while, with an incredible stench coming from his laboratory, the urine began to glow.

He had discovered phosphorus. A Swedish scientist worked out how to manufacture phosphorus without the smell of urine. Phosphorus catches fire easily. That is why most matches these days are manufactured in Sweden.

Thinking scientifically

Figure 1 Discussing fireworks

ACTIVITY

Now look at Figure 1 with your scientific brain.

- Fireworks must be safe to light. Therefore you need a fuse that will last long enough to give you time to get well out of the way.
- Fuses can be made by dipping a special type of cotton into a mixture of two chemicals. One chemical (A) reacts by burning, the other (B) doesn't.
- The chemicals stick to the cotton. Once it is lit, the cotton will continue to burn, setting the firework off. The concentrations of the two chemicals will affect how quickly the fuse burns.

In groups discuss how you could work out the correct concentrations of the chemicals to use. You want the fuse to last long enough to get out of the way, but not to burn so long that we all get bored waiting for the firework to go off!

You can use the following headings to discuss your investigation. One person should be writing your ideas down, so that you can discuss them with the rest of the class.

- What prediction can you make about the concentration of the two chemicals (A and B) and the fuse?
- What would be your independent variable?
- What would be your dependent variable?
- What would you have to control?
- Write a plan for your investigation.
- How could you make sure your results were reliable?

Fundamental ideas about how science works

LEARNING OBJECTIVES

1 How do you spot when a person has an opinion that is not based on good science?
2 What is the importance of continuous, ordered and categoric variables?
3 What is meant by reliable evidence and valid evidence?
4 How can two sets of data be linked?

NEXT TIME YOU...

... read a newspaper article or watch the news on TV ask yourself if that research is valid and reliable. (See page 5.) Ask yourself if you can trust the opinion of that person.

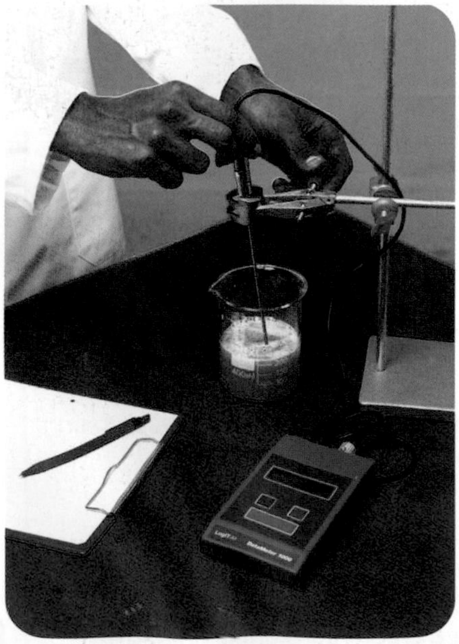

Figure 1 Student recording a range of temperatures

Science is too important for us to get it wrong

Sometimes it is easy to spot when people try to use science poorly. Sometimes it can be funny. You might have seen adverts claiming to give your hair 'body' or sprays that give your feet 'lift'!

On the other hand, poor scientific practice can cost lives.

Some years ago a company sold the drug thalidomide to people as a sleeping pill. Research was carried out on animals to see if it was safe. The research did not include work on pregnant animals. The opinion of the people in charge was that the animal research showed the drug could be used safely with humans.

Then the drug was also found to help ease morning sickness in pregnant women. Unfortunately, doctors prescribed it to many women, resulting in thousands of babies being born with deformed limbs. It was far from safe.

These are very difficult decisions to make. You need to be absolutely certain of what the science is telling you.

a) Why was the opinion of the people in charge of developing thalidomide based on poor science?

Deciding on what to measure

You know that you have an independent and a dependent variable in an investigation. These variables can be one of four different types:

- A **categoric variable** is one that is best described by a label (usually a word). The colour of eyes is a categoric variable, e.g. blue or brown eyes.

- A **discrete variable** is one that you describe in whole numbers. The number of leaves on different plants is a discrete variable.

- An **ordered variable** is one where you can put the data into order, but not give it an actual number. The height of plants compared to each other is an ordered variable, e.g. the plants growing in the woodland are taller than those on the open field.

- A **continuous variable** is one that we measure. Therefore its value could be any number. Temperature (as measured by a thermometer or temperature sensor) is a continuous variable, e.g. 37.6°C, 45.2°C.

When designing your investigation you should always try to measure continuous data whenever you can. This is not always possible, so you should then try to use ordered data. If there is no other way to measure your variable then you have to use a label (categoric variable).

b) Imagine you were growing seedlings in different volumes of water. Would it be better to say that some were tall and some were short; or some were taller than others; or to measure the heights of all of the seedlings?

Making your investigation reliable and valid

When you are designing an investigation you must make sure that others can get the same results as you – this makes it **reliable**.

You must also make sure you are measuring the actual thing you want to measure. If you don't, your data can't be used to answer your original question. This seems very obvious but it is not always quite so easy. You need to make sure that you have **controlled** as many other variables as you can, so that no-one can say that your investigation is not **valid**. A valid investigation should be reliable *and* answer the original question.

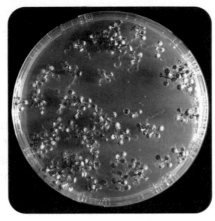

Figure 2 Cress seedlings growing in a petri dish

DID YOU KNOW?

Aristotle, a brilliant Greek scientist, once proclaimed that men had more teeth than women! Do you think that his data collection was reliable?

c) State one way in which you can show that your results are valid.

How might an independent variable be linked to a dependent variable?

Variables can be linked together for one of three reasons:

- It could be because one variable has caused a change in the other, e.g. the more plants there are in a pond, the more oxygen there is in the water. This is a **causal link**.
- It could be because a third variable has caused changes in the two variables you have investigated, e.g. fields that have more grass also have more dandelions in them. There is an *association* between the two variables. This is caused by a third variable – how many sheep there are in the field!
- It could be due simply to *chance*, e.g. the type of weeds growing in different parts of your garden!

d) Describe a causal link that you have seen in biology.

Figure 3 Sheep grazing in a field

SUMMARY QUESTIONS

1 Name each of the following types of variables described in a), b) and c).

   a) People were asked about how they felt inside a new shopping centre: 'warm', 'hot', 'quite warm', 'cold', 'freezing!'
   b) These people were asked as they entered the new shopping centre: 'Warmer than I did outside'; 'Colder than my shed!'
   c) These people had their body temperature measured using a clinical thermometer: 37.1°C; 37.3°C; 36.8°C; 37.0°C; 37.5°C.

2 A researcher claimed that the metal tungsten 'alters the growth of leukaemia cells' in laboratory tests. A newspaper wrote that they would 'wait until other scientists had reviewed the research before giving their opinion.' Why is this a good idea?

KEY POINTS

1 Be on the lookout for non-scientific opinions.
2 Continuous data give more information than other types of data.
3 Check that evidence is reliable and valid.
4 Be aware that just because two variables are related it does not mean that there is a causal link between them.

Starting an investigation

LEARNING OBJECTIVES

1 How can you use your scientific knowledge to observe the world around you?
2 How can you use your observations to make a hypothesis?
3 How can you make predictions and start to design an investigation?

Figure 1 Plant showing positive phototropism

DID YOU KNOW?

Some biologists think that we still have about one hundred million species of insects to discover – plenty to go for then! Of course, observing one is the easy part – knowing that it is undiscovered is the difficult bit!

Observation

As humans we are sensitive to the world around us. We can use our many senses to detect what is happening. As scientists we use observations to ask questions. We can only ask useful questions if we know something about the observed event. We will not have all of the answers, but we know enough to start asking the correct questions.

If we observe that the weather has been hot today, we would not ask if it was due to global warming. If the weather was hotter than normal for several years then we could ask that question. We know that global warming takes many years to show its effect.

When you are designing an investigation you have to observe carefully which variables are likely to have an effect.

a) Would it be reasonable to ask if the plant in Figure 1 is 'growing towards the glass'? Explain your answer.

A farmer noticed that her corn was much smaller at the edge of the field than in the middle (observation). She noticed that the trees were quite large on that side of the field. She came up with the following ideas that might explain why this was happening:

- The trees at the edge of the field were blocking out the light.
- The trees were taking too many nutrients out of the soil.
- The leaves from the tree had covered the young corn plants in the spring.
- The trees had taken too much water out of the soil.
- The seeds at the edge of the field were genetically small plants.
- The drill had planted fewer seeds on that side of the field.
- The fertiliser spray had not reached the side of the field.
- The wind had been too strong over winter and had moved the roots of the plants.
- The plants at the edge of the field had a disease.

b) Discuss each of these ideas and use your knowledge of science to decide which four are the most likely to have caused the poor growth of the corn.

Observations, backed up by really creative thinking and good scientific knowledge can lead to a **hypothesis**.

What is a hypothesis?

A hypothesis is a 'great idea'. Why is it so great? – well because it is a great observation that has some really good science to try to explain it.

For example, you observe that small, thinly sliced chips cook faster than large, fat chips. Your hypothesis could be that the small chips cook faster because the heat from the oil has a shorter distance to travel before it gets to the potato in the centre of the chips.

c) Check out the photograph in Figure 2 and spot anything that you find interesting. Use your knowledge and some creative thought to suggest a hypothesis based on your observations.

When making hypotheses you can be very imaginative with your ideas. However, you should have some scientific reasoning behind those ideas so that they are not totally bizarre.

Remember, your explanation might not be correct, but you think it is. The only way you can check out your hypothesis is to make it into a prediction and then test it by carrying out an investigation.

Observation ➕ knowledge ➡ hypothesis ➡

prediction ➡ investigation

Figure 2 Rusting lock

Starting to design a valid investigation

An investigation starts with a prediction. You, as the scientist, predict that there is a relationship between two variables.

- An **independent variable** is one that is changed or selected by you, the investigator.

- A **dependent variable** is measured for each change in your independent variable.

- All other variables become **control variables**, kept constant so that your investigation is a fair test.

If your measurements are going to be accepted by other people then they must be valid. Part of this is making sure that you are really measuring the effect of changing your chosen variable. For example, if other variables aren't controlled properly, they might be affecting the data collected.

d) When investigating his heart rate before and after exercise, Darren got his girlfriend to measure his pulse. Would Darren's investigation be valid? Explain your answer.

Figure 3 Measuring a pulse

SUMMARY QUESTIONS

1 Copy and complete using the words below:

controlled   dependent   hypothesis   independent
knowledge   prediction

Observations when supported by scientific ...... can be used to make a .... This can be the basis for a .... A prediction links an ...... variable to a ...... variable. Other variables need to be ....

2 Explain the difference between a hypothesis and a prediction.

KEY POINTS

1 Observation is often the starting point for an investigation.
2 Hypotheses can lead to predictions and investigations.
3 You must design investigations that produce valid results if you are to be believed.

H4

Building an investigation

LEARNING OBJECTIVES

1 How do you design a fair test?
2 How do you make sure that you choose the best values for your variables?
3 How do you ensure accuracy and precision?

Figure 1 Corn being harvested

Fair testing

A **fair test** is one in which only the independent variable affects the dependent variable. All other variables are controlled, keeping them constant if possible.

This is easy to set up in the laboratory, but almost impossible in fieldwork. Plants and animals do not live in environments that are simple and easy to control. They live complex lives with lots of variables changing constantly.

So how can we set up fieldwork investigations? The best you can do is to make sure that all of the many variables change in much the same way, except for the one you are investigating. Then at least the plants get the same weather, even if it is constantly changing.

a) Imagine you were testing how close together you could plant corn to get the most cobs. You would plant five different plots, with different numbers of plants in each plot. List some of the variables that you could not control.

If you are investigating two variables in a large population then you will need to do a survey. Again it is impossible to control all of the variables.

Imagine you were investigating the effect of diet on diabetes. You would have to choose people of the same age and same family history to test. The larger the sample size you test, the more reliable your results will be.

Control groups are used in investigations to try to make sure that you are measuring the variable that you intend to measure. When investigating the effects of a new drug, the control group will be given a placebo.

The control group think they are taking a drug but the placebo does not contain the drug. This way you can control the variable of '**thinking** that the drug is working' and separate out the effect of the actual drug.

Choosing values of a variable

Trial runs will tell you a lot about how your early thoughts are going to work out.

Do you have the correct conditions?
A photosynthesis investigation that produces tiny amounts of oxygen might not have enough:

- light, • pondweed, • carbon dioxide, or
- the temperature might not be high enough.

Have you chosen a sensible range?
If there is enough oxygen produced, but the results are all very similar:

- you might not have chosen a wide enough range of light intensities.

Have you got enough readings that are close together?
If the results are very different from each other:

- you might not see a pattern if you have large gaps between readings over the important part of the range.

Accuracy

Accurate measurements are very close to the *true value*.

Your investigation should provide data that is accurate enough to answer your original question.

However, it is not always possible to know what that true value is.

How do you get accurate data?
- You can repeat your measurements and your mean is more likely to be accurate.
- Try repeating your measurements with a different instrument and see if you get the same readings.
- Use high quality instruments that measure accurately.
- The more carefully you use the measuring instruments, the more accuracy you will get.

Precision and reliability

If your repeated measurements are closely grouped together then you have precision and you have improved the reliability of your data.

Your investigation must provide data with sufficient precision. It's no use measuring a person's reaction time using the seconds hand on a clock! If there are big differences within sets of repeat readings, you will not be able to make a valid conclusion. You won't be able to trust your data!

How do you get precise and reliable data?

- You have to use measuring instruments with sufficiently small scale divisions.
- You have to repeat your tests as often as necessary.
- You have to repeat your tests in exactly the same way each time.

A word of caution!

Be careful though – just because your results show precision does not mean your results are accurate. Look at the box opposite.

b) Draw a thermometer scale showing 4 results that are both accurate and precise.

The difference between accurate and precise results

Imagine measuring the temperature after a set time when a fuel is used to heat a fixed volume of water. Two students repeated this experiment, four times each. Their results are marked on the thermometer scales below:

Student A — Precise (but not accurate)
Student B — Accurate (but not precise)

- A precise set of repeat readings will be grouped closely together.
- An accurate set of readings will have a mean (average) close to the true value.

SUMMARY QUESTIONS

1 Copy and complete using the following terms:

   range   repeat   conditions   readings

   Trial runs give you a good idea of whether you have the correct ......; whether you have chosen the correct ......; whether you have enough ......; if you need to do ...... readings.

2 Use an example to explain how a set of repeat measurements could be accurate, but not precise.

3 Briefly describe how you would go about setting up a fair test in a laboratory investigation. Give your answer as general advice.

KEY POINTS

1 Care must be taken to ensure fair testing – as far as is possible.
2 You can use a trial run to make sure that you choose the best values for your variables.
3 Careful use of the correct equipment can improve accuracy.
4 If you repeat your results carefully they are likely to become more reliable.

Making measurements

LEARNING OBJECTIVES

1 Why do results always vary?
2 How do you choose instruments that will give you accurate results?
3 What do we mean by the sensitivity of an instrument?
4 How does human error affect results and what do you do with anomalies?

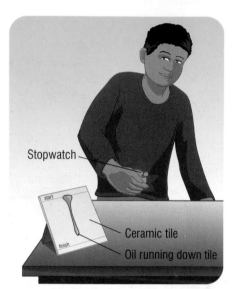

Stopwatch

start

finish

Ceramic tile

Oil running down tile

Figure 1 Student reading the arrival of the oil

DID YOU KNOW?

Professor Hough was investigating possible uses of sucrose (ordinary sugar) in industry. He had created a molecule of sucrose with three atoms of chlorine in it. He asked his new assistant to 'test' it. His assistant thought he had said 'taste' it. Fortunately for his assistant it did him no harm, but he noticed how incredibly sweet it was – a thousand times sweeter than sugar!

Using instruments

Do not panic! You cannot expect perfect results.

Try measuring the temperature of a beaker of water using a digital thermometer. Do you always get the same result? Probably not. So can we say that any measurement is absolutely correct?

In any experiment there will be doubts about actual measurements.

a) Look at Figure 1. Suppose, like this student, you tested the time it takes for one type of oil to flow down the tile. It is unlikely that you would get two readings exactly the same. Discuss all the possible reasons why.

When you choose an instrument you need to know that it will give you the accuracy that you want. That is, it will give you a true reading.

If you have used an electric water bath, would you trust the temperature on the dial? How do you know it is the true temperature? You could use a very expensive thermometer to calibrate your water bath. The expensive thermometer is more likely to show the true temperature. But can you really be sure it is accurate?

You also need to be able to use an instrument properly.

b) In Figure 1 the student is measuring the time it takes for the oil to reach the line. Why is the student unlikely to get a true measurement?

When you choose an instrument you have to decide how accurate you need it to be. Instruments that measure the same thing can have different sensitivities. The **sensitivity** of an instrument refers to the smallest change in a value that can be detected. This determines the precision of your measurements.

Choosing the wrong scale can cause you to miss important data or make silly conclusions, for example 'The amount of gold was the same in the two rings – they both weighed 5 grams.'

c) Match the following weighing machines to their best use:

Used to measure	Sensitivity of weighing machine
Cornflakes delivered to a supermarket	milligrams
Carbohydrate in a packet of cornflakes	grams
Vitamin D in a packet of cornflakes	micrograms
Sodium chloride in a packet of cornflakes	kilograms

Errors

Even when an instrument is used correctly, the results can still show differences.

Results may differ because of **random error**. This is most likely to be due to a poor measurement being made. It could be due to not carrying out the method consistently.

The error might be a **systematic error**. This means that the method was carried out consistently but an error was being repeated.

Check out these two sets of data that were taken from the investigation that Mark did. He tested 5 different oils. The third line is the time calculated from knowing the viscosity of the different oils:

Type of oil used	a	b	c	d	e
Time taken to flow down tile (seconds)	23.2	45.9	49.5	62.7	75.9
	24.1	36.4	48.7	61.5	76.1
Calculated time (seconds)	18.2	30.4	42.5	55.6	70.7

d) Discuss whether there is any evidence for random error in these results.
e) Discuss whether there is any evidence for systematic error in these results.

Anomalies

Anomalous results are clearly out of line. They are not those that are due to the natural variation you get from any measurement. These should be looked at carefully. There might be a very interesting reason why they are so different. If they are simply due to a random error, then they should be discarded (rejected).

If anomalies can be identified while you are doing an investigation, then it is best to repeat that part of the investigation.

If you find anomalies after you have finished collecting data for an investigation, then they must be discarded.

SUMMARY QUESTIONS

1 Copy and complete using the words below:

accurate   discarded   random   sensitivity   systematic
use   variation

There will always be some …… in results. You should always choose the best instruments that you can to get the most …… results. You must know how to …… the instrument properly. The …… of an instrument refers to the smallest change that can be detected. There are two types of error – …… and ……. Anomalies due to random error should be …… .

2 Which of the following will lead to a systematic error and which to a random error?
a) Using a weighing machine, which has something stuck to the pan on the top.
b) Forgetting to re-zero the weighing machine.

DID YOU KNOW?

Sir Alexander Fleming was showing his research assistant some plates on which he had grown bacteria. He noticed an anomaly. There was some mould growing on one of the plates and around it there were no bacteria. He investigated further and grew the mould, identifying it as *Penicillium rubrum*.

He persuaded an assistant to taste it and he said it tasted like Stilton cheese! He later injected the assistant with it – and he didn't die!

Only because Fleming checked out his anomaly did it lead to the discovery of penicillin. Oh, and Fleming also let his nose dribble onto one plate and he discovered lysozyme!!

KEY POINTS

1 Results will nearly always vary.
2 Better instruments give more accurate results.
3 Sensitivity of an instrument refers to the smallest change that it can detect.
4 Human error can produce random and systematic errors.
5 We examine anomalies; they might give us some interesting ideas. If they are due to a random error, we repeat the measurements. If there is no time to repeat them, we discard them.

Presenting data

H6

LEARNING OBJECTIVES

1 What do we mean by the 'range' and the 'mean' of the data?
2 How do you use tables of results?
3 How do you display your data?

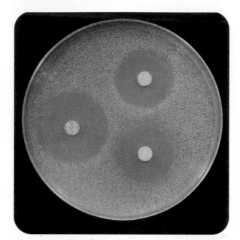

Figure 1 Petri dish with discs showing growth inhibition

For this section you will be working with data from this investigation:

Mel spread some bacteria onto a dish containing nutrient jelly. She also placed some discs onto the jelly. The discs contained different concentrations of an antibiotic. The dish was sealed and then left for a couple of days.

Then she measured the diameter of the clear part around each disc. The clear part is where the bacteria have not been able to grow. The bacteria grew all over the rest of the dish.

Tables

Tables are really good for getting your results down quickly and clearly. You should design your table **before** you start your investigation.

Your table should be constructed to fit in all the data to be collected. It should be fully labelled, including units.

In some investigations, particularly fieldwork, it is useful to have an extra column for any notes you might want to make as you work.

While filling in your table of results you should be constantly looking for anomalies.

● Check to see if a repeat is sufficiently close to the first reading.
● Check to see if the pattern you are getting as you change the independent variable is what you expected.

Remember a result that looks anomalous should be checked out to see if it really is a poor reading or if it might suggest a different hypothesis.

Planning your table

Mel knew the values for her independent variable. We always put these in the first column of a table. The dependent variable goes in the second column. Mel will find its values as she carries out the investigation.

So she could plan a table like this:

Concentration of antibiotic (μg/ml)	Size of clear zone (mm)
4	
8	
16	
32	
64	

Or like this:

Concentration of antibiotic (μg/ml)	4	8	16	32	64
Size of clear zone (mm)					

All she had to do in the investigation was to write the correct numbers in the second column to complete the top table.

Mel's results are shown in the alternative format in the table below:

Concentration of antibiotic (μg/ml)	4	8	16	32	64
Size of clear zone (mm)	4	16	22	26	28

The range of the data

Pick out the maximum and the minimum values and you have the range. You should always quote these two numbers when asked for a range. For example, the range is between .... (the lowest value) and .... (the highest value) – and don't forget to include the units!

a) What is the range for the dependent variable in Mel's set of data?

The mean of the data

Often you have to find the mean of each repeated set of measurements.

You add up the measurements in the set and divide by how many there are. Miss out any anomalies you find.

The repeat values and mean can be recorded as shown below:

Concentration of antibiotic (μg/ml)	Size of clear zone (mm)			
	1st test	2nd test	3rd test	Mean

Displaying your results

Bar charts

If you have a categoric or an ordered independent variable and a continuous dependent variable then you should use a bar chart.

Line graphs

If you have a continuous independent and a continuous dependent variable then a line graph should be used.

Scatter graphs (or scattergrams)

Scatter graphs are used in much the same way as line graphs, but you might not expect to be able to draw such a clear line of best fit. For example, if you wanted to see if people's lung capacity was related to how long they could hold their breath, you would draw a scatter graph with your results.

SUMMARY QUESTIONS

1 Copy and complete using the words below:

   categoric   continuous   mean   range

   The maximum and minimum values show the ...... of the data. The sum of all the values divided by the total number of the values gives the .... Bar charts are used when you have a ...... independent variable and a continuous dependent variable.
   Line graphs are used when you have ...... independent and dependent variables.

2 Draw a graph of Mel's results from the top of this page.

NEXT TIME YOU...

... make a table for your results remember to include:
- headings,
- units,
- a title.

... draw a line graph remember to include:
- the independent variable on the x-axis,
- the dependent variable on the y-axis,
- a line of best fit,
- labels, units and a title.

GET IT RIGHT!

Marks are often dropped in the ISA by candidates plotting points incorrectly. Also use a **line of best fit** where appropriate – don't just join the points 'dot-to-dot'!

KEY POINTS

1 The range states the maximum and the minimum values.

2 The mean is the sum of the values divided by how many values there are.

3 Tables are best used during an investigation to record results.

4 Bar charts are used when you have a categoric or an ordered independent variable and a continuous dependent variable.

5 Line graphs are used to display data that are continuous.

H7 Using data to draw conclusions

Identifying patterns and relationships

Now that you have a bar chart or a graph of your results you can begin to look for patterns. You must have an open mind at this point.

Firstly, there could still be some anomalous results. You might not have picked these out earlier. How do you spot an anomaly? It must be a significant distance away from the pattern, not just within normal variation.

A line of best fit will help to identify any anomalies at this stage. Ask yourself – do the anomalies represent something important or were they just a mistake?

Secondly, remember a line of best fit can be a straight line or it can be a curve – you have to decide from your results.

The line of best fit will also lead you into thinking what the relationship is between your two variables. You need to consider whether your graph shows a **linear** relationship. This simply means, can you be confident about drawing a straight line of best fit on your graph? If the answer is yes – then is this line positive or negative?

a) Say whether graphs (i) and (ii) in Figure 1 show a positive or a negative linear relationship.

Look at the graph in Figure 2. It shows a positive linear relationship. It also goes through the origin (0,0). We call this a **directly proportional** relationship.

Your results might also show a curved line of best fit. These can be predictable, complex or very complex! Look at Figure 3 below.

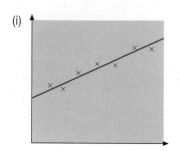

(i)

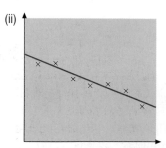

(ii)

Figure 1 Graphs showing linear relationships

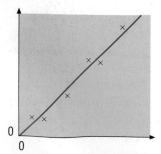

Figure 2 Graph showing a directly proportional relationship

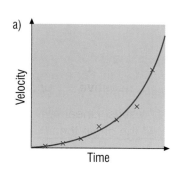

a) Velocity / Time

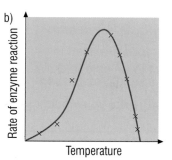

b) Rate of enzyme reaction / Temperature

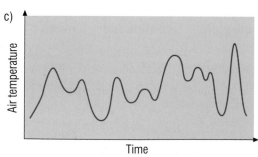

c) Air temperature / Time

Figure 3 a) Graph showing predictable results. b) Graph showing complex results. c) Graph showing very complex results.

Drawing conclusions

Your graphs are designed to show the relationship between your two chosen variables. You need to consider what that relationship means for your conclusion.

There are three possible links between variables. (See page 5.) They can be:

- causal,

- due to association, or

- due to chance.

You must decide which is the most likely. Remember a positive relationship does not always mean a causal link between the two variables.

Poor science can often happen if a wrong decision is made here. Newspapers have said that living near electricity sub-stations can cause cancer. All that scientists would say is that there is possibly an association. Getting the correct conclusion is very important.

You will have made a prediction. This could be supported by your results. It might not be supported or it could be partly supported. Your results might suggest some other hypothesis to you.

Your conclusion must go no further than the evidence that you have. For example, an ultrasound device designed for cleaning metal parts might not clean plastic coated metals.

Evaluation

If you are still uncertain about a conclusion, it might be down to the reliability and the validity of the results. You could check these by:

- looking for other similar work on the Internet or from others in your class,

- getting somebody else to re-do your investigation, or

- trying an alternative method to see if you get the same results.

NEXT TIME YOU…

… read scientific claims, think carefully about the evidence that should be there to back up the claim.

DID YOU KNOW?

Pythagoras of Samos declared that 'Everything is number'. He believed that everything in the Universe can be explained by simple mathematical relationships. He went on to discover the relationship between the length of a string and the sound it produces when it vibrates.

He developed this idea into a theory that the Sun, the Moon and the planets produced a sort of music that kept them in their orbits!

KEY POINTS

1 Drawing lines of best fit help us to study the relationship between variables.

2 The possible relationships are linear, positive and negative; directly proportional; predictable and complex curves.

3 Conclusions must go no further than the data available.

4 The reliability and validity of data can be checked by looking at other similar work done by others, perhaps on the Internet. It can also be checked by using a different method or by others checking your method.

SUMMARY QUESTIONS

1 Copy and complete using the words below:

    anomalous   complex   directly   negative   positive

Lines of best fit can be used to identify …… results. Linear relationships can be …… or ……. If a graph goes through the origin then the relationship could be …… proportional. Often a line of best fit is a curve which can be predictable or …… .

2 Nasma knew about the possible link between cancer and living near to electricity sub-stations. She found a quote from a National Grid Company survey of sub-stations:

'Measurements of the magnetic field were taken at 0.5 metre above ground level within 1 metre of fences and revealed 1.9 microteslas. After 5 metres this dropped to the normal levels measured in any house.'

Discuss the type of experiment and the data you would expect to see to support a conclusion that it is safe to build houses over 5 metres from an electricity sub-station.

H8

Scientific evidence and society

MOBILE PHONE TUMOUR RISK?

Swedish researchers found that the risk of developing an ear tumour increased if you used a mobile phone. The study was of 750 people. This type of tumour affects one in 100,000 people and the risk increased four times if you used the phone for more than 10 years.

Now you have reached a conclusion about a piece of scientific research. So what is next? If it is pure research then your fellow scientists will want to look at it very carefully. If it affects the lives of ordinary people then society will also want to examine it closely.

You can help your cause by giving a balanced account of what you have found out. It is much the same as any argument you might have. If you make ridiculous claims then nobody will believe anything you have to say.

Be open and honest. If you only tell part of the story then someone will want to know why! Equally, if somebody is only telling you part of the truth, you cannot be confident with anything they say.

a) 'X-rays are safe, but should be limited' is the headline in an American newspaper. What information is missing? Is it important?

You must be on the lookout for people who might be biased when representing scientific evidence. Some scientists are paid by companies to do research. When you are told that a certain product is harmless, just check out who is telling you.

b) Suppose you wanted to know about safe levels of noise at work. Would you ask the scientist who helped to develop the machinery or a scientist working in the local university? What questions would you ask, so that you could make a valid judgement?

We also have to be very careful in reaching judgements according to who is presenting scientific evidence to us. For example, if the evidence might provoke public or political problems, then it might be played down.

Equally others might want to exaggerate the findings. They might make more of the results than the evidence suggests. Take as an example the siting of mobile phone masts. Local people may well present the same data in a totally different way from those with a wider view of the need for mobile phones.

c) Check out some websites on mobile phone masts. Get the opinions of people who think they are dangerous and those who believe they are safe. Try to identify any political bias there might be in their opinions.

The status of the experimenter may place more weight on evidence. Suppose an electricity company wants to convince an inquiry that it is perfectly reasonable to site a wind turbine in remote moorland in the UK. The company will choose the most eminent scientist in that field who is likely to support them. The small local community might not be able to afford an eminent scientist. The inquiry needs to be carried out very carefully to make a balanced judgement.

VILLAGERS PROTEST AGAINST WIND FARM

There was considerable local opposition from local villagers to building a wind farm near the A14 road in Cambridgeshire. Planners turned down the application after seven months of protests by local residents. Some described it as being like 16 football pitches rotating in the sky. Others were concerned at the effect on the value of their houses. Friends of the Earth were, in principle, in favour. The wind farm company said that it would provide energy for 20,000 homes.

SUMMARY QUESTIONS

1 Copy and complete using the words below:

status   balanced   bias   political

Evidence from scientific investigations should be given in a ...... way. It must be checked for any ...... from the experimenter.
Evidence can be given too little or too much weight if it is of ...... significance.
The ...... of the experimenter is likely to influence people in their judgement of the evidence.

2 Collect some newspaper articles to show how scientific evidence is used. Discuss in groups whether these articles are honest and fair representations of the science. Consider whether they carry any bias.

3 Extract from BBC website about Sizewell nuclear power station:

'A radioactive leak can have devastating results but one small pill could protect you. "Inside out" reveals how for the first time these life-saving pills will be available to families living close to the Sizewell nuclear power station.'

Suppose you were living near Sizewell power station. Who would you trust to tell you whether these pills would protect you from radiation? Who wouldn't you trust?

KEY POINTS

1 Scientific evidence must be presented in a balanced way that points out clearly how reliable and valid the evidence is.
2 The evidence must not contain any bias from the experimenter.
3 The evidence must be checked to appreciate whether there has been any political influence.
4 The status of the experimenter can influence the weight placed on the evidence.

How is science used for everybody's benefit?

The development of oral contraceptives shows how science can be used for technological development. There were many unscientific ways in which women would attempt contraception.

In Europe, women wore the foot of a weasel around their neck to prevent them becoming pregnant. In North Africa the flower silphium was thought to be an oral contraceptive. It became very expensive to buy and eventually was used so much that it became extinct.

These ideas might have had some basis, because some plants do contain human sex hormones. Some plants such as black kohosh, are used today as relief from problems related to the menopause.

Figure 1 A yam plant

Yam was used as a pain relief. A Japanese scientist found diosgenin in yam plants. Diosgenin was used as the starting point for an investigation that led to the development of the hormone progesterone. This led to the development of the first contraceptive pill. It was some unscientific thinking that encouraged people to eat yam plants as a natural contraceptive – they aren't!

Frank Colton developed Enovid, one of the first oral contraceptives, in 1960. By 1961 the birth control pill was available to 'everyone'. Some doctors were in a dilemma for social as well as medical reasons. They would not prescribe it to unmarried women, because it 'encourages sex outside marriage'.

The UK Government said it couldn't afford the cost. They said the pill could have long-term effects. A woman's body was likened to a clock; 'Whilst it was running well it should be left alone,' said Sir Charles Dodds, a leading expert on drugs. The pill allowed women to take control of their own fertility.

Figure 2 Frank Colton

It was known that there was a risk of cardiovascular disease and stroke. However, new developments have reduced the dose and therefore this risk.

Today there are hundreds of different contraceptive pills. There are even male contraceptive pills. The morning-after pill has raised problems for some people who consider it is a form of abortion.

Some of these hormones are now in such a high concentration in river water that it affects the ability of some fish to reproduce.

Figure 3 Contraceptive pills

The contraceptive pill still raises social, ethical, economic and even environmental issues.

There are many questions left for science to answer. For example, how to develop an oral contraceptive that is 100% safe for all people. However, science cannot answer questions about whether or not we should use contraception.

KEY POINTS

1 Scientific knowledge can be used to develop technologies.
2 People can exploit scientific and technological developments to suit their own purposes.
3 The uses of science and technology can raise ethical, social, economic and environmental issues.
4 These issues are decided upon by individuals and by society.
5 There are many questions left for science to answer. But science cannot answer questions that start with 'Should we ....?'

SUMMARY QUESTIONS

Use the account of the development of contraceptive pills to answer these questions.

1 What scientific knowledge was available to Frank Colton that enabled him to develop Enovid?

2 How did different groups of people react to the development of Enovid?

3 a) Identify some of these issues raised by the development of contraceptive pills: i) ethical, ii) social, iii) economic, iv) environmental.
   b) Which of these issues are decided by individuals and which by society?

SUMMARY QUESTIONS

1 Fit these words into order. They should be in the order in which you might use them in an investigation.

design; prediction; conclusion; method; repeat; controls; graph; results; table; improve; safety

2 a) How would you tell the difference between an opinion that was scientific and a prejudiced opinion?

b) Suppose you were investigating the amount of gas produced in a reaction. Would you choose to investigate a categoric, continuous or ordered variable? Explain why.

c) Explain the difference between a causal link between two variables and one which is due to association.

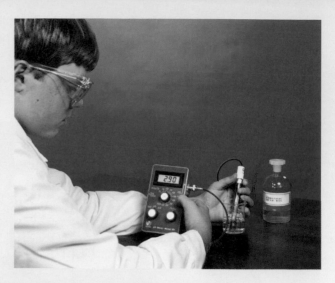

3 You might have observed that marble statues weather badly where there is air pollution. You ask the question why. You use some accepted theory to try to answer the question.

a) Explain what you understand by a hypothesis.

b) Sulfur dioxide in the air forms acids that attack the statues. This is a hypothesis. Develop this into a prediction.

c) Explain why a prediction is more useful than a hypothesis.

d) Suppose you have tested your prediction and have some data. What might this do for your hypothesis?

e) Suppose the data does not support the hypothesis. What should you do to the theory that gave you the hypothesis?

4 a) What do you understand by a fair test?

b) Suppose you were carrying out an investigation into what effect diluting acid had on its pH. You would need to carry out a trial. Describe what a trial would tell you about how to plan your method.

c) How could you decide if your results were reliable?

d) It is possible to calculate the effect of dilution on the pH of an acid. How could you use this to check on the accuracy of your results?

5 Suppose you were watching a friend carry out an investigation using the equipment shown on page 10. You have to mark your friend on how accurately he is making his measurements. Make a list of points that you would be looking for.

6 a) How do you decide on the range of a set of data?

b) How do you calculate the mean?

c) When should you use a bar chart?

d) When should you use a line graph?

7 a) What should happen to anomalous results?

b) What does a line of best fit allow you to do?

c) When making a conclusion, what must you take into consideration?

d) How can you check on the reliability of your results?

8 a) Why is it important when reporting science to 'tell the truth, the whole truth and nothing but the truth'?

b) Why might some people be tempted not to be completely fair when reporting their opinions on scientific data?

9 a) 'Science can advance technology and technology can advance science.' What do you think is meant by this statement?

b) Who answers the questions that start with 'Should we . . . '?

10 Wind turbines are an increasingly popular way of generating electricity. It is very important that they are sited in the best place to maximise energy output. Clearly they need to be where there is plenty of wind. Energy companies have to be confident that they get value for money. Therefore they must consider the most economic height to build them. Put them too high and they might not get enough extra energy to justify the extra cost of the turbine. Before deciding finally on a site they will carry out an investigation to decide the best height.

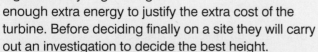

The prediction is that increasing the height will increase the power output of the wind turbine.

A test platform was erected and the turbine placed on it. The lowest height that would allow the turbines to move was 32 metres. The correct weather conditions were waited for and the turbine began turning and the power output was measured in kilowatts.

The results are in the table.

Height of turbine (m)	Power output 1 (kW)	Power output 2 (kW)
32	162	139
40	192	195
50	223	219
60	248	245
70	278	270
80	302	304
85	315	312

a) What was the prediction for this test?

b) What was the independent variable?

c) What was the dependent variable?

d) What is the range of the heights for the turbine?

e) Suggest a control variable that should have been used.

f) This is a fieldwork investigation. Is it possible to control all of the variables? If not, say what you think the scientist should have done to produce more accurate results.

g) Is there any evidence for a random error in this investigation?

h) Was the precision of the power output measurement satisfactory? Provide some evidence for your answer from the data in the table.

i) Draw a graph of the results for the second test.

j) Draw a line of best fit.

k) Describe the pattern in these results.

l) What conclusion can you make?

m) How might this data be of use to people who might want to stop a wind farm being built?

n) Who should carry out these tests for those who might object?

B1a | Human biology

What you already know

Here is a quick reminder of previous work that you will find useful in this unit:

- When you reach puberty your body changes so that you can reproduce.

- Girls have a regular menstrual cycle when their body prepares for pregnancy. Boys begin making sperm which can fertilise an egg.

- Eating a balanced diet is an important part of keeping healthy.

- A balanced diet will include carbohydrates, proteins, fats, minerals, vitamins, fibre and water.

- The food you eat is used as a fuel during respiration. It provides energy for the cells of your body. Your food also gives you the raw materials you need to grow and to repair worn out cells.

- Your food needs breaking down (digesting) before it is any use in your body.

- Diseases caused by bacteria and viruses can affect your health.

- Your body can defend itself against disease-causing microorganisms. However sometimes you give it a helping hand by taking medicines or being immunised.

- Both legal and illegal drugs can damage your health if you abuse them. It is against the law to use illegal drugs.

RECAP QUESTIONS

1 What are the main food groups you need to eat to have a balanced diet?

2 How does your body use the food that you eat?

3 What are the main changes which take place in:

a) boys and

b) girls

when they go through puberty?

4 How often is an egg produced in the menstrual cycle?

5 a) Which types of microorganism can cause disease?

b) What is an *infectious* disease?

c) Give the names of three infectious diseases.

6 How does your body defend itself against disease?

7 Explain how medicines and immunisation can help to keep you healthy.

Making connections

Becki and Sam want to have a baby but although they've been trying for over a year, Becki still isn't pregnant. She might not be making enough hormones to release an egg each month, so we are doing some blood tests to find out what is happening. If lack of hormones really is the problem, we can soon sort things out for them.

Sara makes sure baby Jaz gets his injections right on time. Years ago, lots of children died young from infectious diseases. By immunising babies like Jaz against many of the most dangerous diseases, we can help today's children grow up as healthily as possible.

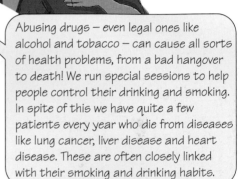

Keeping well is important – if you are healthy you can enjoy life to the full at work, at school and at home. As doctors, we want to help people to be as healthy as possible, but usually we only see people when they are not feeling at their best. Meet some of our patients!

Abusing drugs – even legal ones like alcohol and tobacco – can cause all sorts of health problems, from a bad hangover to death! We run special sessions to help people control their drinking and smoking. In spite of this we have quite a few patients every year who die from diseases like lung cancer, liver disease and heart disease. These are often closely linked with their smoking and drinking habits.

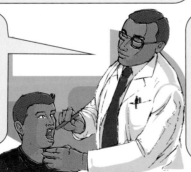

Liam's got a temperature, a headache and a terribly sore throat. When we looked into his throat it was covered in yellow pus. He's got a bad case of tonsillitis. A course of antibiotics will soon have him feeling much better and ready to go back to school!

Being overweight isn't good for your joints or your heart. Some of our overweight patients find it hard to join in activities and they often get teased about being so big. We help them to lose weight and take more exercise, which makes them feel fitter – and better.

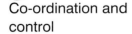

ACTIVITY

Many people think that doctors are only there to make them better when they feel ill. As you can see, most doctors do much more than this!

Design a poster to encourage people to use their GP to help them get healthy and stay healthy.

Your poster is going to be put up in all sorts of places – schools, libraries, shopping centres – so make sure it is clear, easy to read and gets the message across.

Chapters in this unit

Co-ordination and control   Healthy eating   Drug abuse   Controlling infectious disease

B1a 1.1

Responding to change

LEARNING OBJECTIVES

1 How is your body controlled?
2 What is the difference between your nervous system and your hormones?
3 How do you respond to changes in your surroundings?

You need to know what is going on in the world around you. Your **nervous system** makes this possible. It enables you to react to your surroundings and co-ordinate your behaviour.

Your nervous system carries electrical signals (*impulses*) which travel fast – from 1 to 120 metres per second. This means you can react to changes in your surroundings very quickly indeed.

Figure 1 Your body is made up of millions of cells which have to work together. Whatever you do with your body – whether it's winning a race or playing on the computer – your movements need to be co-ordinated. The conditions inside your body must also be controlled.

a) What is the main job of the nervous system?

Hormones are chemical substances. They control many of the processes going on inside your body. Special **glands** make and release (**secrete**) these hormones into your body. Then the hormones are carried around your body to their target organs in the bloodstream. They can act very quickly, but often their effects are quite slow and long lasting.

b) What type of messengers are hormones?

The nervous system

Like all living things, you need to avoid danger, find food and – eventually – find a mate! This is where your nervous system comes into its own. Your body is particularly sensitive to changes in the world around you. Any changes (known as **stimuli**) are picked up by cells called **receptors**.

These receptors are usually found clustered together in special **sense organs**, such as your eyes and your skin. You have many different types of sensory receptors (see Figure 3).

c) Where would you find receptors which would respond to i) a loud noise, ii) touching a hot oven, iii) a strong perfume?

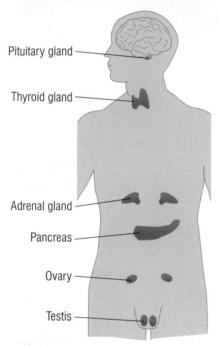

Pituitary gland
Thyroid gland
Adrenal gland
Pancreas
Ovary
Testis

Figure 2 Hormones act as chemical messengers. They are made in glands in one part of your body but having an effect somewhere else entirely

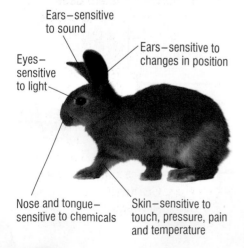

Ears – sensitive to sound
Ears – sensitive to changes in position
Eyes – sensitive to light
Nose and tongue – sensitive to chemicals
Skin – sensitive to touch, pressure, pain and temperature

Figure 3 Look at this rabbit. Being able to detect changes in the environment is important. It can often be a matter of life and death.

How your nervous system works

Once a sensory receptor detects a stimulus, the information (sent as an electrical impulse) passes along special cells called **neurones**. These are usually found in bundles of hundreds or even thousands of neurones known as *nerves*.

The impulse travels along the neurone until it reaches the **central nervous system** or **CNS**. The CNS is made up of the brain and the spinal cord. The cells which carry impulses from your sense organs to your central nervous system are called *sensory neurones*.

> d) What is the difference between a neurone and a nerve?

Your brain gets huge amounts of information from all the sensory receptors in your body. It co-ordinates the information and sends impulses out along special cells. These cells carry information from the CNS to the rest of your body. The cells are called *motor neurones*. They carry impulses to make the right bits of your body – the **effector organs** – respond.

Effector organs are muscles or glands. Your muscles respond to the arrival of impulses by contracting. Your glands respond by releasing (**secreting**) chemical substances.

The way your nervous system works can be summed up as:

receptor   sensory neurone   co-ordinator   motor neurone   effector
(CNS)

> e) What is the difference between a sensory neurone and a motor neurone?

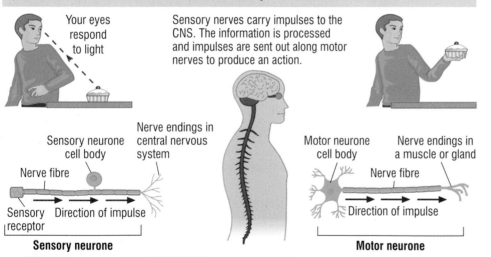

Your eyes respond to light

Sensory nerves carry impulses to the CNS. The information is processed and impulses are sent out along motor nerves to produce an action.

Sensory neurone cell body

Nerve endings in central nervous system

Nerve fibre

Sensory receptor   Direction of impulse

Sensory neurone

Motor neurone cell body

Nerve endings in a muscle or gland

Nerve fibre

Direction of impulse

Motor neurone

DID YOU KNOW?

Some male moths are so sensitive to chemicals that they can detect the scent of a female several kilometres away. What's more they can follow the scent trail and find her!

GET IT RIGHT!

Be careful to use the terms **neurone** and **nerve** correctly. Talk about **impulses** (*not* **messages**) travelling along a neurone.

Figure 4 The rapid responses of our nervous system allow us to respond to our surroundings quickly – and in the right way!

KEY POINTS

1 Hormones, secreted by special glands, are chemicals that help control and co-ordinate processes in your body.
2 The nervous system uses electrical impulses to enable you to react to your surroundings and co-ordinate what you do.
3 Cells called receptors detect stimuli (changes in the environment).
4 Impulses from receptors pass along sensory neurones to the brain. Impulses are sent from the brain to the effector organs along motor neurones.

SUMMARY QUESTIONS

1 Copy and complete using the words below:

  blood   chemical   electrical   glands   nervous

  Your ...... system carries fast ...... impulses. Your hormones are ...... messengers secreted by special ...... and carried around the body in the ....

2 Make a table to show the different types of sense receptors. For each one, give an example of the sort of things it responds to, e.g. touch receptors respond to an insect crawling on your skin.

3 Explain i) what happens in your nervous system when you see a piece of chocolate, pick it up and eat it, ii) the differences between hormonal and nervous control in your body.

B1a 1.2

Reflex actions

Your nervous system lets you take in information about the world around you and respond in the right way. However some of your responses are so fast that they happen without giving you time to think.

When you touch something hot, or sharp, you pull your hand back before you feel the pain. If something comes near your face, you blink. Automatic responses like these are known as **reflexes**.

What are reflexes for?

Reflexes are very important both for human beings and for other animals. They help you to avoid danger or harm because they happen so fast. There are also lots of reflexes which take care of your basic body functions. These functions include breathing and moving the food through your gut.

It would make life very difficult if you had to think consciously about those things all the time – and could be fatal if you forgot to breathe!

a) Why are reflexes important?

How do reflexes work?

Reflex actions involve just three types of neurone. These are:

● sensory neurones,
● motor neurones, and
● relay neurones which simply connect a sensory neurone and a motor neurone. We find relay neurones in the CNS, often in the spinal cord.

An electrical impulse passes from the sensory receptor along the sensory neurone to the CNS. It then passes along a relay neurone (usually in the spinal cord) and straight back along a motor neurone. From there the impulse arrives at the effector organ (usually a muscle in a reflex). We call this a *reflex arc*.

The key point in a reflex arc is that the impulse bypasses the conscious areas of your brain. The result is that the time between the stimulus and the reflex action is as short as possible. When you put your hand on something hot, you have moved your hand away before you feel the pain!

b) Why is it important that the impulses in a reflex arc do not go to the conscious brain?

How synapses work

Your nerves are not joined up directly to each other. There are junctions between them called **synapses**. The electrical impulses travelling along your neurones have to cross these synapses but they cannot leap the gap. Look at Figure 1 to see what happens next.

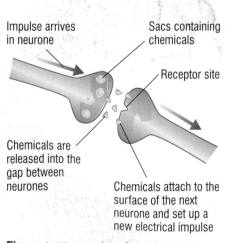

Impulse arrives in neurone

Sacs containing chemicals

Receptor site

Chemicals are released into the gap between neurones

Chemicals attach to the surface of the next neurone and set up a new electrical impulse

Figure 1 When an impulse arrives at the junction between two neurones, chemicals are released which cross the synapse and arrive at **receptor sites** on the next neurone. This starts up an electrical impulse in the next neurone.

The reflex arc in detail

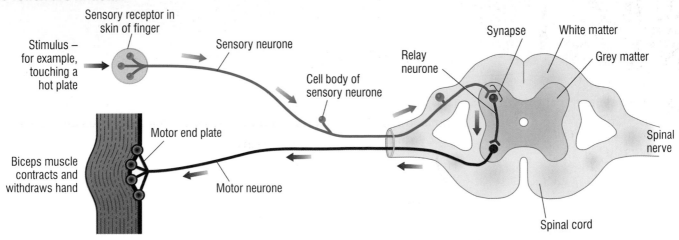

Figure 2 The reflex action which moves your hand away from something hot can save you from a nasty burn!

Look at Figure 2. It shows what would happen if someone touched a hot object.

When they touch it, a receptor in their skin is stimulated. An electrical impulse passes along a sensory neurone to the central nervous system – in this case the spinal cord.

When an impulse from the sensory neurone arrives in the synapse with a relay neurone, a chemical message is released. This crosses the synapse to the relay neurone and sets off an electrical impulse that travels along the relay neurone.

When the impulse reaches the synapse between the relay neurone and a motor neurone returning to the arm, another chemical message is released.

This crosses the synapse and starts an electrical impulse travelling down the motor neurone. When the impulse reaches the organ (effector), it is stimulated to respond. In this example the impulses arrive in the muscles of the arm, causing them to contract and move the hand rapidly away from the source of pain.

Most reflex actions can be shown as follows:

stimulus → receptor → co-ordinator → effector → response

This is not very different from a normal conscious action. However, in a reflex action the co-ordinator is a relay neurone either in the spinal cord or in the unconscious areas of the brain. The whole reflex is very fast indeed.

SUMMARY QUESTIONS

1 Copy and complete using the words below:

**conscious   motor   reflex   relay   response
sensory   stimulus**

In a ...... arc the electrical impulse bypasses the ...... areas of your brain. The time between the ...... and the ...... is as short as possible. Only ...... neurones, ...neurones and ...... neurones are involved.

2 Explain why some actions, such as breathing and swallowing, are reflex actions, while others such as speaking and eating are under your conscious control.

3 Draw a flow chart to explain what happens when you step on a pin. Make sure you include an explanation of how a synapse works.

DID YOU KNOW?

Newborn babies have a number of special reflexes, which disappear as they grow. If something touches the palm of the hand of a newborn baby it will grip on tightly by reflex. In theory the baby would hang from a washing-line! Doctors check for these reflexes to show that a new baby is fit and well.

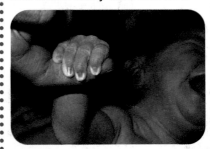

Figure 3 A baby's gripping reflex is very strong

KEY POINTS

1 Some responses to stimuli are automatic and rapid and are called reflex actions.
2 Reflex actions run everyday bodily functions and help you to avoid danger.

B1a 1.3

The menstrual cycle

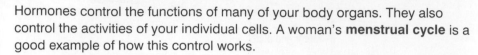

1 How is the menstrual cycle controlled?
2 When is a woman most likely to conceive?

Figure 1 The bodies of young boys and girls work in very similar ways. But once the sex hormones kick in during puberty, some big differences appear in the shape of their bodies and how they work. This shows you the power of the hormones!

Hormones control the functions of many of your body organs. They also control the activities of your individual cells. A woman's **menstrual cycle** is a good example of how this control works.

Hormones made in a woman's brain and in her ovaries control her menstrual cycle. The levels of the different hormones rise and fall in a regular pattern. This affects the way her body works.

What is the menstrual cycle?

The average length of the menstrual cycle is about 28 days. Each month the lining of the womb thickens ready to support a developing baby. At the same time an egg starts maturing in the ovary.

About 14 days after the egg starts maturing it is released from the ovary. This is known as **ovulation**. The lining of the womb stays thick for several days after the egg has been released.

If the egg is fertilised by a sperm, then pregnancy takes place. The lining of the womb provides protection and food for the developing embryo. If the egg is not fertilised, the lining of the womb and the dead egg are shed from the body. This is the monthly bleed or *period*.

All of these changes are brought about by hormones. These are made and released by the **pituitary gland** (a pea sized gland in the brain) and the **ovaries**.

a) What controls the menstrual cycle?
b) Why does the lining of the womb build up each month?

How the menstrual cycle works

Once a month, a surge of hormones from the pituitary gland in the brain starts eggs maturing in the ovaries. The hormones also stimulate the ovaries to produce the female sex hormone *oestrogen*.

● **FSH:** secreted by the pituitary gland. It makes eggs mature in the ovaries. *FSH* also stimulates the ovaries to produce *oestrogen*.

● **Oestrogen:** made and secreted by the ovaries. It stimulates the lining of the womb to build up ready for pregnancy. It also stimulates the pituitary gland to make another hormone known as *LH*.

● **LH:** secreted by the pituitary gland. It stimulates the release of a mature egg from one of the ovaries in the middle of the menstrual cycle.

c) Which hormones are made in the pituitary gland?
d) Which hormone is made by the ovary?

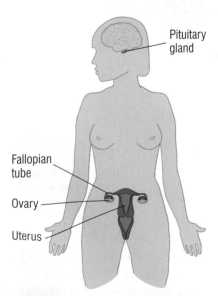

Pituitary gland

Fallopian tube

Ovary

Uterus

Figure 2 Hormones from the pituitary and the ovaries work together to control a woman's fertility

The hormones produced by the pituitary gland and the ovary act together to control what happens in the menstrual cycle. As the oestrogen levels rise they inhibit (slow down) the production of FSH and encourage the production of LH by the pituitary. When LH levels reach a peak in the middle of the cycle, they stimulate the release of a mature egg.

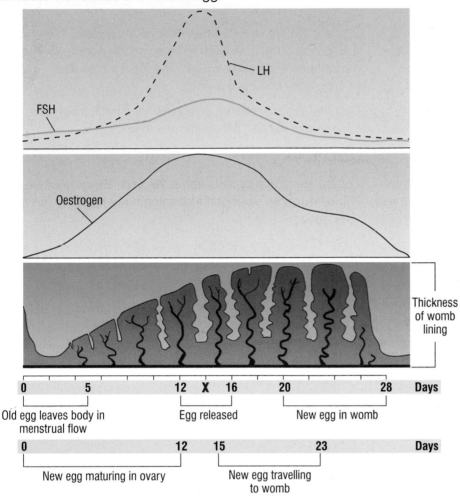

GET IT RIGHT!

Make sure you know the difference between eggs maturing and eggs being released.

Figure 3 The changing levels of the female sex hormones control the different stages of the menstrual cycle

SUMMARY QUESTIONS

1 Copy and complete using the list below:

   28   hormones   FSH   LH   menstrual   oestrogen   ovary

   During the ...... cycle a mature egg is released from the ...... about every ...... days. The cycle is controlled by several ...... including ......, ... and ....

2 Look at Figure 3.

   a) On which day is the woman most likely to get pregnant?
   b) On which days is she having a menstrual period?
   c) On which day is the level of LH highest?
   d) Which hormone controls the build up of the lining of the womb?

3 Produce a leaflet to explain the events of the menstrual cycle to women who are hoping to start a family. You will need to explain the graphs at the top of this page and show when they are most likely to get pregnant.

KEY POINTS

1 Hormones control the release of an egg from the ovary and the build up of the lining of the womb in the menstrual cycle.
2 The main hormones involved are FSH and LH from the pituitary gland and oestrogen from the ovary.

B1a 1.4

The artificial control of fertility

Figure 1 The contraceptive pill contains a mixture of hormones which effectively trick the body into thinking it is already pregnant, so no more eggs are released

For centuries people have tried to control when they have children. They have used substances from camel dung to vinegar to try and stop people having babies. Other people have carved fertility figures, made sacrifices and swallowed horrible herbs to try and have a child.

But it is only in the last fifty years or so that scientists have really been able to help couples control their own fertility, if they choose to do so.

Contraceptive chemicals

In the 21st century it is possible to choose when to have children – and when not to have them. One of the most important and widely used ways of controlling fertility is to use *oral contraceptives* (the *contraceptive pill*).

The pill contains female hormones, particularly oestrogen. The hormones affect your ovaries, preventing the release of any eggs. The pill inhibits (stops) the production of FSH so no eggs mature in the ovaries. Without mature eggs, you can't get pregnant.

Anyone who uses the pill as a contraceptive has to take it very regularly. If they forget to take it, the artificial hormone levels drop. Then their body's own hormones can take over very quickly. This can lead to the unexpected release of an egg – and an unexpected baby!

Fertility treatments

In the UK as many as one couple in six have problems having a family when they want one. There are many reasons for this infertility. It may be linked to a lack of female hormones. Some women want children but simply do not make enough FSH to stimulate the eggs in their ovaries. Fortunately artificial FSH can be used as a fertility drug. It stimulates the eggs in the ovary to mature and also triggers oestrogen production.

Figure 2 Most people who take fertility drugs end up with one or two babies. But the Walton family in the UK had six baby girls who all survived and are now young adults in their own right!

Fertility drugs are also used when a couple is trying to have a baby by IVF (*in vitro* fertilisation). If your fallopian tubes are damaged, eggs cannot reach your womb so you cannot get pregnant naturally.

Fortunately doctors can now help. They remove eggs from the ovary and fertilise them with sperm outside the body. Then they place the tiny developing embryos back into the uterus of the mother, bypassing the faulty tubes.

To produce as many ripe eggs as possible for IVF, the woman is given fertility drugs as part of her treatment. IVF is expensive and not always successful.

Advantages and disadvantages

The use of hormones to control fertility has been a major scientific breakthrough. But like most things there are pros and cons!

In the developed world, using the pill has helped make families much smaller than they used to be. There is less poverty because people have fewer mouths to feed.

The pill has also helped to control population growth in countries such as China, where they find it difficult to feed all their people. In many other countries of the developing world the pill is not available because of a lack of money, education and doctors.

The pill can cause health problems so a doctor always oversees its use.

The use of fertility drugs can also have some health risks for the mother and it can be expensive for society. A large multiple birth can be tragic for the parents if some or all of the babies die. It also costs hospitals a lot of money to keep very small premature babies alive.

Controlling fertility artificially also raises many ethical issues for society and individuals. For example, some religious groups think that preventing conception is denying life and ban the use of the pill.

The mature eggs produced by a woman using fertility drugs may be stored, or fertilised and stored, until she wants to get pregnant later. But what happens if the woman dies, or does not want the eggs or embryos any more?

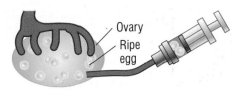
1 Fertility drugs are used to make lots of eggs mature at the same time for collection

2 The eggs are collected and placed in a special solution in a petri dish

3 A sample of semen is collected

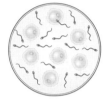
4 The eggs and sperm are mixed in the petri dish

5 The eggs are checked to make sure they have been fertilised and the early embryos are developing properly

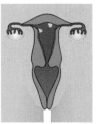
6 When the fertilised eggs have formed tiny balls of cells, 1 or 2 of the tiny embryos are placed in the uterus of the mother. Then, if all goes well, at least one baby will grow and develop successfully.

Figure 3 New reproductive technology using hormones and IVF has helped thousands of infertile couples to have babies

DID YOU KNOW?

In the early days of using fertility drugs there were big problems with the doses used. In 1971 an Italian doctor removed fifteen four-month-old fetuses (ten girls and five boys) from the womb of a 35-year-old woman after treatment with fertility drugs. Not one of them survived.

SUMMARY QUESTIONS

1 Define the following terms: oral contraceptive, fallopian tube, fertility drug, *in vitro* fertilisation.

2 Explain how artificial female hormones can be used to:

   a) prevent unwanted pregnancies,
   b) help people overcome infertility.

3 What, in your opinion, are the main advantages and disadvantages of using artificial hormones to control female fertility?

KEY POINTS

1 Hormones can be used to control fertility.
2 Oral contraceptives contain hormones, which stop FSH production so no eggs mature.
3 FSH can be used as a fertility drug for women, to stimulate eggs to mature in their ovaries. These eggs may be used in IVF treatments.

B1a 1.5

Controlling conditions

LEARNING OBJECTIVES

1 How are conditions inside your body controlled?
2 Why is it so important to control your internal environment?

The conditions inside your body are known as its *internal environment*. Your organs cannot work properly if this keeps changing. Many of the processes which go on inside your body aim to keep everything as constant as possible. This balancing act is called **homeostasis**.

It involves your nervous system, your hormone systems and many of your body organs.

a) Why is homeostasis important?

Controlling water and ions

Water can move in and out of your body cells. How much it moves depends on the concentration of mineral ions (like salt) and the amount of water in your body. If too much water moves into or out of your cells, they can be damaged or destroyed.

You take water and minerals into your body as you eat and drink. You lose water as you breathe out, and in your sweat. You lose salt in your sweat as well. You also lose water and salt in your *urine*, which is made in your *kidneys*.

Your kidneys can change the amount of salt and water lost in your urine, depending on your body conditions. They play an important part in controlling the balance of water and mineral ions in your body. The concentration of the urine produced by your kidneys is controlled by a combination of nerves and hormones.

b) What do your kidneys control?

Figure 1 Running a marathon affects your internal environment

NEXT TIME YOU...

... drink a lot of water all in one go, watch how often you need to go to the toilet afterwards! Your kidneys will remove the extra water from your blood and you will produce lots of pale urine.

PRACTICAL

Helping your body out
When you do a lot of exercise you lose a lot of salt and water from your body as you sweat. It is important to keep your cells hydrated so your body can work properly.

There are lots of special 'sports drinks' you can buy. They claim to rehydrate your body fast, supply you with energy and replace the salt you have lost. Some people think that plain water is just as good! Your kidneys control your internal environment very effectively unless you are exercising really hard for a long time. However, the manufacturers of sports drinks have scientific evidence to back up their claims. You can investigate these claims, and discover just what the drinks contain. See if they help you to perform better!

● How will you carry out your investigation?

Figure 2 A real help in sport – or a good way of making money? Sports drinks are becoming more and more popular, but do most of us really need them?

New reproductive technology using hormones and IVF has made it possible for women in their 50s and 60s to have babies of their own – but is it a good idea?

I married late – I was 40 – and we wanted a family, but my periods stopped when I was 41. Now we have a chance again. I haven't got any eggs so doctors will use FSH as a fertility drug to help them take lots of eggs from my donor (a younger woman). We want this child so much!

We've got three lovely children. I decided to donate some of my eggs to help couples who aren't as lucky as we are. I don't mind the age of the woman who gets my eggs as long as she manages to have a baby and loves it!

I think it is disgraceful and un-natural for women to have babies at this age. We are interfering with nature and with God's will and no good will come of it. The mother might die before the child is an adult!

I can't see anything wrong with older women having babies as long as they are fit and well. I know some people object to it, but some women have babies in their fifties naturally – and lots of men father children in their 60s and even their 70s and no-one objects to that, do they?

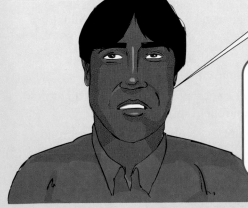

All our evidence shows that infertility treatment is just as successful in older women as it is in younger ones. We have to use artificial hormones to get the womb ready but once the women are pregnant their own hormones take over.

ACTIVITY

There is a lot of debate about the issues explored on these two pages. Use what you have learned in this chapter to help you write a 2–3 minute report for your school radio. It will go out in a regular slot called *Science Issues*. Choose one of these for your report:

- The contraceptive pill – good or bad?
- Older mothers – should science help?

SUMMARY QUESTIONS

1 Match up the following parts of sentences:

a) Many processes in the body	A ...... effector organs.
b) The nervous system allows you	B ...... secreted by glands.
c) The cells which are sensitive to light	C ...... to react to your surroundings and co-ordinate your behaviour.
d) Hormones are chemical substances	D ...... are found in the eyes.
e) Muscles and glands are known as	E ...... are known as nerves.
f) Bundles of neurones	F ...... are controlled by hormones.

2 a) What is the job of your nervous system?

b) Where in your body would you find nervous receptors which respond to:
   i) light?
   ii) sound?
   iii) heat?
   iv) touch?

c) Draw a simple diagram of a reflex arc. Explain carefully how a reflex arc works and why it allows you to respond quickly to danger.

3 a) What is the menstrual cycle?

b) What is the role of:
   i) FSH
   ii) LH
   iii) oestrogen
   in the menstrual cycle?

4 a) Explain carefully the difference between nervous and hormone control of your body.

b) What is a synapse and why are they important in your nervous system?

c) How can hormones be used to control the fertility of a woman?

5 It is very important to keep the conditions inside the body stable.

a) Taking part in school sports on a hot day without a drink bottle for the afternoon would be difficult for your body. Explain how your body would keep the internal environment as stable as possible.

b) Plan an investigation to see whether sports drinks or water are most effective in helping you perform well when you are exercising.

EXAM-STYLE QUESTIONS

1 Oral contraceptives can stop someone becoming pregnant by . . .

A preventing the ovaries from releasing an egg.

B killing eggs that have been released from the ovaries.

C killing sperms before they can reach an egg.

D preventing a fertilised egg from implanting in the uterus lining. (1)

2 A person puts their foot on a sharp object. They automatically lift their foot. The structures involved in this reflex action are shown below:

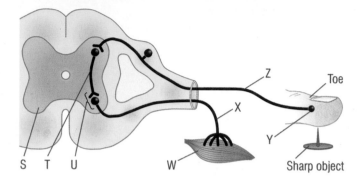

(a) The relay neurone is labelled with the letter

   A T   **B** U
   C X   **D** Z

(b) The motor neurone is labelled with the letter

   A T   **B** X
   C Y   **D** Z

(c) A synapse is labelled with the letter

   A S   **B** U
   C W   **D** Y

(d) The structure labelled W is known as

   A a synapse   **B** a receptor
   C a coordinator   **D** an effector organ

(e) In this reflex action the correct path taken by an impulse is . . .

   A sensory neurone   receptor   coordinator
       motor neurone   effector.

   B effector   coordinator   receptor
       sensory neurone   motor neurone.

   C receptor   sensory neurone   coordinator
       motor neurone   effector.

   D coordinator   receptor   sensory neurone
       effector   motor neurone. (5)

HOW SCIENCE WORKS QUESTIONS

3   The graph below shows the concentrations of three hormones involved in the menstrual cycle:

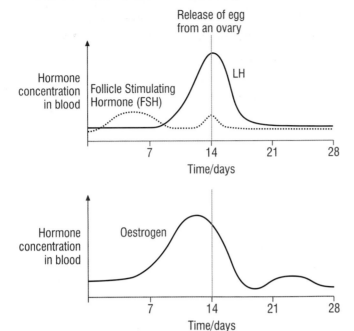

(a) Both FSH and LH are produced in the same gland. What is the name of this gland? (1)

(b) Where in the body is the hormone oestrogen produced? (1)

(c) What is the name given to the release of an egg from the ovary? (1)

(d) The lining of the uterus thickens from around day 5 until some days after the egg has been released. Suggest two purposes for this thickened lining? (2)

(e) Use the information in the graph as well as your knowledge to explain how the concentration of oestrogen affects and controls the release of an egg during the menstrual cycle. (4)

4   Some women are unable to have children naturally. The hormone FSH can sometimes be used to help these women have children.

(a) (i)   Harriet does not produce eggs at ovulation. Explain how FSH could be the cause of the problem. (1)

(ii) Explain how FSH could help Harriet to produce and release an egg. (4)

(b) Sharon has had an infection of her Fallopian tubes that has left them blocked. Although she still produces eggs, they are unable to pass down the Fallopian tubes. Describe a method by which Sharon and her partner John could still have children. Include in your account the role of FSH. (4)

The girls in the class set a challenge to the boys. The girls suggested that they had much better control over their nervous reactions than the boys did. The boys accepted the challenge and agreed to the investigation.

The equipment was set up as you can see in this picture.

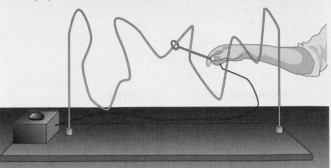

Five girls and five boys took it in turns to move the metal ring along the wire. If anyone touched the ring onto the wire the circuit would be completed and the bell would ring. The teacher counted the number of times the bell rang for each student.

The results are in this table:

Girls' names	Number of touches	Boys' names	Number of touches
Alexandra	6	Arthur	7
Farzana	0	Barnaby	2
Kerry	4	Zahir	4
Summer	3	Jameel	1
Annabel	8	Terry	5

a) In this investigation, which was the dependent variable? (1)

b) Suggest one variable that had not been controlled. (1)

c) Why did the group decide to use the teacher to record the results? (1)

d) Calculate the average for:
i) the boys   ii) the girls. (2)

e) Which of the following words would you use to describe the independent variable?
i) continuous   ii) categoric
iii) ordered   iv) discrete. (1)

f) How might you present these results?
i) bar chart   ii) line graph
iii) scattergraph   iv) use a line of best fit. (1)

g) Do you think that the girls' prediction is supported by the data collected? Explain your answer. (2)

B1a 2.1

Diet and exercise

Figure 1 Everyone needs a source of energy to survive – and your energy source is your food. Whatever food you eat – whether you prefer sushi, dahl, or roast chicken – most people eat a varied diet that includes everything you need to keep your body healthy.

What makes a healthy diet?

A healthy diet contains:

- carbohydrates,
- proteins,
- fats,
- vitamins,
- minerals,
- fibre and
- water

and the energy you need to live, all in the right amounts!

If your diet isn't balanced, you will end up ***malnourished***. If you don't take in enough vitamins and minerals, you will end up with deficiency diseases like scurvy. (Scurvy is caused by a lack of vitamin C.)

Fortunately, in countries like the UK, most of us take in all the minerals and vitamins we need from the food we eat. However, our diet can easily be less well balanced in terms of the energy we take in. If we take in too much energy we get fat – but if we don't eat enough we get too thin.

It isn't always easy to get it right because different people need different amounts of energy.

a) Why do you need to eat food?

How much energy do you need?

The amount of energy you need to live depends on lots of different things. Some of these things you can change and some you can't.

If you are male, you will need to take in more energy than a female of the same age – unless she is pregnant.

If you are a teenager, you will need more energy than if you are in your 70s – and there isn't much you can do about it!

b) Why does a pregnant woman need more energy than a woman who isn't pregnant?

The amount of exercise you do affects the amount of energy you use up. If you do very little exercise, then you don't need much food. The more you exercise, the more food you need to take in. Your food supplies energy to your muscles as they work.

People who exercise regularly are usually much fitter than people who take little exercise. They make bigger muscles – and muscle tissue burns up much more energy than fat. But exercise doesn't always mean time spent training or 'working out' in the gym. Walking to school, running around the house and garden looking after small children or having a physically active job all count as exercise too.

c) Why do athletes need to eat more food than the average person?

Figure 2 Athletes who spend a lot of time training and playing a sport will have a great deal of muscle tissue on their bodies – up to 40% of their body mass. So they have to eat a lot of food to supply the energy they need.

The temperature where you live affects your energy needs as well. The warmer it is, the less energy you need. This is because you have to use less energy keeping your body temperature at a steady level. So you need to take in less food!

Figure 3 If you live somewhere really cold, you need lots of high-energy fats in your diet. You need the energy to keep warm!

The metabolic rate

Imagine two people who are very similar in age, sex and size. However, they may still need quite different amounts of energy in their diet. This is because the rate at which all the chemical reactions in the cells of the body take place (the **metabolic rate**) varies from person to person.

The proportion of muscle to fat in your body affects your metabolic rate. Men generally have a higher proportion of muscle to fat than women, so they have a higher metabolic rate. You can change the proportion of muscle to fat in your body by exercising and building up more muscle.

Your metabolic rate is also affected by the amount of activity you do. Exercise increases your metabolic rate for a time even after you stop exercising.

Finally, scientists think that your basic metabolic rate may be affected by factors you inherit from your parents.

GET IT RIGHT!

Metabolic rate is not the same as heart rate or breathing rate – make sure you know the difference.

SUMMARY QUESTIONS

1 What do we mean by 'a balanced diet'?

2 a) Why does an old person need less energy in their diet than a teenager?
   b) Why does a top footballer need more energy in their diet than you do? Where does the energy in the diet come from?

3 a) What is meant by the 'metabolic rate'?
   b) Explain why some people put on weight more easily than others.

KEY POINTS

1 Most people eat a varied diet, which includes everything needed to keep the body healthy.
2 Different people need different amounts of energy.
3 The metabolic rate varies from person to person.
4 The less exercise you take, the less food you need.

B1a 2.2 Weight problems

Human beings come in all sorts of shapes and sizes. Most people look about right but there will always be extremes. Some people are very overweight and others appear unnaturally thin. Scientists and doctors don't just measure what you weigh. They look at your *body/mass index* or *BMI*. This compares your body mass with your height in a simple formula:

$$BMI = \frac{body\ mass\ in\ kg}{(height\ in\ metres)^2}$$

Most people have a BMI in the range 20–30. But if you have a BMI of below 18.5, or above 35, then you may have some real health problems.

a) What does your body/mass index measure?

Obesity

If you take in more energy than you use, the excess is stored as fat. You need some body fat to cushion your internal organs. Your fat also acts as an energy store for when you don't feel like eating. But if someone eats a lot more food than they need, over a long period of time, they could end up *obese*.

Carrying too much weight is often inconvenient and uncomfortable. Far worse, it can lead to serious health problems. Obese people are more likely to suffer from **arthritis** (worn joints), **diabetes** (high blood sugar levels which are hard to control), *high blood pressure* and *heart disease*. They are more likely to die young than slimmer people.

b) What health problems are linked to obesity?

Losing weight

Many people want to lose weight. This might be for their health or just to look better. You gain fat by taking in more energy than you need, so there are three main ways you can lose it.

- You can reduce the amount of energy you take in by cutting back the amount of food you eat – particularly energy-rich foods like biscuits, crisps and chips.
- You can increase the amount of energy you use up by taking more exercise.
- And the best way to lose weight is to do both – reduce your energy intake and exercise more!

Many people find it easier to lose weight by attending slimming groups. At these weekly meetings they get lots of advice, plus support from other slimmers. All the different slimming programmes involve eating less food and/or taking more exercise!

Increasing your exercise levels can be an important part of losing weight and getting fitter. However, you need to take care. If you suddenly start working out hard in the gym, or taking other vigorous exercise, you can cause other health problems.

Figure 1 In spite of some of the media hype, most people are not obese – but the amount of weight people carry certainly varies a great deal!

Different slimming programmes approach weight loss in different ways. Many simply give advice on healthy living. They advise lots of fruit and vegetables, not too much fat or too many calories and plenty of exercise. Some are more extreme suggesting you cut out almost all of the fat or the carbohydrates from your diet.

Others claim that 'slimming teas' or 'herbal pills' will enable you to eat what you like and still lose weight. What sort of evidence would you look for to decide which approaches worked best?

c) What must you do to lose weight?

Starvation

In some parts of the world obesity is rare, because the biggest problem is lack of food. Civil wars, droughts and pests can destroy local crops so people cannot get enough to eat. Starvation leads to a number of symptoms including:

● You become very thin and your muscles waste away.

● Your immune system can't work properly so you pick up infections.

● If you are female, your periods will become irregular or stop altogether.

These symptoms are also sometimes seen in the developed world in people suffering from the mental disorder called **anorexia** (loss of appetite) **nervosa**.

d) What are the main symptoms of starvation?

Figure 2 Hundreds of thousands of people around the world suffer the symptoms of malnutrition and starvation. There is simply not enough food for them to eat.

GET IT RIGHT!

Make sure you can give specific examples of the problems caused by obesity and starvation.

SUMMARY QUESTIONS

1 Copy and complete using the words below:

energy   fat   less   more   obese

If you take in more ...... than you use, the excess is stored as .... If you eat too much over a long period of time, you will eventually become .... To lose weight you need to eat ...... and exercise ....

2 Plan a simple information sheet about the dangers of being overweight and how to lose weight sensibly.

3 Research the claims of two slimming programmes. Compare and evaluate the claims they make.

KEY POINTS

1 If you take in more energy than you use, you will store the excess as fat.

2 Obese people have more health problems than people of normal weight.

3 People who do not have enough to eat can develop serious health problems.

B1a 2.3

Fast food

LEARNING OBJECTIVES

1 What is cholesterol?
2 Why do your cholesterol levels matter?
3 Is too much salt bad for us?

Figure 1 Fast food tastes good. But it has had lots of things added to make it easy to cook and eat. These often include fat and salt.

People eat fast processed food because it is quick and easy and fits in with their busy lives. But it often contains a lot of fat and salt. These make the food taste good. However, there are some real concerns about the effect that too much fat and salt in your diet can have on your health.

a) What substances do you often find in fast foods?

Cholesterol

Fat is an energy-rich food. So too much fat in your diet can easily make you overweight. But that isn't the only problem with fatty food. The amount and type of fat you eat also seems to affect the levels of **cholesterol** in your blood.

Cholesterol is a substance which you make in your liver. It gets carried around your body in your blood. You need it to make the membranes of your body cells, your sex hormones and the hormones that help your body deal with stress. Without cholesterol, you wouldn't survive. Yet people often talk about cholesterol as if it is a bad thing. Why?

High levels of cholesterol in your blood seem to increase your risk of getting heart disease or diseased blood vessels. The cholesterol builds up in your blood vessels and can even block them. Heart disease is one of the main causes of death in the UK and USA, so no wonder doctors are worried.

b) Why do you need cholesterol in your body?

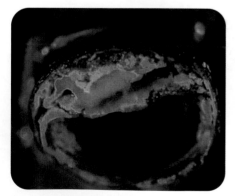

Figure 2 When you get cholesterol building up in the wrong place – like the arteries leading to your heart – it can be very serious indeed

Controlling cholesterol

The amount of cholesterol you have in your blood depends on two things:

- The way your liver works, which is something you inherit from your parents and cannot change.
- The amount of fat in your diet.

Some people have livers that can deal with almost any amount of fat. Their blood cholesterol seems to stay within healthy levels. But for many people the level of cholesterol in their blood is linked to the amount and type of fat they eat.

It isn't just the overall level of cholesterol in your blood which affects your risk of developing heart disease. Cholesterol is carried around your body by two types of *lipoproteins*:

- **Low-density lipoproteins (LDLs)** are known as 'bad' cholesterol. Raised levels of LDLs increase your risk of heart problems.
- **High-density lipoproteins (HDLs)** are known as 'good' cholesterol and they reduce your risk of heart disease.

The balance of LDLs and HDLs in your blood is very important for a healthy heart.

There are three main types of fats in the food you eat and they seem to have different effects on your cholesterol:

- **Saturated fats** increase (raise) blood cholesterol levels. You find them in animal fats like meat, butter and cheese.
- **Mono-unsaturated fats** seem to have two useful effects. They may reduce your overall blood cholesterol levels and improve the balance between LDLs and HDLs in your blood. You find them in foods like olive oil, olives, peanuts and lots of margarines.
- **Polyunsaturated fats** seem to be even better at reducing your blood cholesterol levels and balancing LDLs and HDLs than mono-unsaturates. You find them in foods including corn oil, sunflower oil, many margarines and oily fish.

c) What is the big difference between saturated fats and the other types of fats?
d) Why are raised blood cholesterol levels a worry?

What about salt?

Like fat, salt is vital in your diet. Without it, your nervous system would not work and the chemistry of all your cells would be in chaos. But for about a third of you (30% of the population), too much salt in your diet can lead to high blood pressure. This can damage your heart and kidneys and increase your risk of a stroke.

Many people eat too much salt each day without knowing it. That's because many processed, 'fast' foods contain large amounts of salt. But you can control your salt intake by doing your own cooking – or by reading the labels very carefully when you buy ready-made food!

Oil/fat	% saturated fat	% polyunsaturated fat	% mono-unsaturated fat
Butter	66	4	30
Corn oil	13	62	25
Olive oil	14	12	74
Sunflower oil	11	69	20

Figure 3 Once you start to look at the different fats in the food you are buying, shopping can get very complicated!

SUMMARY QUESTIONS

1 Copy and complete using the words below:

salt   heart   salt   blood pressure   fat   cholesterol

Fast food can contain too much …… and ……. . Raised …… in the blood can lead to …… disease, while too much …… can give some people high …… …… .

2 Look at Figure 3 and use it to help you answer these questions:

a) Which fat or oil has the highest percentage of mono-unsaturates?
b) Which fat or oil has the highest percentage of polyunsaturates?
c) Which fat or oil has the highest percentage of saturated fats?
d) Decide which of these fats or oils would be the best to use for a healthy heart, and which would be the worst. Explain your answer carefully, including the balance of LDLs and HDLs in your blood.

3 Many people want to lower the amount of salt in fast foods.

a) Explain why salt is important in your diet.
b) Why are people worried about high salt levels in food?
c) Would lowering the salt levels in processed foods make everyone healthier? Explain your answer.

KEY POINTS

1 Fast food often contains high proportions of fat and/or salt.
2 Cholesterol is made in the liver and found in the blood. High cholesterol levels have been linked to heart disease.
3 The level and type of cholesterol in your blood is influenced by the type of fat you eat.
4 Too much salt in the diet can lead to raised blood pressure in about a third of the population.

B1a 2.4   Health issues

The Statin Revolution

Doctors have an amazing new weapon against high cholesterol levels and the problems they can bring. They can use a group of drugs called **statins**. Statins stop the liver producing so much cholesterol. Patients need to keep to a relatively low fat diet as well for the best effects.

Here are some different opinions about these exciting new drugs:

Some people just can't get their cholesterol balance right by changing their diet. It doesn't matter how hard they try. I've been very pleased with the results using statins. Almost all my patients have now got healthy cholesterol levels. What's more, we have lost far fewer people to strokes and heart attacks since we started using the drugs.

We are delighted with the results we are getting with statins. We have got data from several really large, powerful research trials involving over 30,000 patients. The trials all show similar results. Using a statin drug can lower your chances of having a heart attack or stroke by 25 to 40% – and we didn't find too many side effects.

The great thing about these new statins that the doctor's given me is that they control my cholesterol for me. It's back to the cream cakes and chips for me – and I won't have to worry about my heart!

I'm so pleased with my new medicine – the pills have brought my cholesterol levels right down and I'm feeling really well

I'm very worried about possible side effects with these new tablets – the leaflet said they can cause liver damage. I know my cholesterol levels were very high without the tablets, but I think I'm going to stop taking them. I don't want my liver to rot!

Scientists wear blinkers – and we pay the price!

For many years now scientists and doctors have been telling us that we are at risk from heart disease because we eat too much animal fat and our blood cholesterol is too high. But a lot of people still die of heart disease. Now it seems that vitamins might be just as important to our hearts as fat. What's more, this idea was first discovered years ago – so why didn't we find out sooner?

Thirty years ago, Kilmer McCully was a young researcher at Harvard University in the USA. He discovered a possible link between an amino acid called homocysteine and changes in the blood supply to the heart which can lead to heart attacks. High levels of homocysteine are linked to low levels of B vitamins in the diet – and these B vitamins are often missing in processed foods!

Changing your diet or taking a cheap supplement of B vitamins lets your body remove the homocysteine and prevents the damage to your heart.

Unfortunately McCully did his research at the same time as many top scientists were supporting the link between fats and heart disease. Time and money had been spent developing anti-cholesterol drugs and low-fat foods. No-one wanted to hear about McCully's cheap and simple solution. He lost his funding at Harvard and his ideas were quashed.

Thirty years on – and in spite of the fact that we have all cut back on our fat levels and taken our anti-cholesterol medicines, deaths from heart disease are still high. Kilmer McCully's work is finally being taken seriously. Major trials on B vitamins and heart disease are taking place around the world. It seems increasingly likely that McCully really has found one of the pieces in the jigsaw which explains heart disease. It is just a pity that no-one would use it for so long! Perhaps scientists need to take the blinkers off and realise that there can be more than one solution to a problem!

ACTIVITY

Write a letter:
Either from the young Kilmer McCully to a friend explaining what you have discovered about a link between B vitamins and heart disease and what it might mean for patients;

Or from a senior scientist who has been working on treatments for high cholesterol and heart disease to one of his colleagues about McCully's work and how you feel about it.

Menu 1

Turkey twizzlers

Chicken nuggets

Pizza

Chips

Spaghetti hoops

Iced bun

Doughnut

Menu 2

Char-grilled chicken

Spaghetti Bolognese

Fish with pesto topping

Baked potato

Fresh fruit

Yoghurt

ACTIVITY

Plan an assembly for the year 7 pupils in your school on the importance of a healthy diet. It should include help with the food they should choose in the school canteen for lunch.

SUMMARY QUESTIONS

1 a) Define the following:
   i) Balanced diet.
   ii) Metabolic rate.

   b) A top athlete needs to eat a lot of food each day. This includes protein and carbohydrate. Explain how they can eat so much without putting on weight.

2 a) What is obesity?

   b) Why is obesity a threat to your health?

   c) Suggest some ways in which an obese person might lose weight.

   d) How do following a slimming diet and suffering from starvation differ?

3 a) What is the link between a high-fat diet, raised blood cholesterol and the LDL/HDL balance in your blood?

   b) Why are doctors concerned if a patient has raised cholesterol levels or a high ratio of LDLs to HDLs?

   c) Fast food is often linked to an unhealthy lifestyle. Explain the problems with fast foods – and why people still eat them.

   d) Recently there has been a lot of media interest in school dinners. People think they contain far too many 'fast foods' and not enough fresh produce, fruit and vegetables.

   Plan a short report for your local radio station on why healthy school meals are important for the future health of the children who eat them.

4 Here are two young people who have written to a lifestyle magazine problem page for advice about their diet and lifestyle. Produce an 'answer page' for the next edition of the magazine.

   a) Melanie: *I'm 16 and I worry about my weight a lot. I'm not really overweight but I want to be thinner. I've tried to diet but I just feel so tired when I do – and then I buy chocolate bars on the way home from school when my friends can't see me! What can I do?*

   b) Jaz: *I'm nearly 17 and I've grown so fast in the last year that I look like a stick! So my clothes look pretty silly. I'm also really good at football, but I don't seem as strong as I was and my legs get really tired by the end of a match. I want to build up a bit more muscle and stamina – but I don't just want to eat so much I end up getting really heavy. What can I do about it?*

EXAM-STYLE QUESTIONS

1 The table is about some conditions that affect the body as a result of certain diets.
Match descriptions **A**, **B**, **C** and **D** with the words 1 to 4 in the table.

A an imbalance of nutrients in the diet

B a severe shortage of food in the diet

C a psychological disorder leading to a dangerously low body mass

D a body/mass index above 30   (4)

	Condition
1	Obesity
2	Anorexia
3	Malnutrition
4	Starvation

2 The body/mass index (BMI) compares body mass to height. The BMI is calculated using the following formula:

$$\text{BMI} = \frac{\text{body mass in kg}}{(\text{height in metres})^2}$$

(a) A woman has a body mass of 60 kg and a height of 1.6 metres. Her BMI is equal to . . .

   A 18.7   **B** 23.4   **C** 29.2   **D** 37.5   (1)

The graph shows the percentage of a population of people in different BMI groups who suffer from a form of arthritis.

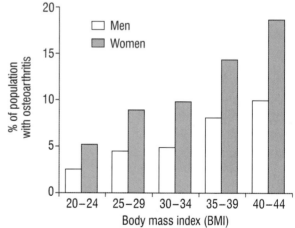

(b) From these data, which group of people have the highest percentage of osteoarthritis?

   A Men with a BMI of 30–34.

   B Men with a BMI of 35–39.

   C Women with a BMI of 35–39.

   D Women with a BMI of 40–44.   (1)

(c) The data were collected by measuring the BMI of a large sample of people. What is the main advantage of using a large sample of people?

   A The data obtained are more reliable.

   B The mean BMI can be calculated.

   C The results obtained are fairer.

   D People of many ages are included.   (1)

(d) What conclusion can be drawn from the data?

A Women with a BMI of 35–39 are the group most likely to suffer from osteoarthritis.

B The higher the BMI in both men and women the greater the risk of suffering from osteoarthritis.

C The lower the BMI in men, the more likely they are to suffer from osteoarthritis.

D Slimming will prevent osteoarthritis. (1)

3 (a) State the seven components that make up a healthy diet. (2)

An investigation was carried out over four different periods to find the energy intake of 14- and 15-year-old girls and boys. The results are shown in the table below.

Period of time	Average energy intake/kJ per day	
	Boys	Girls
1930s	12 873	11 088
1960s	11 739	9 534
1970s	10 962	8 484
1980s	10 478	8 316

(b) Calculate the percentage decrease in energy intake for girls between the 1930s and the 1980s. Show your working. (1)

(c) Explain why the intake of energy for both boys and girls decreased between the 1930s and the 1980s. (2)

(d) If the same study had been carried out with groups of 70-year-olds, how might the results have been different? (1)

(e) Suggest a reason why girls need to take in less energy than boys of the same age. (2)

(f) What would be the best way to display the data from the table above? Explain your choice. (2)

(g) Calculate the mean of the average energy intake of boys from the 1930s to the 1980s. (1)

(h) Who has the larger range of average energy intake between the 1930s and 1980s – boys or girls? Show your working out. (1)

4 What a person eats can affect their health. Explain how doing each of the following might help to keep a person healthy:

(a) Reducing the amount of saturated fat that is eaten. (3)

(b) Eating less salt. (2)

HOW SCIENCE WORKS QUESTIONS

A class of students were asked to test some fruit juices for their vitamin C content. They were given the apparatus set up as in the diagram below. They had to put the sample of fruit juice into the test tube and add the dye (DCPIP) from the burette drop by drop until the mixture retained the blue colour of the dye.

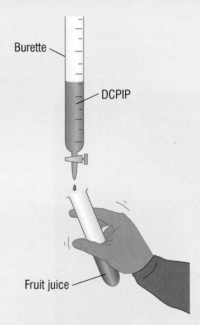

Burette

DCPIP

Fruit juice

a) Name a control variable they would have to use. (1)

b) Describe how they should use the burette to get accurate results. (5)

The class results were as follows:

Juice	Volume of DCPIP added/cm³				
Orange	21.1	26.2	24.8	25.5	25.7
Apple	20.9	19.7	21.3	20.5	21.0

c) Calculate the mean for the amount of DCPIP added by the class to the apple juice. (1)

d) It was suggested that the first result for the orange juice was an anomaly.
Why was this thought to be an anomaly? (1)

e) Calculate the average for the amount of DCPIP added to the orange juice. (1)

f) What conclusion can you make from these results? (1)

g) One student commented that when carrying out the titration with the orange juice it was quite difficult to tell when the DCPIP stayed blue. She thought this was due to the colour of the orange juice. How does this idea affect your conclusion? (2)

B1a 3.1

Drugs

Figure 1 Millions of pounds worth of illegal drugs are brought into the UK every year. It is a constant battle for the police to find and destroy drugs like these.

A drug is a substance that alters the way in which your body works. It can affect your mind, your body or both. In every society there are certain drugs which people use for medicine, and other drugs which they use for pleasure.

Many of the drugs that are used both for medicine and for pleasure come originally from natural substances, often plants. Many of them have been known to and used by indigenous peoples for many years. Usually some of the drugs that are used for pleasure are socially acceptable, while others are illegal.

a) What do we mean by 'indigenous peoples'?

Drugs are everywhere in our society. People drink coffee and tea, smoke cigarettes and have a beer, an alcopop or a glass of wine. They think nothing of it. Yet all of these things contain drugs – caffeine, nicotine and alcohol (the chemical ethanol). These drugs are all legal.

Other drugs, such as cocaine, ecstasy and heroin are illegal. Which drugs are legal and which are not varies from country to country. Alcohol is legal in the UK as long as you are over 18, but it is illegal in many Arab states. Heroin is illegal almost everywhere.

b) Give an example of one drug which is legal and one which is illegal in the UK.

Because drugs affect the chemistry of your body, they can cause great harm. This is even true of drugs we use as medicines. However, because medical drugs make you better, it is usually worth taking the risk.

But legal drugs, such as alcohol and tobacco, and illegal substances, such as solvents, cannabis and cocaine, can cause terrible damage to your body. Yet they offer no long-term benefits to you at all.

What is addiction?

Some drugs change the chemical processes in your body so that you may become addicted to them. You can become dependent on them. If you are addicted to a drug, you cannot manage properly without it.

Once addicted, you generally need more and more of the drug to keep you feeling normal. When addicts try to stop using drugs they usually feel very unwell. They often have aches and pains, sweating, shaking, headaches and cravings for their drug. We call these **withdrawal symptoms**.

c) What do we mean by 'addiction'?

The problems of drug abuse

People take drugs for a reason. Drugs can make you feel very good about yourself. They can make you feel happy and they can make you feel as if your problems no longer matter. Unfortunately, because most recreational drugs are addictive, they can soon become a problem themselves.

No drugs are without a risk. Cannabis is often thought of as a relatively 'soft' – and therefore safe – drug. But evidence is growing which shows that it can cause serious psychological problems to develop in some people.

Hard drugs, such as cocaine and heroin, are extremely addictive. Using them often leads to very severe health problems. Some of these come from the drugs themselves, and some come from the lifestyle which often goes with drugs. Because they are illegal, they are expensive. Young people often end up turning to crime to pay for their drug habit. They don't eat properly or look after themselves. They can also end up with serious illnesses, such as hepatitis, STDs and HIV/AIDS.

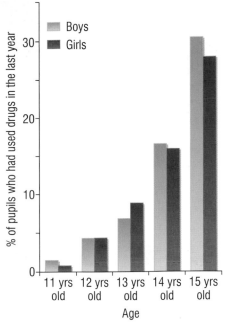

Figure 2 Most of the young people who have used drugs have smoked cannabis – but the number of 15-year-old pupils who have tried drugs is causing a lot of concern

DON'T LET DRUG DEALERS CHANGE THE FACE OF YOUR NEIGHBOURHOOD.
Call Crimestoppers anonymously on 0800 555 111.

Figure 3 Drugs can seem appealing, exciting and fun when you first take them. Many people use them for a while and then leave them behind. But the risks of addiction are high, and no-one can predict who drugs will affect most.

SUMMARY QUESTIONS

1 Copy and complete using the words below:

mind   cocaine   ecstasy   legal   alcohol   drug   body

A ...... alters the way in which your body works. It can affect the ......the ...... or both. Some drugs are ...... e.g. caffeine and .... Other drugs, such as ......, ... and heroin are illegal.

2 a) Why do people often need more and more of a drug?
   b) What happens if you stop taking a drug when you are addicted to it?

3 a) Look at Figure 2. Explain what this tells you about drug-taking in young people.
   b) Why do people take drugs?
   c) Explain some of the problems linked with using cannabis, cocaine and heroin.
   d) Why do you think young people continue to take these drugs when they are well aware of the dangers?
   e) Why does the impact of drug use vary from person to person?

KEY POINTS

1 Drugs change the chemical processes in your body, so you may become addicted to them.
2 Alcohol, tobacco and illegal drugs may harm your body.
3 Smoking cannabis may cause psychological problems.
4 Hard drugs, such as cocaine and heroin, are very addictive and can cause serious health problems.

B1a 3.2

Legal and illegal drugs

Figure 1 Even everyday drugs, like the caffeine we take for granted, have an effect on your nervous system and brain

What is the most widely used drug in the world? It is probably one that most of you will have used at least once today, yet no-one really thinks about. The caffeine in your cup of tea, mug of coffee or can of cola is a drug.

Many people find it hard to get going in the morning without their first mug of coffee – they are probably addicted to the drug! Caffeine is a mild stimulant. It stimulates your brain and increases your heart rate and blood pressure.

a) What makes a can of cola a drug?

How do drugs affect you?

Many of the drugs used for medical treatments have little or no effect on your nervous system. However, all of the drugs which people use for pleasure (see table on page 51) affect the way your brain and nervous system work. It is these changes which people enjoy when they use the drugs. The same changes can cause addiction, so your body doesn't work properly without the drug.

Some drugs like caffeine, nicotine and cocaine speed up the activity of your brain. They make you feel more alert and energetic.

Others like alcohol and cannabis slow down the responses of your brain, making you feel calm and better able to cope. Heroin actually stops impulses travelling in your nervous system, so you don't feel any pain or discomfort. Other drugs like cannabis produce vivid waking dreams. You see or hear things which are not really there.

Why do people use drugs?

People use these drugs for a variety of reasons. For example, people feel that caffeine, nicotine and alcohol help them cope with everyday life. Few people who use these drugs would think of themselves as addicts, yet the chemicals can have a big impact on your brain (see Figure 2).

As for the other recreational drugs – people who try them may be looking for excitement or escape. They might want to be part of the crowd or just want to see what happens. Yet because many of these drugs are addictive, you don't have to try them many times before your body starts to demand a regular fix!

Figure 2 NASA scientists have shown that common house spiders spin their webs very differently when given some of the commonly used drugs shown in the table on page 51. The effect of caffeine on the nervous system of a spider is particularly dramatic!

Marijuana   Benzedrine   Caffeine   Chloral hydrate

Some of these recreational drugs are more harmful than others. Most media reports on the dangers of drugs use focus on illegal drugs (see table below). But in fact the impact of legal drugs on health is much greater than the impact of illegal drugs. That's because far more people take them. Millions of people in the UK smoke – but only a few thousand take heroin.

b) Why do legal drugs cause many more health problems than illegal drugs?

Legal recreational drugs	Illegal recreational drugs
Ethanol (alcoholic drinks)	cannabis
Nicotine (cigarette smoke)	cocaine
Caffeine (coffee, tea, cola etc)	heroin
	ecstasy
	LSD

Drugs in sport

The world of sport has a major problem with the illegal use of drugs. In theory competition in sport is to find the best natural athlete. The only difference between the competitors should be their natural ability and the amount they train. However, there are many drugs which can enhance your performance in sport – and sadly some athletes use them and cheat. Drugs can build up your muscle mass, make you body produce more blood, make you more alert and speed up your reactions.

The sports authorities produce new tests for drugs and run random drugs tests to try and identify the cheats. Athletes are banned from competing if they are discovered using illegal drugs. But competitors are always looking for new ways to get ahead, so the illegal use of drugs in sport continues.

Figure 3 At the 2000 Summer Olympic Games in Sydney, Australia, the Romanian gymnast Andrea Raducan won a gold medal. It was taken away when she tested positive for a banned stimulant.

SCIENCE @ WORK

Many people are involved in developing new drugs tests, carrying out random tests on athletes and analysing the results. You need science qualifications to be involved!

FOUL FACTS

There are drugs that can be given to horses and dogs to make them perform better – or worse – than expected. Some people use these to make a lot of money through betting on a result they already know!

SUMMARY QUESTIONS

1 Copy and complete using the words below:

brain   health   illegal   legal   recreation

Drugs which people use for …… all affect the …… and nervous system. Some of these drugs are legal but some of them are …… More people suffer …… problems caused by the …… drugs than illegal ones.

2 a) What are the main reasons for using illegal drugs?
  b) Plan a TV advert against the use of illegal substances in sport to be shown in the run-up to the 2012 Olympics in the UK.

3 Compare the impact of legal and illegal drugs on individuals and on society.

KEY POINTS

1 Many recreational drugs affect the brain and nervous system.
2 Some recreational drugs are legal and others are illegal.
3 The overall impact of legal drugs on health is much greater than illegal drugs because more people use them.

B1a 3.3   Alcohol – the acceptable drug?

Figure 1 You can buy alcoholic drinks like these legally in the UK once you are 18. But you have to be over 21 in many places in the USA and they are completely illegal in some other countries.

For many people **alcohol** is part of their social life. They like to share a drink with friends or enjoy a glass of wine with a meal. They probably don't think of themselves as drug users.

In small amounts, alcohol makes people feel relaxed and cheerful. It makes you less inhibited. So shy people can feel more confident when they've had an alcoholic drink.

But alcohol has a powerful effect on your body. It is very addictive and it is also very poisonous. Just imagine if alcohol was discovered today. It would almost certainly be illegal and thought of as a very dangerous drug.

In fact, alcohol is one of the most widely used drugs in the UK. Although some religions ban the use of alcohol, it is accepted all over the world. Perhaps this is because alcohol has been around for thousands of years. We also see that many important and famous people like a drink!

a) Why is alcohol described as a drug?

How does alcohol affect your body?

Alcohol is poisonous. However, your liver can usually break it down. Your liver gets rid of the alcohol before it causes permanent damage and death.

When you have an alcoholic drink, the alcohol passes through the wall of your gut and goes into your bloodstream. From your blood, the alcohol passes easily into nearly every tissue of your body. It gets into your nervous system and brain. This slows down your reactions. It can make you lose your self-control.

When you have had too much to drink, you lack judgement. You can end up making stupid or dangerous decisions. Some people end up making mistakes they regret for the rest of their lives.

If you drink large amounts of alcohol, like a whole bottle of spirits, your liver simply cannot cope. You would suffer from alcohol poisoning. This can quickly lead to unconsciousness, coma and death.

b) Give an example of a poor decision that someone under the influence of alcohol might make.

Some people drink heavily for many years, becoming **alcoholics**. They are addicted to the drug. Their liver and brain suffer long-term damage and eventually the drink may kill them.

They may develop **cirrhosis of the liver**. This disease destroys your liver tissue. They can also get *liver cancer*, which spreads quickly and can be fatal. In some heavy drinkers their brain is so damaged (it becomes soft and pulpy) that it can't work any longer. This causes death.

Short bouts of very heavy drinking can cause the same symptoms to develop quite quickly.

c) Which organs are most affected by heavy drinking?

DID YOU KNOW?

It takes your liver about one hour to break down the alcohol in a glass of wine or half a pint of beer (one unit of alcohol).

The effects of drinking on society

Alcohol can also put you at risk because of the way you behave under the influence of the drug. Because alcohol slows down your reactions, you are much more likely to have an accident. This is very dangerous if you drive after drinking. Alcohol is a factor in about 20% of all fatal road accidents in the UK.

Alcohol abuse affects personal lives as well. Domestic violence is often linked to patterns of heavy drinking. Many crimes take place when people are under the influence of alcohol, often mixed with other drugs.

Binge drinking is a recent problem. This often involves young people. They go out and get very drunk several nights a week. They become violent and abusive, damage property and put their own health at risk.

Alcohol related crime in the UK costs us around 20 billion pounds a year and causes great unhappiness. Add to this the medical costs of alcohol abuse and you see that we all pay a high price for this socially acceptable drug!

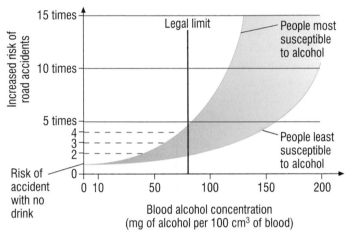

Figure 3 Alcohol affects your brain and your reactions – it is not surprising that if you drink alcohol and then drive a car you are much more likely to have an accident

A healthy liver

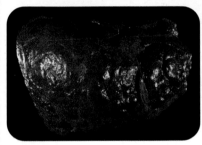

Diseased liver from a heavy drinker with cirrhosis

Figure 2 Your liver deals with all the poisons you put into your body. But if you drink too much alcohol, your liver may not be able to cope. The difference between the healthy liver and the liver with cirrhosis shows just why people are warned against heavy drinking!

SUMMARY QUESTIONS

1 Copy and complete using the words below:

drug   alcohol   brain   alcoholics   liver

…… is a poisonous …… . It is broken down in your …… . It can have a big effect on your …… and your liver.
…… are people who are addicted to alcohol.

2 a) How does alcohol reach your brain after you have had a drink?
 b) What effect does alcohol have on your brain?

3 Look at Figure 3.
 a) What is the approximate legal limit of alcohol that you are allowed in your blood before you drive a car?
 b) Young people are often easily affected by alcohol. If a young person drinks enough to have 125 mg of alcohol per 100 cm³ of their blood and then drives a car, how will this affect their risk of having an accident?
 c) Police advise you not to drink alcohol at all if you are going to drive a car. Based on the evidence of the graph, explain why this is good advice?
 d) Summarise the effects of alcohol on society and discuss banning its use.

KEY POINTS

1 Alcohol affects your nervous system by slowing down your reactions.
2 Alcohol can lead to loss of self-control, unconsciousness, coma and death.
3 Alcohol can cause damage to your liver and brain.

B1a 3.4

Smoking and health

Figure 1 Cigarette smoking increases your risk of developing many serious and fatal diseases. Every packet of cigarettes sold in the UK has to carry a clear health warning. Yet people still buy them in their millions!

Smoking is big business. There are 1.1 billion smokers worldwide, smoking around 6000 billion cigarettes each year!

As a cigarette burns it produces a smoke cocktail of around 4000 chemicals. If you smoke, you breathe this cocktail straight into your lungs. You absorb some of the chemicals into your blood stream. This carries them around your body to your brain.

a) How many chemicals do you inhale in cigarette smoke?

Nicotine is the addictive drug in *tobacco smoke*. It makes people feel calm, contented and able to cope. But you gradually need more and more of the drug to get the same effect. So the number of cigarettes you smoke each day tends to increase over the years. What's more, you feel awful if you don't get your regular dose of nicotine.

b) What is the drug in tobacco smoke?

Smoking related diseases

If you are a non-smoker, small hairs in your breathing system are constantly moving mucus away from your lungs. The mucus traps dirt, dust and bacteria from the air you breathe in. The hairs make sure you get rid of it all.

If you smoke, each cigarette anaesthetises these hairs. They stop working for a time, allowing dirt down into your lungs. This makes you much more likely to suffer from colds and other infections. The mucus also builds up and causes coughing (smoker's cough!).

Tar is a sticky black chemical in tobacco smoke that builds up in your lungs, turning them from pink to grey. It makes smokers much more likely to develop bronchitis. The build-up of tar in your lungs can also lead to the delicate air sacs in the lungs breaking down. We call this *emphysema*. It makes the lungs much less efficient. Your breathing becomes difficult and you can't get enough oxygen.

c) What colour are the lungs of i) a non-smoker? ii) a smoker?
d) What causes the difference in colour in c)ii)?

Tar is also a major **carcinogen** (a cancer causing substance). A build-up of tar can cause lung cancer. We can cure this cancer if it is caught early enough. However, it often grows in the lungs with no obvious symptoms. By the time doctors diagnose it, it has spread to other parts of the body and has become fatal. There are many other carcinogens in tobacco smoke as well.

Many of the chemicals in cigarette smoke are carried right round your body in the blood. Some of them affect your heart and blood vessels. Smoking raises your blood pressure and makes it more likely that your blood vessels will become blocked. This can cause heart attacks, strokes and thrombosis.

NEXT TIME YOU...

. . . see someone lighting up a cigarette, think of the 4000 chemicals they are drawing down into their lungs – and the damage that some of them can do!

FOUL FACTS

Smoking 20 to 60 cigarettes a day will coat your lungs in 1 to 1.5 pounds of tar every year!

Many people want to give up smoking but because nicotine is so addictive, it isn't easy. There are many different ways of giving up. Some are more effective than others. Some are much more expensive than others. The most important thing is always how much you want to give up. Each smoker has to find the method that suits them best and helps them to become a non-smoker!

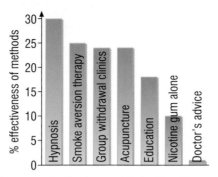

Way of stopping smoking	% effectiveness of method
Hypnosis	30
Smoke aversion therapy	25
Group withdrawal clinics	24
Acupuncture	24
Education	18
Nicotine gum alone	10
Doctor's advice	1

Figure 2 Many people try to give up smoking, but because nicotine is so addictive it isn't easy. There are lots of different methods which can help you – some seem to be more effective than others!

Smoking and pregnancy

Carbon monoxide is a very poisonous gas found in cigarette smoke. It is picked up by your red blood cells. This reduces the amount of oxygen carried in your blood. After smoking a cigarette, up to 10% of a smoker's blood will be carrying carbon monoxide rather than oxygen! This is one reason why smokers often get breathless going up the stairs!

Carbon monoxide in cigarette smoke affects pregnant women in particular. During pregnancy a woman needs oxygen, not just for her own cells but for her developing fetus as well.

If she smokes, the amount of oxygen in her blood will be lower than normal. This means her fetus will be deprived of oxygen. Then it may not grow as well as it should.

Mums who smoke when they are pregnant have a much higher risk of having:

● a premature birth (baby born too early so may struggle to survive),
● a baby with a low birth mass (so it is more at risk of developing problems),
● a stillbirth (where the baby is born dead).

Be specific when you describe the effects of smoking on the body.

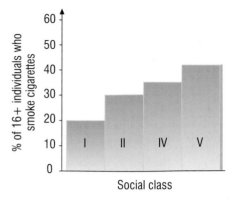

Social class

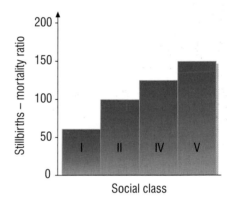

Social class

Figure 3 Data like this show that the people who smoke more are more likely to have stillborn babies. The same pattern is there for babies who are born weighing less than they should.

SUMMARY QUESTIONS

1 Define the following words: a) **nicotine**; b) **tar**; c) **carbon monoxide**; d) **carcinogenic**

2 Smokers are more likely to get infections of their breathing tubes and lungs than non-smokers. Explain why this is the case.

3 Explain how cigarette smoking is linked to an increased risk of getting:

a) lung cancer   b) emphysema   c) heart disease

4 a) Look at Figure 3. What is the evidence that smoking during pregnancy can be dangerous for the developing baby?
b) Explain how smoking causes these problems during pregnancy.
c) How do scientists try to make sure that the correlations they spot in their data are as reliable as possible?
d) Carry out some research to find two ways of helping smokers kick the habit. Compare and contrast each method.

KEY POINTS

1 Tobacco smoke contains substances which can help to cause lung cancer, lung diseases such as emphysema and bronchitis, and diseases of the heart and blood vessels.
2 Tobacco smoke contains carbon monoxide, which reduces the oxygen-carrying capacity of the blood. In pregnant women this can deprive a fetus of oxygen and lead to a low birth mass or death.

B1a 3.5 Lung cancer and smoking

A timeline of evidence

At several stages in its history, smoking was thought to be really good for you. We now think up to 90% of all lung cancers are due to smoking! Here is a brief history of how the evidence against smoking began to build up:

I really like smoking – and I'm really fit. I don't think it does you any harm at all.

1908	The sale of cigarettes to children under the age of sixteen was banned.
1912	Dr Isaac Adler suggested that there was a strong link between lung cancer and smoking based on what he saw among his patients.
1914–18	More people smoked than ever before during the First World War.
1925	The cigarette manufacturers set out to persuade women to smoke.
1930s	Britain had the highest lung cancer rate in the world.
1930s	Scientists in Germany found a strong statistical link between people who smoked and people who got lung cancer.
1939–45	Another war – and smoking continued to rise.
1951	Dr Richard Doll and Professor Austin Hill interviewed 5000 patients in British hospitals. They found that out of 1357 men with lung cancer, 99.5% of them were smokers. This was a very strong statistical link which connected smoking with a high risk of disease.
1953	Dr Ernst Wynder painted cigarette tar on the backs of mice and they developed cancers, showing a biological link between the chemicals in cigarettes and cancer.
1962	The Royal College of Physicians published a report suggesting the restriction of smoking, tax on tobacco products and warning of the dangers of smoking. For the first time in many years, cigarette sales fell.
1964	Doll and Hill published the results of a ten-year study into death rates in relation to smoking. They found a dramatic fall in lung cancer cases in people who had given up smoking compared to those who still smoked.
1997	Hackshaw and colleagues analysed 37 different studies. They found the risk of developing lung cancer in life-long non-smokers who lived with a smoker was 24% higher than if they lived with another non-smoker. They also found carcinogens from tobacco smoke in the blood of the non-smokers. People who lived with heavy smokers were more at risk than people living with light smokers. They concluded that '*The epidemiological and biochemical evidence . . . provides compelling confirmation that breathing other people's tobacco smoke is a cause of lung cancer.*'

It is a free country – if people want to smoke they have every right to do so.

My gran smoked all her life – and she lived until she was ninety!

We're going to sue the tobacco company. Dad couldn't stop smoking, and it killed him. Yet they knew the risks all along!

We need to keep them hooked. We can do this by increasing the amount of nicotine. This won't affect the tar figures, so we can still advertise them as low tar.

We found out years ago that nicotine is addictive. However, we didn't publish our results in any scientific journals.

A change of image?

Tobacco arrived in Britain in the 16th century, when it was seen as a new and exciting thing. Since then it has been seen as a foul habit, the height of fashion and a calming influence during the wars. Now in the developed world, scientific evidence shows that smoking is a serious health risk. But in the developing world – and for many young people even in the UK – smoking is still seen as a glamorous and desirable habit . . .

ACTIVITY

Discuss the evidence shown here in small groups.

Deaths from lung cancer and smoking

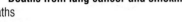
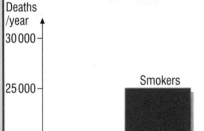

Cigarette consumption and risk of lung cancer death

Number of cigarettes smoked per day	Annual death rate per 100 000	Relative risk
0	14	–
1–14	105	8
15–24	208	15
25+	355	25

Death rates from lung cancer in men by age group: England and Wales 1974–92

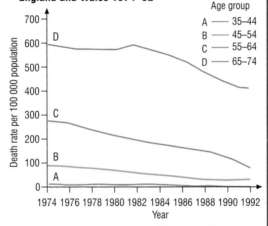

Age group
A — 35–44
B — 45–54
C — 55–64
D — 65–74

Age-adjusted death rates for lung cancer and breast cancer among women, United States, 1930–1997

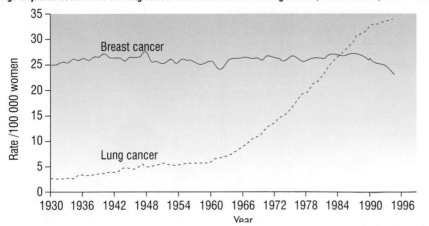

Evidence like this gradually builds up a compelling picture showing that smoking is a major risk factor in the development of lung cancer. Similar evidence can be put together linking smoking with cancers of the throat and tongue and with heart disease.

ACTIVITY

In 21st-century Britain most people accept that there is a link between smoking and lung cancer, heart disease and other health problems – but it doesn't stop them smoking! There has also been a lot in the press about *passive smoking* – breathing in other people's cigarette smoke. In some countries smoking has been banned in almost all public places.

Look at the evidence published by Hackshaw and colleagues in the *British Medical Journal* in 1997 along with all the other evidence here. Use it to carry out one of these two tasks:

Either: Plan an anti-smoking campaign in your school. Target teachers and pupils alike. Plan an article for the school magazine and a presentation which can be used in Citizenship lessons with year 9 pupils.

Or: Plan a campaign to have smoking banned in all public places in your local community – shops, pubs, cafes and restaurants, bars and bowling alleys. Think carefully about how to get the issues across to the general public. You might need posters, leaflets and/or a speech to make at a public meeting.

Whichever task you chose, use plenty of scientific evidence to help make people take notice!

B1a 3.6   Does cannabis lead to hard drugs?

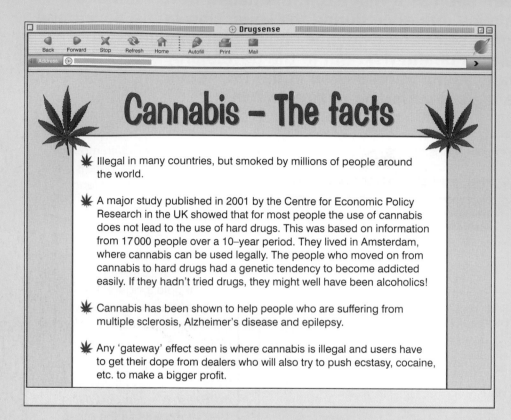

Cannabis – The facts

* Illegal in many countries, but smoked by millions of people around the world.

* A major study published in 2001 by the Centre for Economic Policy Research in the UK showed that for most people the use of cannabis does not lead to the use of hard drugs. This was based on information from 17 000 people over a 10–year period. They lived in Amsterdam, where cannabis can be used legally. The people who moved on from cannabis to hard drugs had a genetic tendency to become addicted easily. If they hadn't tried drugs, they might well have been alcoholics!

* Cannabis has been shown to help people who are suffering from multiple sclerosis, Alzheimer's disease and epilepsy.

* Any 'gateway' effect seen is where cannabis is illegal and users have to get their dope from dealers who will also try to push ecstasy, cocaine, etc. to make a bigger profit.

What You Need To Know

Keeping your children safe!

* Cannabis smoke contains more carcinogens than cigarette smoke.

* Almost everyone who uses heroin started off using cannabis.

* Cannabis can cause complete mental breakdown.

* Cannabis is addictive.

* If your children use cannabis they will be in contact with drug dealers who will try to push other more expensive drugs like cocaine onto them.

In the minds of many people – parents, teachers and politicians – cannabis is a 'gateway' drug. It opens the door to the use of other much harder drugs, such as cocaine and heroin. Your health – and indeed your life itself – is at risk. How accurate is this picture? Let's look at what the scientists say . . .

Scientist A: In some people cannabis use can trigger mental illness, which can be very serious and permanent. The people who are affected have usually been ill or have a family history of mental illness. In fact it seems as if these people are particularly attracted to drug use.

There have been some major studies in New Zealand involving many thousands of young people. These show a definite link between using cannabis and both depression and schizophrenia. In fact one research team calculated that if we stopped people using cannabis in the UK, the number of people with schizophrenia would drop by 13%.

Scientist B: There is very little evidence based on serious scientific research to say that cannabis use affects the long- or short-term memory of users or that cannabis has many of the physically damaging effects often reported in the popular press.

Scientist C: Almost all heroin users were originally cannabis users. This sounds as if cannabis use leads to heroin use. But think again! This is not a case of 'cause and effect' as we call it in science. Almost all cannabis users were originally smokers – but we don't claim that smoking cigarettes leads to cannabis use! In fact the vast majority of smokers do not go on to use cannabis – and the vast majority of cannabis users do not move on to hard drugs like heroin.

Scientist D: The number of people who are damaged or killed as a result of the use of illegal drugs each year is a tiny fraction of the people affected by the legal drugs alcohol and tobacco.

A lot of scientific research has been done into the effects of cannabis on our health, and on the links between cannabis use and addiction to hard drugs.

Unfortunately many of the studies have been quite small. They have not used large sample sizes, so the evidence is not reliable.

Most of the bigger studies show that the effects of cannabis use are not as serious as was thought. They also show that cannabis acts as a 'gateway' to other drugs. That's **not** because it makes people want a stronger drug but because it puts them in touch with illegal dealers.

The UK Government downgraded cannabis in 2004, although it is still an illegal drug. More states in the USA are looking at decriminalising cannabis use as the evidence grows steadily that it is less dangerous than alcohol. Some scientists support the moves, while others feel that cannabis should remain illegal.

ACTIVITY

You are going to set up a classroom debate. The subject is: *'We believe that cannabis should not be made a legal drug.'*
You are going to prepare **two** short speeches – one **for** the idea of legalising cannabis and one **against**.
You can use the information on these pages and also look elsewhere for information – in books and leaflets from PSHE, in the media and on the Internet.
In both of your speeches you must base your arguments on scientific evidence as well as considering the social, moral and ethical implications of any change in the law. You have to be prepared to argue your case (both 'for' and 'against') and answer any questions – so do your research well!

SUMMARY QUESTIONS

1 a) What is a drug?

b) Why can drugs be so harmful?

c) Give two examples of legal drugs and two examples of illegal drugs.

d) What is addiction to a drug?

2 a) How does alcohol affect your body?

b) Why do so many people use alcohol?

c) Alcohol can cause serious damage to your body, both if you take a single big overdose or if you drink too much over many years.
Explain what happens to your body in both cases.

d) Alcohol costs our society millions of pounds in health care and in sorting out the social problems that it causes. It is a very dangerous drug.
Explain why you think it is still legal and easy to get hold of.

3 a) Copy and complete:

Every cigarette contains ...... leaves which burn to produce around ...... chemicals which are breathed into your lungs. Some of those chemicals are ...... into the blood stream to be carried around your body and to your brain. ...... is the......drug found in tobacco smoke. It is absorbed into your .... On the other hand ......stays in your ...... where it can cause cancer.

b) Explain the following facts about smokers.

i) They are more likely than non-smokers to cough.

ii) They are more likely than non-smokers to suffer from lung diseases like emphysema.

iii) They are more likely than non-smokers to suffer from lung cancer.

iv) They are more likely than non-smokers to have a baby which is born dead or has a low birth mass.

4 Use the data on page 57 to help you answer the following questions:

a) What percentage of the people who die of lung cancer are smokers?

b) Draw a bar chart to show the effect of the number of cigarettes you smoke on your relative risk of dying.

c) i) What happened to the death rates from cancer from 1974–1992 in men in England and Wales?

ii) What do you think happened to the numbers of men smoking over the same time period? Explain your answer.

d) The numbers of women smoking increased steadily from the 1950s. How does the evidence suggest that smoking is linked to lung cancer but not to breast cancer.

EXAM-STYLE QUESTIONS

1 The table below contains statements about certain drugs. Match the words **A**, **B**, **C** and **D** with the statements **1** to **4** in the table.

A Alcohol   **B** Nicotine

C Cannabis   **D** Heroin

	Statement
1	May cause cirrhosis of the liver.
2	May cause psychological problems.
3	Is an illegal drug.
4	Acts as a stimulant.

(4)

2 Which of the following substances in tobacco smoke is addictive?

A Tar   **B** Nicotine

C Carbon monoxide   **D** Carbon dioxide   (1)

3 Which of the following is **not** a disease that can be caused by smoking cigarettes?

A Liver damage   **B** Bronchitis

C Emphysema   **D** Lung cancer   (1)

4 Which of the options **A**, **B**, **C** or **D** in the following statement is **not** correct?
Women who smoke during pregnancy have a higher risk of having a:

A longer pregnancy.

B premature birth.

C baby of lower than average birth mass.

D baby that is dead at birth.   (1)

5 The graph shows the number of cigarettes smoked per male per year in the UK from 1910–1980. The dotted line shows the number of male deaths from lung cancer over the same period.

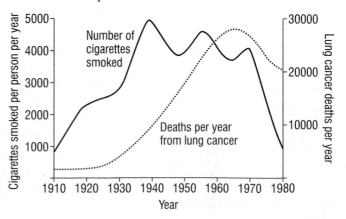

(a) Name the independent variable shown on the graph. (1)

(b) Most graphs display data collected on one dependent variable.
  (i) Why is this graph unusual? (1)
  (ii) Explain why the data has been presented like this. (1)

(c) Use the information in the graph to show that there is a link between cigarette smoking and lung cancer. (2)

(d) Assuming smoking does cause lung cancer, give two reasons why there is a time lag of around 30 years between a high level of cigarette smoking and a high rate of death from lung cancer. (2)

(e) What is the effect on the body of the carbon monoxide in tobacco smoke? (1)

6 The table shows the results of a survey carried out in 1995 to find out how much alcohol youngsters drink in a week.

Amount drunk* (units of alcohol)	Aged 11 (%)	Aged 12 (%)	Aged 13 (%)	Aged 14 (%)	Aged 15 (%)
None	75.7	64.1	56.7	44.7	31.4
1–6	20.1	29.0	29.7	34.2	38.8
7–10	1.7	3.5	5.4	8.8	13.3
11–14	1.6	1.4	3.7	5.2	5.4
15–20	0.7	1.0	2.1	3.8	4.6
21+	0.3	1.0	2.4	3.3	7.5

(* One unit = half a pint of beer/one glass of wine/one measure of spirit).

(a) Calculate the percentage of 14-year-olds who drink more than six units of alcohol in a week. (1)

(b) In a school with 240 pupils aged 15, how many of these pupils does the data suggest drink more than 21 units per week? (1)

(c) Calculate the mean percentage of 11 to 15 year olds who drink 1–6 units of alcohol in a week. (1)

(d) What is the range of the proportion of youngsters aged 11–15 who drink 15–20 units of alcohol per week? (1)

(e) Explain why driving a car under the influence of alcohol can be dangerous. (3)

(f) Suggest two other effects on society of alcohol abuse apart from drink-driving. (2)

HOW SCIENCE WORKS QUESTIONS

Some students decided to test whether drinking coffee could affect heart rate. They asked the class to help them with their investigation. They divided the class into two groups. Both groups had their pulses taken. They gave one group a drink of coffee. They waited for 10 minutes and then took their pulses again. They then followed the same procedure with the second group.

a) What do you think the second group were asked to drink? (1)

b) State a control variable that should have been used. (1)

c) Explain why it would have been a good idea not to tell the two groups exactly what they were drinking. (1)

d) Which of the following best describes the type of dependent variable being measured?
  i) continuous
  ii) discrete
  iii) categoric
  iv) ordered. (1)

e) Study this table of results that they produced.

Group	Increase in pulse rates (beats per min.)
With caffeine	12, 15, 13, 10, 15, 16, 10, 15, 16, 21, 14, 13, 16
Without caffeine	4, 3, 4, 5, 7, 5, 7, 4, 2, 6, 5, 4, 7

Can you detect any evidence for systematic error in these results? If so, describe this evidence. (2)

f) Is there any evidence for a random error in these results? If so, describe this evidence. (2)

g) What is the range for the increase in pulse rates without caffeine? (1)

h) What is the mean (or average) increase in pulse rate:
  i) with caffeine
  ii) without caffeine. (2)

i) What conclusion, if any, does the data collected suggest? (1)

j) How could you make your data more reliable? (1)

B1a 4.1   Pathogens

LEARNING OBJECTIVES

1 What are the differences between bacteria and viruses?
2 How do pathogens cause disease?
3 How did Ignaz Semmelweiss change the way we look at disease?

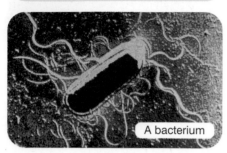
A bacterium

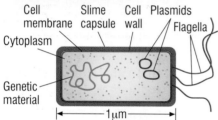

Cell membrane   Slime capsule   Cell wall   Plasmids

Cytoplasm   Flagella

Genetic material

—1µm—

Figure 1 Bacteria come in a variety of shapes and sizes, which help us to identify them under the microscope, but they all have the same basic structure

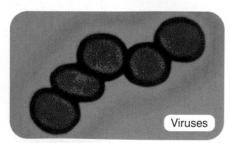
Viruses

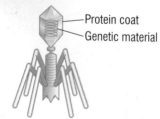
Protein coat
Genetic material

Figure 2 Viruses are really tiny with a very simple structure. Scientists are still arguing about whether they are actually living organisms or not.

Infectious diseases are found all over the world, in every country. The diseases range from relatively mild ones, such as the common cold and tonsillitis, through to known killers, such as tetanus, influenza and HIV/AIDS.

An infectious disease is caused by a **microorganism** entering and attacking your body. People can pass these microorganisms from one person to another. This is what we mean by *infectious*.

Microorganisms which cause disease are called **pathogens**. Common pathogens are bacteria and viruses.

a) What causes infectious diseases?

The differences between bacteria and viruses

Bacteria are single celled living organisms that are much smaller than animal and plant cells.

A bacterium is a single cell. It is made up of cytoplasm surrounded by a membrane and a cell wall. Inside the bacterial cell is the genetic material. Unlike animal and plant cells, this genetic material is not contained in a nucleus.

Although some bacteria cause disease, many are harmless and some are really useful to us. We use them to make food like yoghurt and cheese, in sewage treatment and to make medicines.

b) How are bacteria different from animal and plant cells?

Viruses are even smaller than bacteria. They usually have regular shapes. A virus is made up of a protein coat surrounding simple genetic material. They do not carry out any of the functions of normal living organisms except reproduction. But they can only reproduce by taking over another living cell. As far as we know, all naturally occurring viruses cause disease.

c) Give one way in which viruses differ from bacteria?

How pathogens cause disease

Bacteria and viruses cause disease because once they are inside the body they reproduce rapidly. Bacteria simply split in two. They often produce toxins (poisons) which affect your body. Sometimes they directly damage your cells. Viruses take over the cells of your body as they reproduce, damaging and destroying them. They very rarely produce toxins.

Common disease symptoms are a high temperature, headaches and rashes. These are caused by the damage and toxins produced by the pathogens. The symptoms also appear as a result of the way your body responds to the damage and toxins.

d) How do pathogens make you feel ill?

The work of Ignaz Semmelweiss

When Ignaz Phillipp Semmelweiss was a doctor in the mid-1850s, many women who gave birth in hospital died a few days later. They died from childbed fever, but no-one knew what caused it.

Semmelweiss realised that his medical students were going straight from dissecting a dead body to delivering a baby without washing their hands. He wondered if they were carrying the cause of disease from the corpses to their patients.

Then another doctor cut himself while working on a body and died from symptoms which were identical to childbed fever. Now Semmelweiss was sure the fever was caused by an infectious agent.

He insisted that his medical students wash their hands before delivering babies. Immediately, fewer mothers died.

Getting his ideas accepted

Semmelweiss presented his findings to other doctors. He thought his evidence would prove to them that childbed fever was spread by doctors. But his ideas were mocked.

Many doctors thought that childbed fever was God's punishment to women. They didn't want to accept the idea that the disease was caused by something invisible passed from patient to patient. Also it was hard for doctors to admit that they might have spread the disease and killed their patients instead of curing them.

What's more, hand-washing seemed a strange idea at the time. There was no indoor plumbing, the water was cold, and the chemicals used eventually damaged the skin of your hands. It is difficult for us to imagine just how difficult hand-washing must have seemed in the 19th century! It took years for Semmelweiss's ideas to be accepted.

In hospitals today, bacteria such as MRSA, which are resistant to antibiotics, are causing lots of problems (see page 68). Getting doctors, nurses and visitors to wash their hands more often is seen as part of the solution – just as it was in Semmelweiss's time!

GET IT RIGHT!

Make sure you know the differences between bacteria and viruses.

Figure 3 Ignaz Semmelweiss – his battle to persuade medical staff to wash their hands to prevent infections is still going on today!

DID YOU KNOW?

Semmelweiss couldn't bear to think of the thousands of women who died because other doctors ignored his findings. By the 1860s he suffered a major breakdown and in 1865, aged only 47, he died – from an infection picked up from a patient during an operation!

KEY POINTS

1. Infectious diseases are caused by microorganisms such as bacteria and viruses.
2. Microorganisms which cause disease are called pathogens.
3. Bacteria and viruses reproduce rapidly inside your body. They may produce toxins which make you feel ill.
4. Viruses use and damage your cells as they reproduce. This can also make you feel ill.

SUMMARY QUESTIONS

1 Copy and complete using the words below:

**toxins   viruses   microorganisms   reproduce   pathogens
damage   symptoms   bacteria**

The …… which cause infectious diseases are known as ……. Once …… and …… get inside your body they …… rapidly. They …… your tissues and may produce …… which cause the …… of disease.

2 Bacteria and viruses can both cause disease. Make a table which shows how bacteria and viruses are different, and how they are similar.

3 Give five examples of the way we now accept the germ theory of disease in our everyday lives, e.g. washing your hands after using the toilet.

4 Write a letter by Ignaz Semmelweiss to a friend explaining how you formed your ideas and the struggle to get them accepted.

B1a 4.2

Defence mechanisms

LEARNING OBJECTIVES

1 How does your body stop pathogens getting in?
2 How do white blood cells protect us from disease?

Figure 1 Droplets carrying millions of pathogens fly out of your mouth and nose at up to 100 miles an hour when you sneeze!

There are a number of ways in which we can spread pathogens from one person to another. The more pathogens that get into your body, the more likely it is that you will get an infectious disease.

Droplet infection: When you cough, sneeze or talk you expel tiny droplets full of pathogens from your breathing system. Other people breathe in the droplets, along with the pathogens they contain. So they pick up the infection, e.g. 'flu (influenza), tuberculosis or the common cold. (See Figure 1.)

Direct contact: Some diseases are spread by direct contact of the skin, e.g. impetigo and some sexually transmitted diseases like genital herpes.

Contaminated food and drink: Eating raw or undercooked food, or drinking water containing sewage can spread disease, e.g. diarrhoea or salmonellosis. You get these by taking large numbers of microorganisms straight into your gut.

Through a break in your skin: Pathogens can enter your body through cuts, scratches and needle punctures, e.g. HIV/AIDS or hepatitis.

When people live in crowded conditions, with no sewage treatment, infectious diseases can spread very rapidly.

a) What are the four main ways in which infectious diseases are spread?

Preventing microbes getting into your body

Each day you come across millions of disease-causing microorganisms. Fortunately your body has a number of ways of stopping these pathogens getting inside.

Your skin covers your body and acts as a barrier. It prevents bacteria and viruses from reaching the vulnerable tissues underneath.

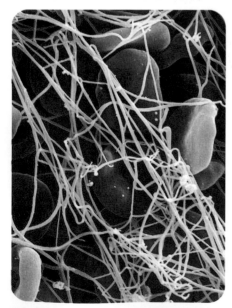

Figure 2 When you get a cut, the platelets in your blood set up a chain of events to form a clot which dries to a scab. This stops pathogens from getting into your body. It also stops you bleeding to death as well!

If you damage or cut your skin in any way you bleed. Platelets in your blood quickly help to form a clot which dries into a scab. The scab forms a seal over the cut, stopping pathogens getting in through the wound. (See Figure 2.)

Your breathing system could be a weak link in your body defences. That's because every time you breathe you draw air loaded with pathogens right inside your body. However, your breathing organs produce a sticky liquid, called mucus, which covers the lining of your lungs and tubes. It traps the pathogens. The mucus is then moved out of your body or swallowed down into your gut. Then the acid in your stomach destroys the microorganisms. In the same way, the stomach acid destroys most of the pathogens you take in through your mouth.

b) What are the three main ways in which your body prevents pathogens from getting in?

How white blood cells protect you from disease

In spite of your various defence mechanisms, some pathogens still manage to get inside your body. Once there, they will meet your second line of defence – the *white blood cells* of your **immune system**. The white blood cells help to defend your body against pathogens in several ways:

Role of white blood cell	How it protects you against disease
Ingesting microorganisms	Some white blood cells ingest (take in) pathogens, destroying them and preventing them from causing disease.
Producing antibodies Antibody, Antigen, Bacterium, White blood cell, Antibody attached to bacterium	Some white blood cells produce special chemicals called **antibodies**. These target particular bacteria or viruses and destroy them. You need a unique antibody for each type of bacterium or virus. Once your white blood cells have produced antibodies against a particular pathogen, they can produce them again very rapidly if that pathogen invades again.
Producing antitoxins Antitoxin molecule, Toxin and antitoxin joined together, Toxin molecule, Bacterium	Some white blood cells produce antitoxins. These counteract the toxins (poisons) released by pathogens.

Figure 3 Ways in which your white blood cells destroy pathogens and protect you against disease

FOUL FACTS

Have you ever wondered why the mucus produced from your nose turns green when you have a cold? Some white blood cells containing green coloured enzymes are secreted in your mucus to destroy the cold viruses – and any bacteria which decide to infect the mucus at the same time. The dead white blood cells along with the dead bacteria and viruses are removed in the mucus, making it look green.

GET IT RIGHT!

Avoid using words like 'battle' and 'fight' when you explain how antibodies work. Such words suggest that the white blood cells think about what they are doing.

SUMMARY QUESTIONS

1 Explain how diseases are spread by:

  a) droplet infection   c) contaminated food and drink
  b) direct contact   d) through a cut in the skin.

2 Certain diseases mean you cannot fight infections very well. Explain why the following symptoms would make you less able to cope with pathogens.

  a) Your blood won't clot properly.
  b) The number of white cells in your blood falls.
  c) Your skin is damaged to expose a large area of raw tissue underneath.

3 Here are four common things we do. Explain carefully how each one helps to prevent the spread of disease.

  a) Washing your hands before preparing a salad.
  b) Throwing away tissues after you have blown your nose.
  c) Making sure that sewage does not get into drinking water.
  d) Putting your hand in front of your mouth when you cough or sneeze.

4 Explain in detail how the white blood cells in your body work.

KEY POINTS

1 Your body has several methods of defending itself against the entry of pathogens using the skin, the mucus of the breathing system and the clotting of the blood.
2 Your white blood cells help to defend you against pathogens by ingesting them, making antibodies and making antitoxins.

B1a 4.3 Using drugs to treat disease

When you have an infectious disease, you often take medicines which contain useful drugs. Often the medicine has no effect at all on the pathogen that is causing the problems. It just eases the symptoms and makes you feel better.

For example, drugs like aspirin and paracetamol are very useful as painkillers. When you have a cold they will help relieve your headache and sore throat. On the other hand, they will have no effect on the virus which has invaded your tissues and made you feel ill!

Many of the medicines you can buy at a chemist's are like this. They are symptom relievers rather than pathogen killers. They do not make you better any faster. You have to wait for your immune system to overcome the invading microorganisms.

a) Why don't medicines like aspirin actually cure your illness?

Antibiotics

While drugs which make us feel better are useful, what we really need are drugs that can *cure* us. We use antiseptics and disinfectants to kill bacteria outside the body. But they are far too poisonous to use inside you. They would kill you and your pathogens at the same time!

The drugs which have really changed the way we treat infectious diseases are **antibiotics**. These are medicines which can kill disease-causing bacteria inside your body.

b) What is an antibiotic?

Figure 1 Taking paracetamol will make this child feel better, but she will not actually get well any faster as a result.

Alexander Fleming was a scientist who studied bacteria and was keen to find ways of killing them. In 1928, he was growing lots of bacteria on agar plates. Alexander was a rather sloppy scientist, and his lab was quite untidy. So he often left the lids off his plates for a long time. He also often forgot about cultures he had set up!

When he came back from a holiday, Fleming noticed that lots of his culture plates had mould growing on them. Just before he put the plates in the washing up bowl, he noticed a clear ring in the jelly around some of the spots of mould. Something had killed the bacteria covering the jelly.

There and then Fleming saw how important this was. He worked hard on the mould, extracting a 'juice' which he called **penicillin**. But he couldn't get much penicillin from the mould. He couldn't make it keep, even in the fridge. So he couldn't prove it would actually kill bacteria and make people better. By 1934 he gave up on penicillin and went on to do different work!

About 10 years after penicillin was first discovered, Ernst Chain and Howard Florey set about trying to use it on people. Eventually they managed to make penicillin on an industrial scale. The process was able to produce enough penicillin to supply the demands of the Second World War. We have used it as a medicine ever since.

DID YOU KNOW?

A patient called Albert Alexander was dying of a blood infection when Florey and Chain gave him some of their experimental penicillin for five days. The effect was almost miraculous and Albert recovered. But then the penicillin ran out. Florey and Chain even tried to collect spare penicillin from Albert's urine, but it was no good. The infection came back and sadly Albert died.

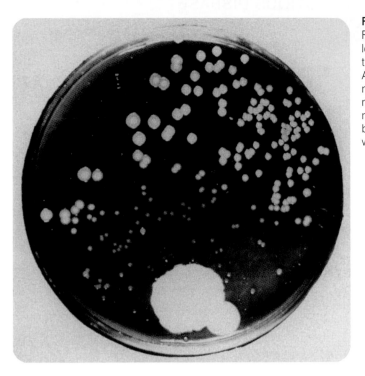

Figure 2 Alexander Fleming was on the lookout for something that would kill bacteria. As a result of him noticing the effect of this mould on his cultures, millions of lives have been saved around the world.

c) Who was the first person to discover penicillin?

How antibiotics work

Antibiotics, such as penicillin, work by killing the bacteria which cause disease while they are inside your body. They damage the bacterial cells without harming your own cells. They have had an enormous effect, because we can now cure bacterial diseases such as plague and TB. These same diseases killed millions of people in years gone by.

Unfortunately antibiotics have not been a complete answer to the problem of infectious diseases. Antibiotics have no effect on diseases caused by viruses. What's more, developing drugs which do have an effect on viral diseases is proving very difficult indeed.

The problem with viral pathogens is that they reproduce inside the cells of your body. It is extremely difficult to develop drugs which kill the viruses without damaging the cells and tissues of your body at the same time.

d) How do antibiotics work?

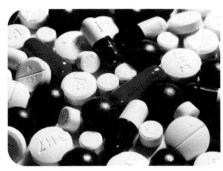

Figure 3 Penicillin was the first antibiotic. Now we have many different ones which kill different types of bacteria. In spite of this, scientists are always on the look out for new antibiotics to keep us ahead in the battle against the pathogens.

SUMMARY QUESTIONS

1 What is the main difference between drugs like paracetamol and drugs such as penicillin?

2 a) How did Alexander Fleming discover penicillin?
  b) Why was it so difficult to make a medicine out of penicillin?
  c) Who developed the industrial process which made it possible to mass-produce penicillin?

3 Explain why it is so much more difficult to develop medicines against viruses than it has been to develop anti-bacterial drugs.

KEY POINTS

1 Some medicines relieve the symptoms of disease but do not kill the pathogens which cause it.

2 Antibiotics cure bacterial diseases by killing the bacteria inside your body.

3 Antibiotics do not destroy viruses because viruses reproduce inside the cells. It is difficult to develop drugs that can destroy viruses without damaging your body cells.

B1a 4.4 | Changing pathogens

If you are given an antibiotic and use it properly, the bacteria which have made you ill are almost all killed. The ones that are left are the ones which have a natural **mutation** which means they are not affected by the antibiotic. They are resistant to it – but your body finishes them off!

Antibiotic-resistant bacteria

If antibiotics are used too often, or you don't take the full course of medicine prescribed by your doctor, more of the resistant bacteria survive. If they go on to make someone else ill, they will not be killed by the original antibiotic. They are **resistant** to that antibiotic. This resistance is the result of a process of **natural selection**. (See Figure 1.)

As more types of bacteria become resistant to more antibiotics, so diseases caused by bacteria are becoming harder and harder to treat.

To prevent more resistant strains of bacteria appearing it is important not to over-use antibiotics. It's best to only use them when you really need them. It is also very important that people finish their course of medicine every time.

a) Why is it important not to use antibiotics too frequently?

The MRSA story

Hospitals treat many patients with infectious diseases. They use large amounts of antibiotics. As a result of natural selection, hospitals often contain a number of bacteria which are not affected by most of the commonly used antibiotics. This is what has happened with **MRSA** (the bacterium **methicillin resistant Staphylococcus aureus**).

In hospitals, where doctors and nurses move from patient to patient, these antibiotic-resistant bacteria are spread easily. MRSA alone now causes around a thousand deaths every year in hospital patients who, although ill, might otherwise have recovered.

How can we control the spread of these antibiotic-resistant bacteria in hospitals? There are a number of simple steps which can have a big effect on the spread of microorganisms such as MRSA. We have known some of them since the time of Semmelweiss, but they sometimes get forgotten!

- Doctors, nurses and other medical staff wash their hands between patients.
- Visitors wash their hands as they come into and leave the hospital.
- Look after patients infected with the bacteria in isolation from other patients.
- Keep hospitals clean – high standards of hygiene.
- Medical staff wear either disposable clothing or clothing which is regularly sterilised.

b) Is MRSA a bacterium or a virus?

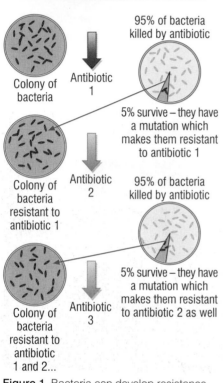

Figure 1 Bacteria can develop resistance to many different antibiotics in a process of natural selection

Colony of bacteria → *Antibiotic 1* → 95% of bacteria killed by antibiotic → 5% survive – they have a mutation which makes them resistant to antibiotic 1

Colony of bacteria resistant to antibiotic 1 → *Antibiotic 2* → 95% of bacteria killed by antibiotic → 5% survive – they have a mutation which makes them resistant to antibiotic 2 as well

Colony of bacteria resistant to antibiotic 1 and 2... → *Antibiotic 3*

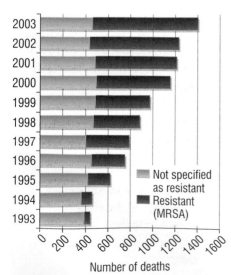

Number of deaths

Not specified as resistant
Resistant (MRSA)

Source: National Statistics Office

Figure 2 The growing impact of MRSA in our hospitals can be seen from this data

Mutation and pandemics

Another problem caused by the mutation of bacteria and particularly viruses is that new forms of diseases can appear. Because no-one is immune to them, they can cause widespread illness. A good example of this in action is influenza, commonly known as 'flu.

'Flu is caused by a virus which mutates easily, so every year new strains appear which can fool your immune system. These new strains are usually quite similar to the old 'flu but just occasionally a very different form of 'flu virus appears.

These usually cause a 'flu *epidemic* (in one country) or even a *pandemic* (across several countries). In 1918–19 a new strain of 'flu emerged which spread rapidly around the world and killed between 20 and 40 million people.

Animals such as chickens and pigs get 'flu-like diseases too. Sometimes an animal virus will mutate so it can infect people. For example, the 1918–19 'flu pandemic may have been linked to bird 'flu.

Many scientists think that a new and serious form of human 'flu is likely to be linked to one of the bird influenzas which keep appearing in China, Thailand and Asia.

We are trying to prevent a new pandemic using research to discover what makes certain types of 'flu so dangerous. The World Health Organisation is monitoring all 'flu outbreaks. Scientists are also trying to produce different *vaccines* to protect us against new and dangerous forms of the disease.

There is no 'antibiotic for viruses' but there are several drugs which can reduce the length of time you suffer with 'flu. This lessens the chance of it spreading.

Finally, many countries have plans to restrict travel and even put people in isolation if there is a 'flu outbreak. This worked very well in 2003 when SARS, a 'flu-like illness, appeared for the first time in China.

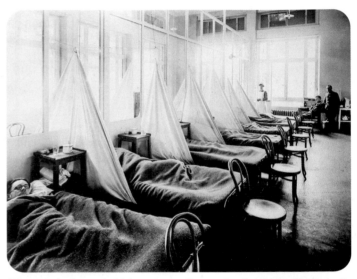

Figure 3 In the early 20th century a 'flu pandemic killed millions. In the 21st century a new form of 'flu could sweep the world even more quickly because of the way we all travel around on aircraft for business and holidays.

SUMMARY QUESTIONS

1 Copy and complete using the words below:

| antibiotics | bacterium | better | disease |
| mutation | mutate | resistant | virus |

If bacteria change or …… they may become …… to …… This means the medicine no longer makes you …… . A …… in a …… or ……can also lead to a new form of …… .

2 Make a flow chart to show how bacteria develop resistance to antibiotics.

3 A new strain of bird 'flu has been discovered in Asia and it is in the national news. Write a piece for the science pages of your local paper. Explain what people can do to protect themselves and what problems they might face preventing the new disease spreading.

KEY POINTS

1 Many types of bacteria have developed antibiotic resistance as a result of natural selection. To prevent the problem getting worse we must not over-use antibiotics.

2 If bacteria or viruses mutate, new strains of a disease can appear which spread rapidly to cause epidemics and pandemics.

B1a 4.5 Developing new medicines

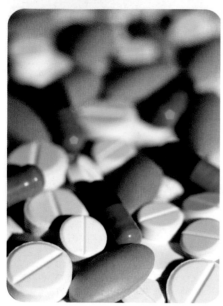

Figure 1 No matter how many medicines we have, there is always room for more as we tackle new diseases!

We are developing new medicines all the time, as scientists and doctors try to find ways of curing more diseases. Every new medical treatment has to be extensively tested and trialled. This process makes sure that it works well and is as safe as possible.

A good medicine is:

● **Effective** – it must prevent or cure the disease it is aimed at, or at least make you feel better.
● **Safe** – the drug must not be toxic (poisonous) and there must be no unacceptable side effects.
● **Stable** – you need to be able to use the medicine under normal conditions and store it for some time.
● **Successfully taken into and removed from your body** – a medicine is no use unless it can reach its target in your body. Then your body must be able to remove the medicine once it has done its work.

Developing and testing a new drug

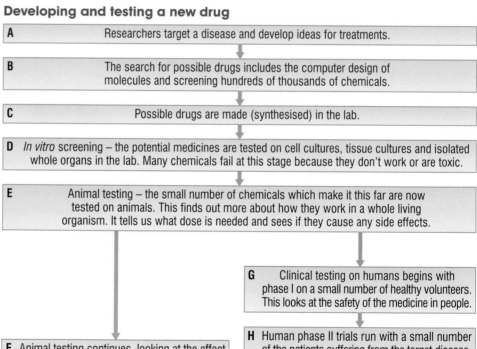

A   Researchers target a disease and develop ideas for treatments.

B   The search for possible drugs includes the computer design of molecules and screening hundreds of thousands of chemicals.

C   Possible drugs are made (synthesised) in the lab.

D   *In vitro* screening – the potential medicines are tested on cell cultures, tissue cultures and isolated whole organs in the lab. Many chemicals fail at this stage because they don't work or are toxic.

E   Animal testing – the small number of chemicals which make it this far are now tested on animals. This finds out more about how they work in a whole living organism. It tells us what dose is needed and sees if they cause any side effects.

F   Animal testing continues, looking at the effect of longer term use of the medicine.

G   Clinical testing on humans begins with phase I on a small number of healthy volunteers. This looks at the safety of the medicine in people.

H   Human phase II trials run with a small number of the patients suffering from the target disease. This is where scientists can really begin to see if the drug will be safe and effective.

I   Human phase III trials continue with a larger number of patients.

J   When the medicine has passed all the tests set down in law, it will be granted a licence. Now your doctor can use the new medicine to treat your illness.

K   Once the medicine is in use, phase IV trials continue. The medicine will be monitored for as long as patients use it. This makes sure it works and is as safe as possible.

Figure 2 It takes a long time and a lot of work to develop a successful new drug

When scientists research a new medicine they have to make sure that all these conditions are met. This is why it takes a very long time. It can take up to 12 years to bring a new medicine into your doctor's surgery. It can also cost a lot of money, up to about £350 million!

a) What are the important properties of a good new medicine?

Testing drugs

We test new medicines in the laboratory. This is to find out if they are toxic and if they seem to do their job. We also trial new medicines on human volunteers. This is to discover if they have any side effects.

Take a look at all the stages a new drug has to go through – no wonder it is a slow process! (See Figure 2.)

Why do we test new medicines so thoroughly?

Thalidomide is a medicine which was developed in the 1950s as a sleeping pill. This was before we had agreed standards for studying the effects of new medicines. In particular, the specific animal tests on pregnant animals which are now known to be essential were not carried out.

Then it was discovered that thalidomide stopped sickness in pregnancy. Because thalidomide seemed very safe for adults, it was assumed that it was also safe for unborn children. Doctors gave it to pregnant women to relieve their morning sickness.

Tragically, thalidomide was **not** safe for developing fetuses. It affected many of the women who took the drug in the early stages of pregnancy. They went on to give birth to babies with severe limb deformities.

The thalidomide tragedy led to a new law which set standards for the testing of all new medicines. Since the Medicines Act 1968, new medicines **must** be tested on animals to see if they have an effect on developing fetuses.

There is another twist in the thalidomide story. Although thalidomide is never given to anyone who is or might become pregnant, doctors are finding more and more uses for the drug. They can use it to treat leprosy and autoimmune diseases (where the body attacks itself). There have also been some very exciting results using thalidomide to treat certain types of cancer.

b) Why was thalidomide prescribed to pregnant women?

Figure 3 This man has limb deformities because his mother took thalidomide during her pregnancy. He was just one of thousands of people affected by the thalidomide tragedy, many of whom have gone on to live full and active lives.

KEY POINTS

1 When we develop new medicines they have to be tested and trialled extensively before we can use them.
2 Drugs are tested to see if they work well. We also make sure they are not toxic and have no unacceptable side effects.
3 Thalidomide was developed as a sleeping pill and was found to prevent morning sickness in early pregnancy. It had not been fully tested and it caused birth defects. Thalidomide is now used to treat leprosy and other diseases.

SUMMARY QUESTIONS

1 Copy and complete using the words below:

   effective   trialled   safe   medicine   stable   tested

   Every new ...... has to be extensively ...... and ...... before you can use it to make sure that it works well. A good medicine can be taken into and removed from your body, and it is ......, ... and ....

2 a) Testing a new medicine costs a lot of money and can take up to 12 years. Explain the main stages in testing new drugs.
   b) What were the flaws in the original development of Thalidomide?
   c) Comment on the benefits and drawbacks of using Thalidomide to treat leprosy and cancer.

B1a 4.6   Immunity

Figure 1 No-one likes having a vaccination very much – but they save millions of lives!

Every cell has unique proteins on its surface called **antigens**. The antigens on the microorganisms which get into your body are different to the ones on your own cells. Your immune system recognises they are different.

Your white blood cells then make antibodies to attack the antigens. This destroys the pathogens. (See page 65.)

Your white blood cells seem to 'remember' the right antibody needed to tackle a particular pathogen. If you meet that pathogen again, they can make the same antibody very quickly. So you become *immune* to that disease.

The first time you meet a new pathogen you get ill. That's because there is a delay while your body sorts out the right antibody needed. The next time, you completely destroy the invaders before they have time to make you feel unwell.

a) How does your immune system work?

Vaccination

Some pathogens can make you seriously ill very quickly. In fact you can die before your body manages to make the right antibodies. Fortunately, you can be protected against many of these diseases by **immunisation** (also known as *vaccination*).

Immunisation involves giving you a *vaccine*. A vaccine is usually made of a dead or weakened form of the disease-causing microorganism. It works by triggering your natural immune response to invading pathogens.

A small amount of dead or inactive pathogen is introduced into your body. This gives your white blood cells the chance to develop the right antibodies against the pathogen **without** you getting ill.

Then if you meet the live pathogens, your white blood cells can respond rapidly. They can make the right antibodies just as if you had already had the disease. This is how vaccination protects you against disease.

b) How do vaccines work?

Figure 2 This is how vaccines protect you against dangerous infectious diseases

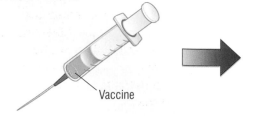

Small amounts of dead or inactive pathogen are put into your body, often by injection.

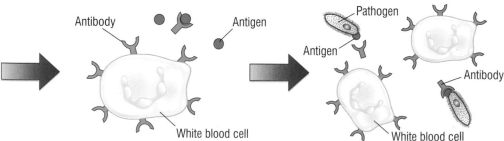

The antigens in the vaccine stimulate your white blood cells into making antibodies. The antibodies destroy the antigens without any risk of you getting the disease.

You are immune to future infections by the pathogen. That's because your body can respond rapidly and make the correct antibody as if you had already had the disease.

We use vaccines to protect us against both bacterial diseases (e.g. tetanus and diphtheria) and viral diseases (e.g. polio, measles, mumps and rubella). Vaccines have saved millions of lives around the world. One disease – smallpox – has been completely wiped out by vaccinations. It also looks as if polio will disappear as well in the next few years.

c) Give an example of one bacterial and one viral disease which you can be immunised against.

The vaccine debate

No medicines are completely risk free. Very rarely, a child will react badly to a vaccine with tragic results. Making the decision to have your baby immunised can be difficult.

Society as a whole needs as many people as possible to be immunised against as many diseases as possible. This keeps the pool of infection in the population as low as we can get it. On the other hand, by taking your healthy child along for a vaccination, you know there is a remote chance that something will go wrong.

Because vaccines are so successful, we never see the terrible diseases they protect us against. We forget that 60 years ago in the UK thousands of children died every year from infectious diseases. Many more were left permanently damaged. So parents today are often aware of the very small risks from vaccination – but sometimes forget about the terrible dangers of the diseases we vaccinate against.

If you are a parent it can be difficult to find unbiased advice to help you make a decision. The media emphasise scare stories which make good headlines. The pharmaceutical companies want to sell vaccines. Doctors and health visitors can weigh up all the information, but they have vaccination targets set by the government.

Most people agree that vaccination is a good thing both for you and for society. The great majority of parents still choose to protect their children – but it is not always an easy choice. (See pages 74 and 75.)

DID YOU KNOW…

… in the first 10 years of the 20th century, nearly 50% of *all* deaths in people aged up to 44 years old were caused by infectious diseases? The development of antibiotics and vaccines means that now only 0.5% of all deaths in the same age group are due to infectious disease!

GET IT RIGHT!

High levels of antibodies do not stay in your blood forever – immunity is the ability of your white blood cells to produce them quickly if you are re-infected by a disease.

KEY POINTS

1 You can be immunised against a disease by introducing small amounts of dead or inactive pathogens into your body.
2 Your white blood cells produce antibodies to destroy the pathogens. Then your body will respond rapidly to future infections by the same pathogen, by making the correct antibody. You become immune to the disease.
3 We can use vaccination to protect against both bacterial and viral pathogens.

SUMMARY QUESTIONS

1 Copy and complete using the words below:

**antibodies   pathogen   immunised   dead   immune
inactive   white blood**

People can be …… against a disease by introducing small quantities of …… or …… forms of a …… into your body. They stimulate the …… …… cells to produce …… to destroy the pathogen. This makes you …… to the disease in future.

2 Explain carefully, using diagrams if they help you:

a) how the immune system of your body works,
b) how vaccines use your natural immune system to protect you against serious diseases.

3 Make a table to show the advantages and disadvantages of giving your child the MMR vaccine. What would you choose to do?

SUMMARY QUESTIONS

1 a) Define the following terms:
   i) infectious disease
   ii) microorganism
   iii) pathogen
   iv) toxin.

   b) What are the main differences between bacteria and viruses?

   c) How do tiny organisms like bacteria and viruses make a large person like you ill?

2 There is going to be a campaign to try and stop the spread of colds in Year 7 of your school. There is going to be a poster and a simple PowerPoint® presentation.
   Make a list of all the important things that the Year 7 children need to know about how diseases are spread. Also cover how to reduce the spread of infectious diseases from one person to another.

3 a) What is the difference between a medicine which makes you feel better and a medicine which actually makes you better?

   b) What is an antibiotic?

   c) What are the limitations of antibiotics?

   d) Where have antibiotic-resistant bacteria (like MRSA found in hospitals) actually come from?

   e) What can we do to prevent the problem of antibiotic resistance getting worse?

4 a) Why do new medicines need to be tested and trialled before doctors can use them to treat their patients?

   b) Why is the development of a new medicine so expensive?

   c) Do you think it would ever be acceptable to use a new medicine before all the trials had been completed?

5 a) Draw a labelled flow diagram to show how your immune system works.

   b) Explain why thousands of children in the UK no longer die of diphtheria or become paralysed by polio.

   c) Make a table to show the risks and benefits of having children vaccinated against serious diseases.

   d) The media like a good story. Many people read the papers and watch television.
   Explain why it is important that stories about medical issues like vaccination should be reported very carefully in the media.

EXAM-STYLE QUESTIONS

1 The table below contains statements about chemicals that act against pathogens.

   Match the list of chemicals **A**, **B**, **C** and **D** with the statements **1** to **4** in the table.

   A Antibodies   B Antitoxins
   C Antibiotics   D Antiseptics

	Statement
1	Counteract poisons released by pathogens
2	Produced by the white blood cells to destroy particular bacteria or viruses
3	Kill infective bacteria outside the body
4	Drugs that kill infective bacteria inside the body

   (4)

2 Which one of the following is **not** used by white blood cells to protect us against pathogens?

   A Production of antibodies to destroy bacteria

   B Sealing of wounds to prevent infection

   C Production of antitoxins to counteract poisons released by pathogens

   D Ingestion of pathogens.   (1)

3 MRSA is a bacterium that kills around a thousand hospital patients each year. Which one of the following has led to these MRSA infections?

   A Ineffective vaccines

   B Overcrowding in hospitals

   C Antibiotic resistance

   D Shortage of hospital equipment   (1)

4 Which of the following would **not** help control the spread of MRSA?

   A Medical staff having regular health checks

   B Medical staff washing their hands between seeing patients

   C Visitors washing their hands as they enter and leave the hospital

   D Cleaning hospitals with antiseptics   (1)

5 New drugs are thoroughly trialled and tested to ensure they have certain features that make them good medicines. When tested on human volunteers, it is not possible to keep every control variable constant. So how can researchers make their data reliable?

   A Use only animals instead of humans in tests.

   B Only test people over the age of 65.

C Use as large a sample of people as possible.

D Calculate the mean age of the people tested. (1)

6 Measles is an extremely infectious disease that is caused by a virus. Measles can cause brain damage and death. To try to prevent epidemics of measles, the MMR vaccine was developed.

(a) Which other two diseases does the MMR vaccine protect against? (2)

(b) The MMR vaccine contains the virus that causes measles. Why does this virus not cause measles when the MMR vaccine is injected into a child? (1)

(c) Name the type of chemical released by certain white blood cells to destroy viruses. (1)

(d) Explain how vaccinating a child with the MMR vaccine makes them immune to measles. (5)

(e) A child that has not been immunised with the MMR vaccine develops measles. Suggest a reason why antibiotics will not cure the child of measles. (1)

(f) In the UK, by 2001, the uptake of the MMR vaccine had fallen from 92% to 75%. Suggest one reason for this decrease. (1)

7 (a) A health centre gives the following advice about antibiotics to its patients. In each case suggest **one** reason why the advice is given.

(i) Do not take an antibiotic for a viral infection like a cold or the flu.

(ii) Do not take an antibiotic that has been prescribed for someone else.

(iii) Ask whether an antibiotic is likely to help your illness or whether there are alternatives.

(iv) Do not stop taking the antibiotic once you feel better – always complete the course of drugs. (4)

(b) Describe how antibiotic resistance arises. (5)

HOW SCIENCE WORKS QUESTIONS

Look at the newspaper advert below and answer the following questions:

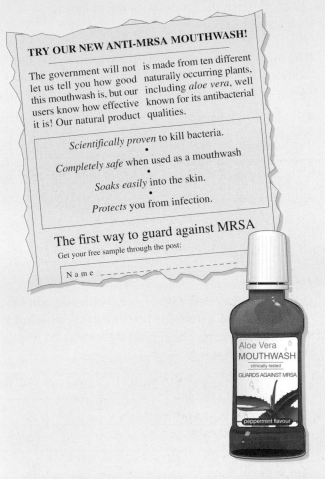

TRY OUR NEW ANTI-MRSA MOUTHWASH!

The government will not let us tell you how good this mouthwash is, but our users know how effective it is! Our natural product is made from ten different naturally occurring plants, including *aloe vera*, well known for its antibacterial qualities.

Scientifically proven to kill bacteria.
•
Completely safe when used as a mouthwash
•
Soaks easily into the skin.
•
Protects you from infection.

The first way to guard against MRSA

Get your free sample through the post:

N a m e

Aloe Vera
MOUTHWASH
clinically tested
GUARDS AGAINST MRSA

peppermint flavour

a) Why does the government want to prevent medical claims for this mouthwash? (1)

b) Describe the tests that you think should have been carried out to make the claim that it is 'Scientifically proven to kill bacteria'. (4)

c) Why might you mistrust any claims made by this company? (2)

d) Mr Skeptic commented that 'it must be crazy to think that a plant could kill something as difficult as MRSA'. What do you think? (1)

e) Describe the sort of investigation needed to be able to say that the mouthwash is 'completely safe'. (3)

EXAMINATION-STYLE QUESTIONS

Science A
In question **1** match the letters with the numbers.
Use **each** answer only **once**.

1 The menstrual cycle in women is controlled by hormones.
Match words **A**, **B**, **C** and **D** with the spaces in **1** to **4** in the sentences.

A	LH	**B**	Pituitary gland
C	FSH	**D**	Oestrogen

Each month, the hormone ......**1**...... is produced from the ......**2**....... and causes an egg to mature in a woman's ovaries. The ovaries, in turn, produce a hormone called ......**3**...... that stimulates the production of another hormone called ......**4**...... which causes the mature egg to be released. *(4 marks)*

Question 2
In **each** part choose only **one** answer.

2 Cigarette smoking has been shown to increase the chances of early death due to diseases such as lung cancer, bronchitis and emphysema.
The table below shows the percentage of the adult UK population by age group who were smokers. The figures were taken at five different times over a 25-year period.

	Age 16–19	Age 20–24	Age 25–34	Age 35–49	Age 50–59	Age 60+
1978	34	44	45	45	45	30
1988	28	37	36	36	33	23
1998	31	40	35	30	27	16
2000	29	35	35	29	27	16
2003	26	36	34	30	25	16

Figures show the percentage UK population who were smokers.

(a) Which of the following statements is the only one supported by the data in the table?

    A Fewer individuals smoked in 2003 that in 1978.

    B A greater percentage of the population smoked in 1978 than in 2003.

    C In 1978 a greater percentage of 20–24-year olds smoked than 25–34-year olds.

    D From 16 years old, the proportion of the population who smoke declines steadily with age. *(1 mark)*

(b) What would be the most reliable way of collecting the data in the table?

    A Samples of people were asked about their smoking habits.

    B Samples of people had their blood tested for nicotine.

    C Tobacconists were asked about the people who buy cigarettes.

    D Researchers recorded the number of people smoking in various places. *(1 mark)*

See pages 28–9

GET IT RIGHT!

Start with what you know.
In question 1 you have to match four letters with four numbers. Read the **whole** question first and then match up the pairs that you are certain about. If you can match all four, all well and good. If however, you are only sure about three of them, do not worry as the last one can be paired by a process of elimination. Not the best examination technique (far better to know all four) but preferable to leaving blank spaces.

See page 5

Science B

1 In 2003 a health survey was carried out on over 8 000 men and women over 16 to find out their blood cholesterol levels. From this survey a table of results was made. It shows . . .

1 The mean level of cholesterol for each group.

2 The percentage of the group with a cholesterol level above the recommended healthy limit of 5.0 mmol/dm³.

	Age 25–34 %	Age 35–44 %	Age 45–54 %	Age 55–64 %	Age 65–74 %
Men					
Group size	718	789	675	585	401
Mean (mmol/dm³)	5.3	5.8	5.9	5.8	5.5
% 5.0 and above	59.8	76.9	81.0	79.7	67.4
Women					
Group size	717	794	674	603	455
Mean (mmol/dm³)	5.0	5.4	5.8	6.3	6.2
% 5.0 and above	54.9	69.3	79.3	83.7	77.1

(a) The sample groups were of different sizes. Suggest which one provides the most reliable result and why. *(2 marks)*

(b) To remain healthy a maximum level of 5.0 mmol/dm³ for blood cholesterol is often recommended.

 (i) Which group is most at risk from ill-health due to their cholesterol levels? *(1 mark)*

 (ii) Which group is the second most at risk from ill-health due to their cholesterol levels? *(1 mark)*

 (iii) What in particular is the health risk from having a high blood cholesterol level? *(2 marks)*

(c) In which organ of the body is cholesterol made? *(1 mark)*

(d) What a person eats affects how much cholesterol the body makes. What other factor also affects how much is made? *(1 mark)*

(e) The amount and type of fat in the diet affects the level of cholesterol in the blood. For each of the following, state whether their presence in the diet increases, decreases a lot or decreases a little the blood cholesterol level.

 (i) saturated fats

 (ii) mono-unsaturated fats

 (iii) polyunsaturated fats *(3 marks)*

See page 8

See pages 42–3

GET IT RIGHT!

Check the mark allowance. The number of marks allowed for a question gives you valuable information on the extent and detail of an answer. For example, in question 1 part (b) (iii), there are 2 marks compared to only 1 mark for the other sections of part (b). This suggests that two health risks are needed. (Think about which parts of the body are affected.)

B1b | Evolution and environment

Our environment is precious and needs protecting

What you already know

Here is a quick reminder of previous work that you will find useful in this unit:

- In sexual reproduction, fertilisation happens when a male and a female sex cell join together.

- There are variations between members of the same species. Because of these variations, some individuals are more successful than others. In harsh conditions, the fittest survive.

- In sexual reproduction, information from two parents is mixed to make a new plan for the offspring. This leads to variation between members of a species.

- Variation between organisms of the same species has *environmental* as well as *inherited causes*.

- You can produce animals and plants with the features you want by a process called **selective breeding**.

- There are ways in which we can protect living organisms and the environment they live in.

- Sustainable development (when we replace the plants and animals we use) is becoming more and more important.

- Competition for the resources available is one thing which affects the size of populations of animals and plants.

- Toxic (poisonous) materials that we produce can build up in food chains and cause big problems.

RECAP QUESTIONS

1 a) How does sexual reproduction lead to variation within a species?

   b) How can the environment cause variation in a species?

2 What do we mean by 'the fittest survive'?

3 Imagine that you want to produce a type of dog with a very curly tail using selective breeding. Explain how you would do this. You can use diagrams if you think they will be helpful.

4 Think of as many ways as you can in which we might protect the environment and the living things in it. Explain why it is important to do this.

5 a) What do we mean by a 'population' of animals or plants?

   b) What sort of things do animals compete for? List as many things as you can which might affect the numbers of animals in a population.

   c) There is competition in a culture of bacteria. What sort of things might bacteria compete for?

6 a) Give two examples of food chains.

   b) Give an example of the way toxins can build up in a food chain.

   c) Explain how it is the carnivores at the end of the chain which are most likely to be affected.

Making connections

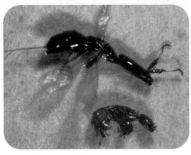

A female (top) and a male (bottom) fig wasp

There are about 700 different species of fig trees. Each one has its own species of pollinating wasps, without which it will die! The fig flowers of the trees are specially adapted so that they attract the right wasps.

Male fig wasps vary. Some species can fly but others are adapted to live in a fig fruit all their life. If they are lucky, a female wasp will arrive in the flower. Then the male will fertilise her. After this, he digs an escape tunnel for the female through the fruit and dies himself! The male wasp has special adaptations, such as a loss of his wings and very small eyes. These adaptations help him move around inside the fig fruit to find a female.

Female fig wasps have specially shaped heads for getting into fig flowers. They also have ovipositors. These allow them to place their eggs deep in the flowers of the fig tree.

Dr James Cook

If a fig tree cannot attract the right species of wasp, it will never be able to reproduce. In fact in some areas the trees are in danger of extinction because the wasp populations are being wiped out. Dr James Cook and his team at Imperial College, London are looking at the adaptations of the different wasps and their genetic material. They are trying to work out the relationships between all the different species.

ACTIVITY

Fig wasps are very strange animals. They have lots of adaptations which help them to reproduce sexually in just one species of tree.

Make a list of as many different types of animals or plants as you can that have strange ways of reproducing. Use a large piece of paper or a white board to record your ideas. You can include animals or plants which depend on just one other type of organism to be successful in life!

Chapters in this unit

Adaptation for survival   Variation   Evolution   How people affect the planet

B1b 5.1

Adaptation in animals

1 How can hair help animals survive in very cold climates?
2 What are the advantages – and disadvantages – of lots of body fat?
3 What is your surface area : volume ratio?

The variety of conditions on the surface of the Earth is huge. If you are a living organism, you could find yourself living in the dry heat of a desert or in wastelands of ice and snow. Fortunately, living organisms have special features (known as **adaptations**). These make it possible for them to survive in their particular habitat – however extreme the conditions might be!

Animals in cold climates

To survive in a cold environment you must be able to keep yourself warm. Arctic animals are adapted to reduce the heat they lose from their bodies as much as possible. You lose body heat through your body surface (mainly your skin). The amount of heat you lose is closely linked to your surface area : volume (SA/V) ratio.

Look at Figure 2. This explains why so many Arctic mammals, such as seals, walruses, whales, and polar bears, are relatively large. It keeps their surface area : volume ratio as small as possible and so helps them hold on to their body heat.

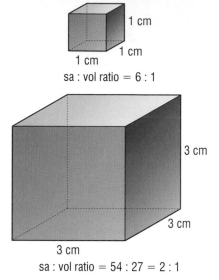

sa : vol ratio = 6 : 1

sa : vol ratio = 54 : 27 = 2 : 1

Figure 2 The ratio of surface area to volume falls as objects get bigger. You can see this clearly in the diagram. This is very important when you look at the adaptations of animals which live in cold climates.

a) Why are so many Arctic animals large?

Figure 1 The Arctic is a cold and bleak environment. However, the animals which live there are well adapted for survival. Notice the large size, small ears, thick coat and white camouflage of this polar bear.

Animals in very cold climates often have other adaptations too. The surface area of the thin skinned areas of their bodies – like their ears – is usually very small. This reduces their heat loss – look at the ears of the polar bear in Figure 1.

Many Arctic mammals have plenty of insulation, both inside and out. Blubber – a thick layer of fat that builds up under the skin – and a thick fur coat on the outside will insulate an animal very effectively. They really reduce the amount of heat lost through their skin.

The fat layer also provides a food supply. Animals often build up their blubber in the summer. Then they can live off their body fat through the winter when there is almost no food.

b) List three ways in which Arctic animals keep warm in winter.

Camouflage is important both to predators (so their prey doesn't see them coming) and to prey (so they can't be seen). The colours which would camouflage an Arctic animal in summer against plants would stand out against the snow in winter. Many Arctic animals, including the Arctic fox, the Arctic hare and the stoat, exchange the greys and browns of their summer coats for pure white in the winter.

... that polar bears stay white all year round? The white colour makes them less visible to the seals they hunt. Adult polar bears don't have any natural predators on the land – who would dare to attack a polar bear?

Surviving in dry climates

Dry climates are often also hot climates – like deserts! Deserts are very difficult places for animals to live. There is scorching heat during the day, followed by bitter cold at night, while water is in short supply.

The biggest challenges if you live in a desert are:

- coping with the lack of water, and
- stopping your body temperature from getting too high.

Many desert animals are adapted to need little or no drink. They get the water they need from the food they eat.

Mammals keep their body temperature the same all the time. So as the environment gets hotter, they have to find ways of keeping cool. Most mammals sweat to help them cool down. But this means they lose water, which is not easy to replace in the desert.

c) Why do mammals try to lose heat without sweating in hot, dry conditions?

Desert animals have other adaptations for cooling down. They are often most active in the early morning and late evening, when the temperature is comfortable. During the cold nights and the heat of the day they rest. You find them in burrows well below the surface, where the temperature doesn't change much.

Many desert animals are quite small, so their surface area is large compared to their volume. This helps them to lose heat through their skin. They often have large, thin ears as well to increase their surface area for losing heat.

Another adaptation of many desert animals is that they don't have much fur. Any fur they do have is fine and silky. They also have relatively little body fat stored under the skin. Both of these features make it easier for them to lose heat through the surface of the skin. The animals keep warm during the cold nights by retreating into their burrows.

Figure 3 Animals like this fennec fox have many adaptations to help them cope with the hot dry conditions. How many can you spot?

Figure 4 An elephant is pretty big but it lives in hot, dry climates. Its huge wrinkled skin would cover an animal which was much bigger still. The wrinkles increase the surface area to aid heat loss.

GET IT RIGHT!

Remember, the **larger** the animal, the **smaller** the surface area : volume (SA/V) ratio.
Animals often have **increased** surface areas in **hot** climates, and **decreased** surface areas in **cold** climates.

FOUL FACTS

Animals from the deep oceans are adapted to cope with enormous pressure, no light and very cold water. But if these deep-water organisms are brought up to the surface too quickly, they explode because of the rapid change in pressure.

KEY POINTS

1 All living things have adaptations which help them to survive in the conditions where they live.
2 Animals which are adapted for cold environments are often large, with a small surface area : volume (SA/V) ratio. They have thick insulating layers of fat and fur.
3 Changing coat colour in the different seasons gives animals year-round camouflage.
4 Adaptations for hot, dry environments include a large SA/V ratio, thin fur, little body fat and behaviour patterns that avoid the heat of the day.

SUMMARY QUESTIONS

1 a) List the main problems which face animals living in cold conditions like the Arctic.
   b) List the main problems which face animals living in the desert.
2 Give three ways in which animals that stay in the Arctic throughout the winter keep warm. Explain how these adaptations work.
3 Give three ways in which animals which live in a desert manage to keep cool without sweating so they don't lose water.
4 Explain why being quite large helps many Arctic animals to keep warm.

B1b 5.2   Adaptation in plants

LEARNING OBJECTIVES

1 How are plants adapted to live in dry conditions?
2 How do plants store water?

There are some places where plants simply cannot grow. In deep oceans no light penetrates, and no plants can grow. In the icy wastes of the Antarctic, no plants grow.

Almost everywhere else, including the hot, dry areas of the world, we can find plants growing. Without them there would be no food for the animals. But plants need water both for photosynthesis and to keep their tissues upright. If a plant does not get the water it needs, it wilts and eventually dies.

a) Why do plants need water?

Plants take in water through their roots in the soil. It moves up through the plant and is lost through the leaves in the **transpiration stream**. Plants lose water all the time through their leaves. There are small openings called **stomata** in the leaves of a plant. These open to allow gases in and out for photosynthesis and respiration. But at the same time water is lost by evaporation.

The rate at which a plant loses water is linked to the conditions it is growing in. When it is hot and dry, photosynthesis and respiration take place quickly. As a result, plants also lose water very fast. So how do plants that live in dry conditions cope? Most of them either reduce their surface area so they lose less water or they store water in their tissues. Some do both!

b) How do plants lose water from their leaves?

Changing surface area

When it comes to stopping water loss through the leaves, the surface area : volume ratio (see page 82) is very important to plants. There are a few desert plants which have broad leaves with a large surface area. These leaves collect the dew that forms in the cold evenings. They then funnel the water towards their shallow roots

Some plants in dry environments have curled leaves. This reduces the surface area of the leaf. It also traps a layer of moist air around the leaf which really cuts back the amount of water they lose by evaporation.

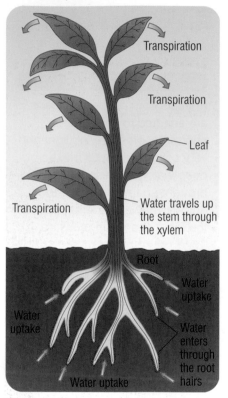

Figure 1 The transpiration stream means plants are losing water all the time by evaporation from their leaves. When the conditions are hot and dry, they lose water very quickly.

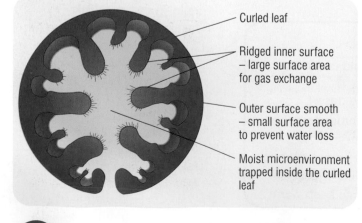

Curled leaf

Ridged inner surface – large surface area for gas exchange

Outer surface smooth – small surface area to prevent water loss

Moist microenvironment trapped inside the curled leaf

Figure 2 Plants that live on sand dunes near the sea have to survive very dry conditions. This marram grass, which you can find all around the British coast, has tightly curled leaves, which reduce the surface area available for water loss.

However, most plants which live in dry conditions have reduced the surface area of their leaves. This cuts down the area from which water can be lost. Some desert plants have small fleshy leaves with a thick cuticle to keep water loss down. The cuticle is a waxy covering on the leaf that stops water evaporating away.

The best-known desert plants are the cacti. Their leaves have been reduced to spines with a very small surface area indeed. This means the cactus only loses a tiny amount of water – and the spines put animals off eating the cactus as well!

c) Why do plants often reduce the surface area of their leaves to help them prevent water loss?

Storing water

Plants can also cope with dry conditions by storing water in their tissues. When there is plenty of water available after a period of rain, the plant stores it. Plants which store water in their fleshy leaves, stems or roots are known as **succulents**.

Cacti don't just rely on their spiny leaves to help them survive in dry conditions. They are succulents as well. The fat green body of a cactus is its stem, which is full of water-storing tissue. All these adaptations make cacti the most successful plants in a hot dry climate.

d) In which parts can a plant store its water?

Keep away!

One of the biggest problems for plants is being eaten by animals. Plants have a wide variety of adaptations designed to deal with this. Vicious thorns, unpleasant tastes and poisonous chemicals can all put animals off!

We have made use of some of these adaptations. For example, we use the bitter chemical in the bark of the cinchona tree to make quinine. This helps relieve the symptoms of malaria. What's more, the poison digitalis from foxgloves is used as a very effective heart medicine.

GET IT RIGHT!

Remember that plants need their stomata open for photosynthesis and respiration. This is why they lose water by evaporation from their leaves.

DID YOU KNOW?

An apple tree in the UK can lose a whole bath of water from its leaves every day. A large saguaro cactus in the desert loses less than one glass of water in the same amount of time!

Figure 3 Cacti are well adapted to survive in desert conditions

SUMMARY QUESTIONS

1 Copy and complete using the words below:

  adaptations   desert   plants   spiny   stem   water

  Cacti are …… which live in the …… . They have two main …… to help them survive. Their leaves have become …… and they store …… in their …… .

2 a) Explain why plants lose water through their leaves all the time.
  b) Why does this make living in a dry place such a problem?

3 Explain three adaptations which help plants living in dry conditions to reduce water loss from their leaves.

KEY POINTS

1 Plants lose water all the time by evaporation from their leaves.
2 Plants which live in dry places have adaptations which help to reduce water loss. These adaptations may often include reduced surface area of their leaves and/or water-storage tissues.

B1b 5.3

Competition in animals

Figure 1 Some herbivores, like these silk worms eating their mulberry leaves and the panda with its bamboo, only feed on one particular plant. They are very open to competition from other animals or to a disease that damages their food plant.

Animals and plants grow alongside lots of other living things, some from the same species and others completely different. In any area there will only be a limited amount of food, water and space, and a limited number of mates. As a result, living organisms spend their time competing for the things they need.

The best adapted organisms are most likely to be the winners of the **competition** for resources. They will be most likely to survive and produce healthy offspring.

a) Why do living organisms compete?

What do animals compete for?

Animals compete for many things, including:

- water
- territory
- mates

Competition for food is very common. Herbivores (animals which eat only plants) sometimes feed on many types of plant, and sometimes on only one or two different sorts. Many different species of herbivores will all eat the same plants. Just think how many types of animals eat grass!

The animals which eat a wide range of plants are most likely to be successful. If you are a picky eater, you risk dying out if anything happens to your only food source. An animal with wider tastes will just eat something else for a while!

Competition is common among carnivores (animals which eat only meat). They compete for prey. Small mammals like mice are eaten by animals like foxes, owls, hawks and domestic cats. The different types of animals all hunt the same mice. So the animals which are best adapted to the area will be most successful.

Carnivores have to compete with other members of their own species for their prey as well as with members of different species. Successful predators are adapted to have long legs for running fast and sharp eyes to spot prey. These features will be passed on to their offspring.

Animals often avoid direct competition with members of other species when they can. It is the competition between members of the same species which is most intense!

Prey animals compete with each other too – to be the one which **isn't** caught! Adaptations like camouflage colouring, so you don't get seen, and good hearing, so you pick up a predator approaching, are important for success.

b) Give one adaptation which would be useful to a plant-eater and one which would be helpful to a carnivore.

Competition for mates can be fierce. In many species the male animal puts a lot of effort into impressing the females. The males compete in different ways to win the privilege of mating with her.

In some species – like deer and lions – the males fight between themselves. Then the winner gets the females.

Many male animals display to the female to get her attention. Some birds have spectacular adaptations to help them stand out. Male peacocks have the most amazing tail feathers. They use them for displaying to other males (to warn them off) and to females (to attract them).

What makes a successful competitor?

A successful competitor is an animal which is adapted to be better at finding food or a mate than the other members of its own species. It also needs to be better at finding food and water than the members of other local species. What is more, it must also breed successfully.

Many animals are successful because they avoid competition with other species as much as possible. They feed in a way that no other local animals do, or they eat a type of food that other animals avoid. For example, one plant can feed many animals without direct competition. While caterpillars eat the leaves, greenfly drink the sap, butterflies suck nectar from the flowers and beetles feed on pollen.

It is much harder to avoid competition within the same species, but many animals try to do just that. They may set up and defend a **territory** – an area where they live and feed. This is a common way of making sure that they will be able to find enough food for themselves and for their young when they breed.

Figure 2 The spectacular display of a male peacock certainly attracts the attention of the females. And unlike animals which fight for their mates, the peacock doesn't risk getting hurt when he tries to win over the females.

FOUL FACTS

Different types of African dung beetles avoid competition with each other by attacking the same pile of dung at different times of day and in different ways. The most active beetles work in the heat of the day and make balls of dung which they roll away. The quieter tunnellers and the beetles that actually live in the dung heaps work as dusk is falling.

SUMMARY QUESTIONS

1 Match the following words to their definitions:

a) Competition	A An animal which eats plants.
b) Carnivore	B An area where an animal lives and feeds.
c) Herbivore	C An animal which eats meat.
d) Territory	D The way animals compete with each other for food, water, space and mates.

2 a) Give an example of animals competing with members of other species for food.
  b) Give an example of animals competing with members of the same species for food.
  c) Why can animals which rely on a single type of food be killed off so easily?

3 a) Give two ways in which animals compete for mates.
  b) What sort of adaptations would be needed to be successful in the two types of competitions in a)?

4 Explain the adaptations would you expect to find in:

  a) an animal which hunts mice?
  b) an animal which eats grass?
  c) a fish which feeds on other fish?
  d) an animal which feeds on the tender leaves at the top of trees?

GET IT RIGHT!

Learn to look at an animal and spot the adaptations which make it a successful competitor!

KEY POINTS

1 Animals often compete with each other for food and territories.
2 Animals compete for mates.
3 Animals have adaptations which make them good competitors.

B1b 5.4 Competition in plants

LEARNING OBJECTIVES

1 What do plants compete for?
2 Why do plants need light?

EXPERIMENTAL DATA

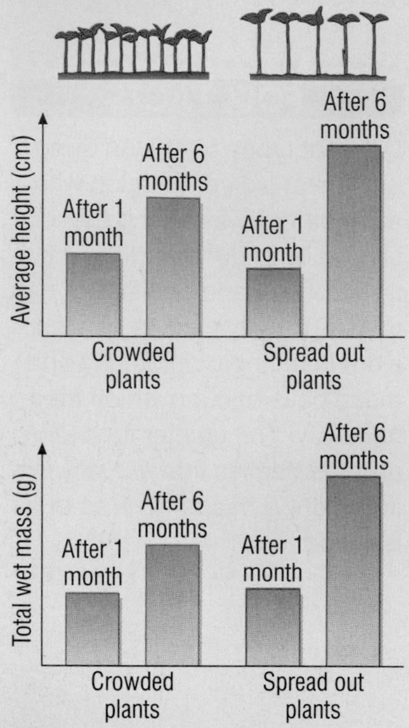

Figure 1 Experiments like this can be carried out to show the effect of competition on plants. All the conditions – light level, the amount of water and minerals available and the temperature were kept exactly the same for both sets of plants. The differences in their growth were the result of overcrowding and competition for resources in one of the groups.

Plants might look like peaceful organisms, growing silently in your local park. But the world of plants is full of cut-throat competition. They compete with each other for light, for water and for nutrients (minerals) from the soil.

They need light for photosynthesis, to make food using energy from sunlight. They need water for photosynthesis and to keep their tissues rigid and supported. And plants need minerals so they can make all the chemicals they need in their cells.

a) What do plants compete with each other for?

Why do plants compete?

Just like animals, plants are in competition both with other species of plants and with their own species. Big, tall plants like trees take up a lot of water and minerals from the soil and prevent light from reaching the plants beneath them. So the plants around them need adaptations to help them to survive.

If a plant sheds its seeds and they land nearby, the parent plant will be in direct competition with its own seedlings. Because the parent plant is large and settled, it will take most of the water, minerals and light. So the plant will deprive its own offspring of everything they need to grow successfully!

If the seeds from a plant all land close together – even if they are a long way from their parent – they will then compete with each other as they grow. So many plants have special adaptations which help them to spread their seeds over a wide area.

b) Why is it important that seeds are spread as far as possible from the parent plant?

Coping with competition

When plants are growing close to other species they often have adaptations which help them to avoid competition.

Small plants found in woodlands often grow and flower very early in the year. Although it is cold, plenty of light gets through the bare branches of the trees. The dormant trees take very little water out of the soil. The leaves shed the previous autumn have rotted down to provide minerals in the soil.

Plants like snowdrops, anemones and bluebells are all adapted to take advantage of these things. They flower, set seeds and die back again before the trees are in full leaf.

Another way plants compete successfully is by having different types of roots. Some plants have shallow roots taking water and minerals from near the surface of the soil. Others have long, deep roots, which go far underground. Both compete successfully for what they need without affecting the other.

If one plant is growing in the shade of another, it may grow taller to reach the light. It may also grow leaves with a bigger surface area to take advantage of all the light it does get.

c) How can short roots help a plant to compete successfully?

Spreading the seeds

To compete successfully, a plant has to avoid competition with its own seedlings. Usually, the most important adaptation for success is the way they shed their seeds.

Many plants use the wind to help them. Some produce seeds which are so small that they are carried easily by air currents. Many others produce fruits with special adaptations for flight to carry their seeds as far away as possible. Examples include the parachutes of the dandelion 'clock' and the winged seeds of the sycamore.

d) How do the fluffy parachutes of dandelion seeds help the seeds spread out?

Some plants use mini-explosions to spread their seeds. The pods dry out, twist and pop, flinging the seeds out and away.

Juicy berries, fruits and nuts are produced by plants to tempt animals to eat them. Once the fruit gets into the animal's gut, the tough seeds travel right through. They are deposited with the waste material in their own little pile of fertiliser, often miles from where they were eaten!

Fruits which are sticky or covered in hooks get caught up in the fur or feathers of a passing animal. They are carried around until they fall off hours or even days later.

Sometimes the seeds of different plants land on the soil and start to grow together. The plants which grow fastest will compete successfully against the slower-growing plants. For example:

- the plants which get their roots into the soil first will get most of the available water and minerals;
- the plants which open their leaves fastest will be able to photosynthesise and grow faster still, depriving the competition of light.

Plants compete at all levels, from spreading their seeds to the height they grow and how early they flower each year. The winners of the competitions are the ones we see. The losers just don't make it!

Figure 2 Coconuts will float for weeks or even months on ocean currents which can carry them hundreds of miles from their parents – and any other coconuts!

SUMMARY QUESTIONS

1. a) Give two ways in which plants can overcome the problems of growing in the shade of another plant.
    b) Explain how a primrose plant manages to grow and flower successfully in spite of living under a large oak tree.

2. a) Why is it so important that plants spread their seeds successfully?
    b) Give three examples of successful adaptations for spreading seeds.

3. The dandelion is a successful weed. Carry out some research and evaluate the adaptations that make it a better competitor than other plants on a school field.

KEY POINTS

1. Plants often compete with each other for light, for water and for nutrients (minerals) from the soil.
2. Plants have many adaptations, which make them good competitors.

B1b 5.5 How do *you* survive?

The most amazing plants in the world?

Most plants die without water – but not the resurrection plants. They can survive massive water losses. They don't prevent water loss or store water – they have adapted to cope with water loss when it happens. These amazing plants can lose up to 95% of their water content without suffering permanent damage.

When conditions get dry, the plants lose more and more water until all that is left are the small, shrivelled remains. The plant looks dead. It can last like this for weeks – but within about 24 hours of watering the tissues fill up with water again (rehydrate). The plant looks as good as new!

Dr Peter Scott and his team at the University of Sussex are trying to find out just how this survival mechanism works, because resurrection plants aren't just a fascinating fact. All over the world crops fail every year because conditions are too dry, and so millions of people don't get enough food. If scientists can find a way to produce 'resurrection crops', then starvation might become a thing of the past.

The difference 24 hours and some water can make to a resurrection plant!

ACTIVITIES

a) How might 'resurrection crops' prevent starvation in the world?

b) You want to get money for some research into the adaptations of a very unusual animal or plant (you can make one up!). Write a brief application for funding for your project. Using the example of the resurrection plants, explain how information about an unusual adaptation might lead to great benefits for people. Use this to back up your claim for money!

The fastest predator in the world?

It takes you about 650 milliseconds to react to a crisis. But the star-nosed mole takes only 230 milliseconds from the moment it first touches its prey to gulping it down. That's faster than the human eye can see!

What makes this even more amazing is that star-nosed moles live underground and are almost totally blind. Their main sense organ is a crown of fleshy tendrils around the nose – incredibly sensitive to touch and smell but very odd to look at!

It seems likely that they have adapted to react so quickly because they can't see what is going on. They need to grab their prey as soon as possible after they touch it. If they don't it might move away or try to avoid them. Then they wouldn't know where it had gone.

When you've got ultra-sensitive tendrils which can try out 13 possible targets every second, who needs eyes?!

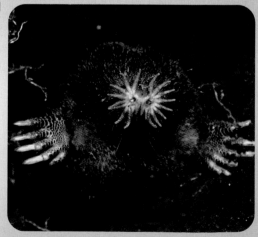

The star-nosed mole

Death by infection

The Komodo dragon – largest lizard in the world

The Komodo dragon is the largest lizard in the world. A big male can be over three metres long! They live in Indonesia and their colour varies depending on which island they make their home. They have long, forked tongues which give them an excellent sense of smell. They can smell rotting meat five miles away!

The Komodo dragon eats carrion (dead animals) but it is also a predator. But the dragons are reptiles. They cannot run fast for long or pounce on their prey, yet they can kill a huge water buffalo. How do they do it?

The dragons have 52 sharp teeth, but it is not the sharpness which makes them deadly, it is the bacteria which grow on them. A dragon will lie in wait for a water buffalo, and then rush out and grab one of the hind legs. It tears the leg, and the 15 different species of bacteria growing on its teeth get straight into the buffalo's blood stream. Within a couple of days, the buffalo will be dead from the lethal infection. The dragon just follows after its prey until the buffalo dies.

The dragon then eats almost 80% of its own weight in buffalo meat before resting quietly for a very long time!

A carnivorous plant

The Venus flytrap is a plant that grows on bogs. Bogs are wet and their peaty soil has very few minerals in it. This makes them a difficult place for plants to live.

Venus flytraps have special 'traps' which contain a sweet-smelling nectar. They sit wide open, displaying their red insides. Insects are attracted to the colour and the smell.

Inside the trap are many small, sensitive hairs. As the insect moves about to find the nectar, it will brush against these hairs. Once the hairs have been touched, the trap is triggered. It shuts, trapping the insect inside. Special enzymes then digest the insect inside the trap.

The Venus flytrap – an insect-eating plant!

The Venus flytrap uses the minerals from the digested bodies of its victims in place of the minerals it can't get from the bog soil. After the insect has been digested, the trap reopens ready for its next victim.

ACTIVITIES

c) Look carefully at the information you have been given about resurrection plants, star-nosed moles, Komodo dragons and Venus flytraps. For each one make a list of the adaptations which help them to survive.

d) There are so many different living organisms, each with their own adaptations. Choose three organisms that you know something about – or find out about three organisms which interest you. Make your own fact file on their adaptations and how these adaptations help them to compete successfully. Choose one organism that has an adaptation which has been used by people in some way. Try to include at least one plant!

SUMMARY QUESTIONS

1 Cold-blooded animals like reptiles and snakes take their body temperature from their surroundings and cannot move until they are warm.

a) Why do you think that there are no reptiles and snakes in the Arctic?

b) What problems do reptiles face in desert conditions and how do they cope with them?

c) Most desert animals are quite small. How does this help them survive in the heat?

2 a) What are the main problems for plants living in a hot, dry climate?

b) Why does reducing the surface area of their leaves help plants to reduce water loss?

c) Describe two ways in which plants can reduce the surface area of their leaves.

d) How else are some plants adapted to cope with hot, dry conditions?

e) Why are cacti such perfect desert plants?

3 a) How does marking out and defending a territory help an animal to compete successfully?

b) Bamboo plants all tend to flower and die at the same time. Why is this such bad news for pandas, but doesn't affect most other animals?

4 Why is competition between animals of the same species so much more intense than the competition between different species?

5 Use Figure 1 on page 88 to answer these questions.

a) Describe what happens to the height of both sets of seedlings over the first six months and explain why the changes take place.

b) The total wet mass of the seedlings after one month was the same whether or not they were crowded. After six months there was a big difference.

   i) Why do you think both types of seedling had the same mass after one month?

   ii) Explain why the seedlings that were more spread out each had more wet mass after six months?

c) When scientists carry out experiments such as the one described on page 88, they try to use large sample sizes. Why?

d) i) Name the control variable mentioned in the caption to Figure 1.

   ii) Why were these variables kept constant?

EXAM-STYLE QUESTIONS

1 Which of the following adaptations would **not** reduce the rate of transpiration in a plant?

  A Fleshy succulent leaves   **B** Thick waxy cuticle

  C Few stomata   **D** Reduced leaf area   (1)

2 Which of the following will **not** help an arctic mammal, such as a polar bear, to survive cold temperatures?

  A Thick fur coat

  B Thick layer of fat beneath the skin

  C Small extremities such as ears

  D A large surface area : volume ratio   (1)

3 To investigate competition between species, a series of 20 enclosures were set up to look at the effect of kangaroo rats on other, smaller rodents. Kangaroo rats were removed from half of the enclosures. Over three years the numbers of the smaller rodents were counted in both sets of enclosures.
The results are shown on the graph.

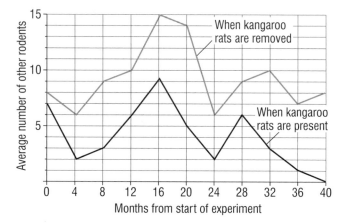

(a) In the enclosure with kangaroo rats, how many other rodents, on average, were present after 12 months?

  A 4   **B** 6   **C** 8   **D** 10   (1)

(b) After how many months was there the biggest difference between the average number of other rodents in the two sets of enclosures?

  A 16   **B** 20   **C** 32   **D** 40   (1)

(c) The conclusion that can be drawn from the experiment is that:

  A kangaroo rats eat other rodents.

  B kangaroo rats compete with other rodents for food.

  C kangaroo rats compete with other rodents and limit their population size.

  D kangaroo rats make it more difficult for other rodents to breed.   (1)

(d) How might the reliability of the results of this experiment be improved?

    A By increasing the number of enclosures of each type.

    B By using the same number of enclosures but increasing the number of kangaroo rats and other rodents in each type.

    C By using just one enclosure of each type rather than ten of each.

    D By having a third set of ten enclosures with only kangaroo rats present. (1)

4 Animals that live in the arctic have a range of adaptations that allow them to survive.

(a) Explain why the coat of an arctic fox is brown in summer and white in winter. (3)

(b) If the arctic fox adapts by changing its coat colour between summer and winter, why then does the polar bear remain white throughout the year? (1)

5 The gemsbok is a large herbivore living in dry desert regions of South Africa. It feeds on grasses that are adapted to the dry conditions by obtaining moisture from the air as it cools at night.

The table below shows the water content of these grasses and the feeding activity of the gemsbok over a 24-hour period.

Time of day	% water content of grasses	% of gemsboks feeding
03.00	18	40
06.00	23	60
09.00	25	20
12.00	8	17
15.00	6	16
18.00	5	19
21.00	7	30
24.00	14	50

(a) (i) Name the independent variable investigated. (1)

    (ii) Is this a categoric, ordered, discrete or continuous variable? (1)

(b) How does the water content of the grasses change throughout the 24 hour period? (1)

(c) Between which recorded times are more than 30% of the gemsboks feeding? (1)

(d) Suggest three reasons why the gemsboks benefit from feeding at this time. (3)

HOW SCIENCE WORKS QUESTIONS

Maize is a very important crop plant. Amongst many other foods, it is made into cornflakes. It is also grown for animal feed. The most important part of the plant is the cob which fetches the most money. In an experiment to find the best growing conditions, three plots of land were used. The young maize plants were grown in different densities in the three plots.

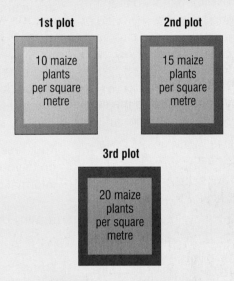

1st plot 10 maize plants per square metre

2nd plot 15 maize plants per square metre

3rd plot 20 maize plants per square metre

The results were as follows:

	Planting density (plants/m²)		
	10	15	20
Dry mass of shoots (kg/m²)	9.7	11.6	13.5
Dry mass of cobs (kg/m²)	6.1	4.4	2.8

a) What was the independent variable in this investigation? (1)

b) Draw a graph to show the effect of the planting density on the mass of the cobs grown. (3)

c) What is the pattern shown in your graph? (1)

d) This was a fieldwork investigation. What would the experimenter have taken into account when choosing the location of the three plots? (2)

e) Did the experimenter choose enough plots? Explain your answer. (1)

f) What is the relationship between the mass of cobs and the mass of shoots at different planting densities? (1)

g) The experimenter concluded that the best density for planting the maize is ten plants per m². Do you agree with this as a conclusion? Explain your answer. (2)

B1b 6.1

Inheritance

Figure 1 This picture shows a mother pig and her offspring. They aren't exactly the same as each other, but they are obviously related!

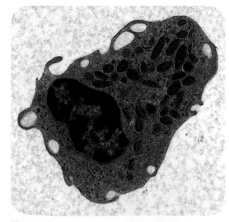

Figure 2 This micrograph shows a highly magnified human cell. In fact the nucleus of the cell would only measure about 0.005 mm! All the instructions for making you and keeping you going are inside this microscopic package. It seems amazing that they work!

Young animals and plants resemble their parents. Horses have foals and people have babies. Chestnut trees produce conkers which grow into little chestnut trees. Many of the smallest organisms that live in the world around us are actually identical to their parents. So what makes us the way we are?

Why do we resemble our parents?

Most families have characteristics which we can see clearly from generation to generation. People find it funny and interesting when one member of a family looks very much like another. Characteristics like nose shape, eye colour and dimples are *inherited* (passed on to you from your parents).

Your resemblance to your parents is the result of *genetic information* passed on to you in the sex cells (**gametes**) from which you developed. This genetic information determines what you will be like.

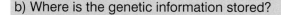

a) Why do you look like your parents?

Genes and chromosomes

The genetic information which is passed from generation to generation during reproduction is carried in the nucleus of your cells. Almost all of the cells of your body contain a nucleus. And it contains all the plans for making and organising a new cell. What's more, the nucleus contains the blueprint for a whole new you!

Imagine the plans for building a car. They would cover many sheets of paper! Yet in every living organism, the nucleus of the cells contains the information to build a whole new animal, plant, bacterium or fungus. A human being is far more complicated than a car. So where does all the information fit in?

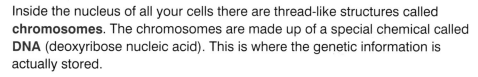

b) Where is the genetic information stored?

Inside the nucleus of all your cells there are thread-like structures called **chromosomes**. The chromosomes are made up of a special chemical called **DNA** (deoxyribose nucleic acid). This is where the genetic information is actually stored.

DNA is a long molecule made up of two strands which are twisted together to make a spiral. This is known as a double helix – imagine a ladder that has been twisted round!

Each different type of organism has a different number of chromosomes in their body cells. Humans have 46 chromosomes while turkeys have 82! You inherit half your chromosomes from your mother and half from your father, so chromosomes come in pairs. You have 23 pairs of chromosomes in all your normal body cells.

Each of your chromosomes contains thousands of **genes** joined together. These are the units of inheritance.

Each gene is a small section of the long DNA molecule. Genes control what an organism is like – its size, its shape and its colour. Each gene affects a different characteristic about you.

Your chromosomes are organised so that both of the chromosomes in a pair carry genes controlling the same things in the same place. This means your genes also come in pairs, one from your father and one from your mother.

c) Where would you find your genes?

Some of your characteristics are decided by a single pair of genes. For example, there is one pair of genes which decides whether or not you will have dimples when you smile! However most of your characteristics are the result of several different genes working together. For example, your hair and eye colour are both the result of several different genes.

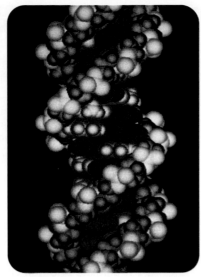

Figure 3 DNA! This huge molecule is actually made up of lots of smaller molecules joined together. Each gene is a small section of the big DNA strand.

Science pioneers: Cracking the code

For a very long time no-one knew how inheritance worked. By the 1940s most scientists thought that DNA was probably the molecule which carried inherited information from one generation to the next.

In the 1950s James Watson (a young American) and Francis Crick (from the UK) were working on the DNA problem at Cambridge. They took all the information they could find on DNA – including X-ray pictures of the molecule taken by another team, Maurice Wilkins and Rosalind Franklin in London.

Watson and Crick tried to build a model of the DNA molecule that would explain everything they knew. When they finally realised that the bases always paired up in the same way they had cracked the code. The now famous DNA double helix was seen for the first time.

SUMMARY QUESTIONS

1 Copy and complete using the words below:

> **chromosomes   DNA   genes   genetic information**
> **gametes   nucleus   resemble**

Offspring …… their parents because of …… …… passed on to them in the …… (sex cells) from which they developed. The information is contained in the ……, made of a chemical called …… found in the …… of the cell. The information is carried in small units called …… .

2 a) What is the basic unit of inheritance?
   b) Offspring inherit information from their parents, but do not look exactly like them – why not?

3 a) Which molecule carries genetic information?
   b) Why do chromosomes come in pairs?
   c) Why do genes come in pairs?
   d) How many genes do scientists think human beings have?

Figure 4 Francis Crick and James Watson – the two men who first showed the world how DNA works. Watson, Crick and Wilkins all received the Nobel Prize for their work. Rosalind Franklin died of cancer before the prizes were awarded.

KEY POINTS

1 Young animals and plants have similar characteristics to their parents. That's because of genetic information passed on to them in the sex cells from which they developed.

2 The nucleus of your cells contains chromosomes. Chromosomes carry the genes that control the characteristics of your body.

B1b 6.2

Types of reproduction

LEARNING OBJECTIVES

1 Why does asexual reproduction result in offspring that are identical to their parents?
2 How does sexual reproduction produce variety?

Reproduction is very important to living things. It is during reproduction that genetic information is passed on from parents to their offspring. There are two very different ways of reproducing – *asexual reproduction* and *sexual reproduction*.

Asexual reproduction

Asexual reproduction only involves one parent. The process produces more organisms completely identical to itself. There is no joining of special sex cells and there is no variety in the offspring.

Asexual reproduction gives rise to offspring known as **clones**. Their genetic material is identical both to the parent and to each other. Although there is no variety, asexual reproduction is very safe – you don't have to worry about finding a partner!

a) Why is there no variety in offspring from asexual reproduction?

Asexual reproduction is very common in the smallest animals and plants and in bacteria. However, many bigger plants like daffodils, strawberries and brambles do it too. Bulbs, corms, tubers (like potatoes), runners and suckers are all ways in which plants reproduce asexually. Asexual reproduction also takes place all the time in your own body, as cells divide to grow and to replace worn-out tissues.

Sexual reproduction

The other way of passing information from parents to their offspring is through sexual reproduction. Sexual reproduction involves the joining of a male sex cell and a female sex cell from two parents. These two special cells (gametes), one from each parent, join together to form a new individual.

Figure 1 A mass of daffodils like this can contain hundreds of identical flowers. This is because they come from bulbs which reproduce asexually.

If you are the result of sexual reproduction, you will inherit genetic information from both parents. You will have some characteristics from both of your parents, but won't be exactly like either of them. This introduces variety. In plants the gametes involved in sexual reproduction are found within ovules and pollen. In animals they are called ova (eggs) and sperm.

Sexual reproduction is more risky than asexual reproduction, because it relies on the sex cells from two individuals meeting. In spite of this, variety is so important to survival that sexual reproduction is seen in species of organisms ranging from bacteria to people!

b) How does sexual reproduction cause variety in the offspring?

Variation

Sexual reproduction involves the joining of different genetic information. This results in offspring which show much more variation than the offspring from asexual reproduction. This is a great advantage in making sure the species survives. That's because the more variety there is in a group of individuals, the more likely it is that at least a few of them will have the ability to survive difficult conditions.

GET IT RIGHT!

Asexual reproduction results in identical genetic information being passed on. Sexual reproduction makes sure the genetic information is mixed so there is variety in the offspring.

If we take a closer look at how sexual reproduction works, it becomes clear how variation appears in the offspring.

Each pair of genes affects a different characteristic about you. However the genes in a pair can come in different forms. These different versions of the same gene are called **alleles**. Most things about you are controlled by lots of different pairs of genes. Luckily some of your characteristics are controlled by one gene with just two possible alleles. For example, there are genes which decide whether:

- your earlobes are attached closely to the side of your head or hang freely,
- your thumb is straight or curved,
- you have dimples when you smile,
- you have hair on the second segment of your ring finger.

We can use these genes to help us understand how inheritance works.

NEXT TIME YOU...

... look at your finger tips, take a close look at your fingerprints. They are one of your few characteristics that aren't controlled completely by your genes! Even identical twins (who have identical genetic information) have different fingerprint patterns. Something that goes on before you are even born must also be involved in making your unique fingerprint patterns.

c) Why is variety important?

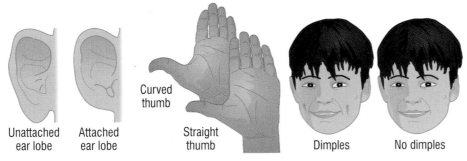

Curved thumb

Straight thumb

Unattached ear lobe

Attached ear lobe

Dimples

No dimples

Figure 2 These are all human characteristics which are controlled by a single pair of genes, so they can be very useful in helping us to understand how sexual reproduction introduces variety and how inheritance works

The gene which controls dimples has two possible forms – an allele for dimples, and an allele for no dimples. The gene for dangly earlobes also has two possible alleles – one for dangly earlobes and one for earlobes which are attached. Some features have lots of different possible alleles.

You will get a random mixture of thousands of alleles from your parents – which is why you don't look exactly like either of them!

Figure 3 Although these young people have some family likenesses, the variety which results from the mixing of their parents' genetic information can clearly be seen!

SUMMARY QUESTIONS

1 Define the following words:

a) asexual reproduction
b) sexual reproduction
c) gamete
d) variety.

2 a) Why is sexual reproduction more risky for individuals than asexual reproduction?
b) What is the big advantage of sexual reproduction over asexual reproduction?

3 A daffodil reproduces asexually using bulbs and sexually using flowers.

a) How does this help to make them very successful plants?
b) Explain the genetic differences between a daffodil's sexually and asexually produced offspring.

KEY POINTS

1 In asexual reproduction there is no joining of gametes and only one parent. There is no genetic variety in the offspring.
2 In sexual reproduction male and female gametes join. The mixture of genetic information from two parents leads to genetic variety.

B1b 6.3

Cloning

LEARNING OBJECTIVES

1 What is a clone?
2 Why do we want to create clones?

Figure1 Simple cloning by taking cuttings is a technique used by gardeners and nurserymen all around the world. It gives us plants like these.

A clone is an individual which has been produced asexually from its parent. It is therefore genetically identical to the parent. Many plants reproduce naturally by cloning, and this has been used by farmers and gardeners for many years.

Cloning plants

Gardeners can produce new plants by taking cuttings from older plants. This is a form of artificial asexual reproduction which has been carried out for hundreds of years. How do you take a cutting? First you remove a small piece of a plant – often part of the stem or sometimes just part of the leaf. If you grow it in the right conditions, new roots and shoots will form to give you a small, complete new plant.

Using this method you can produce new plants quickly and cheaply from old plants. The cuttings will be genetically identical to the parent plants.

Many growers now use hormone rooting powders to encourage the cuttings to grow. They are most likely to develop successfully if you keep them in a moist atmosphere until their roots develop. We produce plants such as orchids and many conifer trees commercially by cloning in this way.

a) Why does a cutting look the same as its parent plant?

Cloning tissue

Taking cuttings is a very old technique. In recent years scientists have come up with a more modern way of cloning plants called *tissue culture*. It is more expensive but it allows you to make thousands of new plants from a tiny piece of plant tissue. If you use the right mixture of plant hormones, you can make a small group of cells from the plant you want produce a big mass of identical plant cells.

Then, using a different mixture of hormones and conditions, you can stimulate each of these cells to form a tiny new plant. This type of cloning guarantees that the plants you grow will have the characteristics you want.

b) What is the advantage of tissue culture over taking cuttings?

Cloning animals

In recent years cloning technology has moved forward even further and now includes animals. In fact cloning animals is now quite common in farming, particularly cloning cattle embryos. Cows normally produce only one or two calves at a time. If you use embryo cloning, your very best cows can produce many more top-quality calves each year.

In embryo cloning, you give a top-quality cow fertility hormones to make her produce a lot of eggs. You then fertilise these eggs using sperm from a really good bull. Often this is done inside the cow, and the embryos which are produced are then gently washed out of her womb. Sometimes the eggs are collected and you add the sperm in a laboratory to produce the embryos.

SCIENCE @ WORK

Cloning cattle embryos and transferring them to host cattle is skilled and expensive work. Teams of scientists, technicians and vets are constantly working to improve the technique even more.

At this very early stage of development every cell of the embryo can still form all of the cells needed for a new cow. They have not become specialised.

1 Divide each embryo into several individual cells.
2 Each cell grows into an indentical embryo in the lab.
3 Transfer embryos into their host mothers, which have been given hormones to get them ready for pregnancy.
4 Identical cloned calves born. They are not biologically related to their mothers.

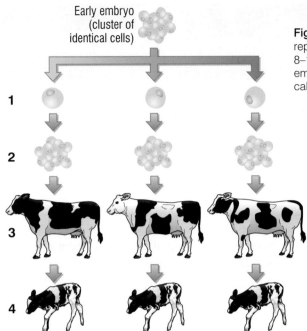

Early embryo (cluster of identical cells)

1

2

3

4

Figure 2 Using normal sexual reproduction, a top cow might produce 8–10 calves during her working life. Using embryo cloning she can produce more calves than that in a single year!

Cloning embryos in this way has made it possible for us to take high-quality embryos all around the world. We can take them to places where cattle with a high milk yield or lots of meat are badly needed for breeding with poor local stock. Embryo cloning is also used to make lots of identical copies of embryos that have been genetically modified to produce medically useful compounds. (See pages 102 and 103.)

GET IT RIGHT!

- Remember clones have identical genetic information.
- Make sure you are clear about the difference between a tissue and an embryo.

SUMMARY QUESTIONS

1 Match up the following definitions:

a) Cuttings	A Splitting cells apart from a developing embryo before they become specialised to produce several identical embryos.
b) Tissue cloning	B Taking a small piece of a stem or leaf and growing it on in the right conditions to produce a new plant.
c) Asexual reproduction	C Getting a few cells from a desirable plant to make a big mass of identical cells each of which can produce a tiny identical plant.
d) Embryo cloning	D Reproduction which involves only one parent, there is no joining of gametes and the offspring are genetically identical to the parent.

2 Tissue cloning and taking cuttings both give you plants which are identical to their parent. How do these two methods of plant cloning differ?

3 a) Why is the ability to clone cattle embryos so useful?
   b) Draw a flow diagram to show the stages of the embryo cloning of cattle.
   c) Comment on the economic and ethical issues involved in embryo cloning in cattle.

KEY POINTS

1 The genetically identical offspring produced by asexual reproduction are known as **clones**.
2 New plants can be produced quickly and cheaply by taking cuttings from older plants. The new plants are genetically identical to the older ones.
3 There are a number of more modern cloning techniques. These include tissue culture of plants and embryo cloning and transfers in animals.

B1b 6.4   New ways of cloning animals

True cloning of animals, without sexual reproduction involved at all, has been a major scientific breakthrough. The basic technique is known as *fusion cell cloning*. It is the most complicated form of asexual reproduction you can find!

Fusion cell cloning

To clone a cell from an adult animal is easy. Asexual reproduction takes place all the time in your body to produce millions of identical cells. But to take a cell from an adult animal and make an embryo or even a complete identical animal is a very different thing.

Here are the steps involved:

● The nucleus is taken from an adult cell.
● At the same time the nucleus is removed from an egg cell from another animal of the same species.
● The nucleus from the original adult cell is placed in the empty egg and the new cell is given a tiny electric shock.
● This fuses the new cell together, and starts the process of cell division.
● An embryo begins to develop which is genetically identical to the original adult animal.

Adult cell cloning

Fusion cell cloning has been used to produce whole animal clones. The first large mammal ever to be cloned from another adult animal was Dolly the sheep, born in 1997. A team of scientists in Edinburgh produced Dolly from the adult cell of another sheep.

When a new animal is produced, this is known as *adult cell* or *reproductive cloning*. It is still relatively rare. You still have to fuse the nucleus of one cell with the empty egg of another animal. Then you have to place the embryo which results into the womb of a third animal. It develops there until it is born.

a) What is the name of the technique which produced Dolly the sheep?

When Dolly was produced she was the only success from hundreds of attempts. The technique is still difficult and unreliable, but scientists hope that it will become easier in future.

Figure 1 Dolly the sheep was the first large mammal to be cloned from another adult animal. Her birth caused great excitement and many scientists have tried to clone other animals since.

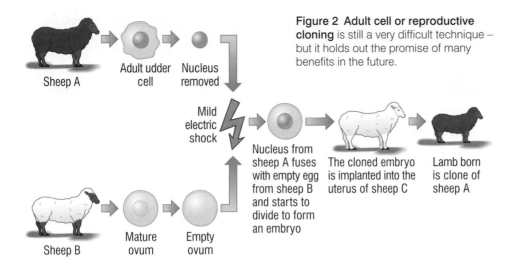

Figure 2 Adult cell or reproductive cloning is still a very difficult technique – but it holds out the promise of many benefits in the future.

Sheep A → Adult udder cell → Nucleus removed

Mild electric shock

Nucleus from sheep A fuses with empty egg from sheep B and starts to divide to form an embryo

The cloned embryo is implanted into the uterus of sheep C

Lamb born is clone of sheep A

Sheep B → Mature ovum → Empty ovum

The benefits and disadvantages of adult cell cloning

One big hope for adult cell cloning is that animals which have been genetically engineered to produce useful proteins in their milk can be cloned. This would give us a good way of producing large numbers of cloned, medically useful animals.

This technique could also be used to help save animals from extinction, or even bring back species of animals which died out years ago. The technique could be used to clone pets or prized animals so that they continue even after the original has died. However, some people are not happy about this idea. (See page 104.)

b) How might adult cell cloning be used to help people?

There are some disadvantages to this exciting science as well. Many people fear that the technique could lead to the cloning of human babies. At the moment this is not possible, but who knows what is in the future?

One big problem is that modern cloning produces lots of plants or animals with identical genes. In other words, cloning reduces variety in a population. This means the population is less able to survive any changes in their environment which might happen in the future. That's because if one of them does not contain a useful mutation, none of them will.

In a more natural population, at least one or two individuals can usually survive change. They go on to reproduce and restock. This could be a problem in the future for cloned crop plants, or for cloned farm animals.

SUMMARY QUESTIONS

1 Copy and complete:

Dolly the sheep was created from the …… cell of another sheep. She was …… to this sheep. This technique is known as …… …… …… .

2 Produce a flow chart to show how fusion cell cloning works.

3 a) List the main advantages of the development of adult cell cloning techniques and the main disadvantages.
   b) Give your own opinion about whether work on the technique should be allowed to continue. Explain your point of view.

KEY POINTS

1 Fusion cell cloning is a form of asexual reproduction
2 In adult cell cloning a whole cloned animal can result. The nucleus from a cell from an adult animal is transferred to an empty egg cell from another animal. A small electric shock fuses the cell and starts embryo development. The embryo is placed in a third animal to develop.

The Sheffield College
Hillsborough LRC

B1b 6.5 Genetic engineering

LEARNING OBJECTIVES

1 What is genetic engineering?
2 How are genes transferred from one organism to another?
3 What are the issues involved in using genetic engineering?

When you clone plants and animals, you are changing the natural processes of reproduction. There is another new technology which takes the changes much further – **genetic engineering** (also known as **genetic modification**). Genetic engineering is used to change an organism and give it new characteristics which we want to see.

What is genetic engineering?

Genetic engineering involves changing the genetic material of an organism. You take a small piece of DNA – a gene – from one organism and transfer it to the genetic material of a completely different organism. So, for example, genes from the chromosomes of one of your human cells can be 'cut out' using enzymes and transferred to the cell of a bacterium. Your gene carries on making a human protein, even though it is now in a bacterium.

a) How is a gene taken out from one organism to be put into another?

If genetically engineered bacteria are cultured on a large scale they will make huge quantities of protein from other organisms. We now use them to make a number of drugs and hormones used as medicines.

Transferring genes to animal and plant cells

There is a limit to the types of proteins bacteria are capable of making. As a result, genetic engineering has moved on. Scientists have found that genes from one organism can be transferred to the cells of another type of animal or plant at an early stage of their development. As the animal or plant grows it develops with the new desired characteristics from the other organism.

b) Why are genes inserted into animals and plants as well as bacteria?

The benefits of genetic engineering

One of the biggest advantages of genetically engineered bacteria is that they can make exactly the protein needed, in exactly the amounts needed and in a very pure form. For example, people with diabetes need supplies of the hormone insulin. It used to be extracted from the pancreases of pigs and cattle but it wasn't quite the same as human insulin, and the supply was quite variable. Both of those problems have been solved by the introduction of genetically engineered human insulin. (See Figure 1.)

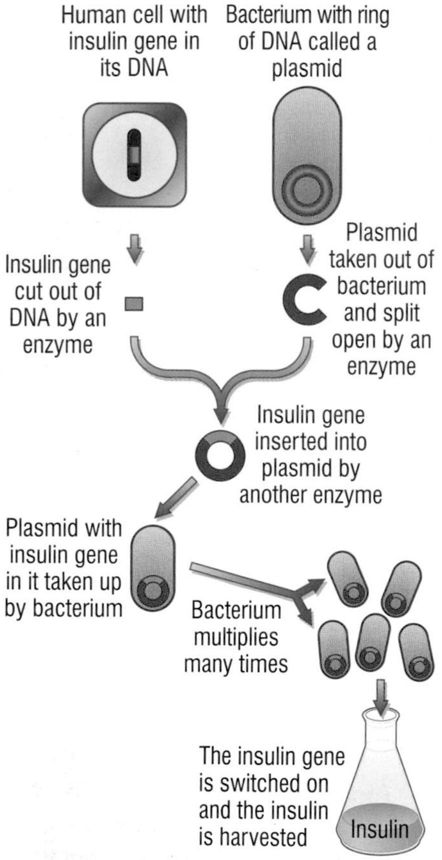

We can use engineered genes to improve the growth rates of plants and animals. They can be used to improve the food value of crops and to reduce the fat levels in meat. They are used to produce plants which make their own pesticide chemicals. GM food lasts longer in the supermarkets. It can also be designed to grow well in dry, hot or cold parts of the world. So it could help to solve the problems of world hunger.

A number of sheep and other mammals have also been engineered to produce life-saving human proteins in their milk. These are much more complex proteins than the ones produced by bacteria. They have the potential to save many lives.

Figure 1 The principles of genetic engineering. A bacterial cell receives a gene from a human being.

Human cell with insulin gene in its DNA

Bacterium with ring of DNA called a plasmid

Insulin gene cut out of DNA by an enzyme

Plasmid taken out of bacterium and split open by an enzyme

Insulin gene inserted into plasmid by another enzyme

Plasmid with insulin gene in it taken up by bacterium

Bacterium multiplies many times

The insulin gene is switched on and the insulin is harvested

Insulin

DID YOU KNOW...

... glowing genes from jellyfish have been used to produce crop plants which give off a blue light when they are attacked by insects. Then the farmer knows when they need spraying!

Human engineering

If there is a mistake in your genetic material, you may have a genetic disease. These can be very serious. Many people hope that genetic engineering can solve the problem.

It might become possible to put 'healthy' DNA into the affected cells by genetic engineering, so they work properly. Perhaps the cells of an early embryo can be engineered so that the individual develops to be a healthy person. If these treatments become possible, many people would have new hope of a normal life for themselves or their children.

Figure 2 These sheep look very normal – but they are a genetically engineered flock producing human proteins in their milk. The proteins are used in life-saving medicines.

c) What do we mean by a 'genetic disease'?

The disadvantages of genetic engineering

Genetic engineering is still a very new science. There are many concerns about it as no-one can yet be completely sure what all of the long-term effects might be. For example, it seems possible that insects may become pesticide-resistant if they eat a constant diet of pesticide-forming plants.

Some people are concerned about the effect of eating genetically modified food on human health. Genes from genetically modified plants and animals might spread into the wildlife of the countryside. Genetically modified crops are often made infertile, which means farmers in poor countries have to buy new seed each year.

And people may want to manipulate the genes of their future children. This might be to make sure they are born healthy, but what if it is to have a child who is clever, or good-looking, or good at sport? The idea of 'designer babies' causes concern for many people. Genetic engineering raises issues for us all to think about. (See page 105.)

Figure 3 You can't tell what is genetically modified and what isn't just by looking at it! In the UK very few genetically modified foods are sold. The ones that are have to be clearly labelled. But many other countries, including the USA, are far less worried and use GM food quite widely.

SCIENCE @ WORK

There are many scientists working in genetic engineering. Some work for pharmaceutical companies developing new medicines. Others are doing medical research and some are involved in agriculture and crop breeding.

KEY POINTS

1 In genetic engineering, genes from the chromosomes of humans and other organisms can be 'cut out' using enzymes and transferred to the cells of bacteria.
2 Genes can also be transferred to the cells of animals and plants at an early stage of their development.
3 There are many potential advantages and disadvantages to the use of genetic engineering.

SUMMARY QUESTIONS

1 Copy and complete using the words below:

cell   engineering   enzymes   gene   genetic   transfer

Genetic ...... involves changing the ...... material of an organism. You cut a ...... from one organism using .... Then ...... it to the ...... of a completely different organism.

2 a) Make a flow diagram that explains the stages of genetic engineering.
   b) Make two lists, one to show the possible advantages of genetic engineering and the other to show the possible disadvantages.
   c) Do you think genetic engineering is a good idea? Should it be allowed? Justify your views.

B1b 6.6   Making choices about technology

Cc – A REAL COPYCAT!

Cc, or Copycat, was the first cloned cat to be produced. Born in 2002, she was a change of direction. Most of the research into cloning had been focused on farm and research animals – but cats are thought of first and foremost as pets.

Cc the cloned cat is unaware of the stir she has caused. Cats like this one are often well-loved pets – but should we really be cloning our old friends?

Much of the funding for cat cloning in the US comes from companies who are hoping to be able to clone people's dying or dead pets for them. It has already been shown that a succesful clone can be produced from a dead animal. Cells from beef from a slaughterhouse were used to create a live cloned calf.

But to make Cc, 188 attempts were made producing 87 cloned embryos, only one of which resulted in a kitten. Cloning your pet won't be easy or cheap. The issue is, should people be cloning their dead pets, or should they be learning to grieve, appreciate the animal they had and give a home to one of the thousands of unwanted cats already in existence? Even if a favourite pet is cloned, it may look nothing like the original because the coat colour of many cats is the result of random gene switching in the skin cells. The markings would never be the same again, even if the DNA was!

THE FOAL WHO COULD CHANGE RACING – FOR GOOD! By David Turf

This little foal with her mum looks just like any other, but in fact she's made history!

The foal in this photo with her mother is no ordinary young horse. Prometea is the first cloned horse – and her surrogate mother is also her identical twin, because the foal is a clone of the mare who gave birth to her. This new technology has in turn led to a breakthrough which could change the breeding of racehorses forever.

Pieraz is a famous Arab horse who has been world endurance racing champion several times. Pieraz 2, a foal born in 2005, is his closest relative. 'So what?', I hear you say. It is very common for successful racehorses to be used for breeding. The difference here is that Peiraz was neutered when he was still a youngster – and Pieraz 2 is not his son, but his clone!

Changing the genes – right or wrong?

Our first daughter has been affected by a dreadful genetic disease. You don't know what it's like until it happens to you. If we had another baby, I'd want them to change the genes if they could when the embryo was really tiny. Then we'd know the baby wasn't going to be ill – and any grandchildren we had in the future would be alright as well.

I think it is wrong to interfere with nature. It must be awful if your family is affected, but if they start fiddling about with the genes of a tiny embryo where will it all end? It'll be designer babies next, you mark my words. If they can change the genes that can make you ill, you can't tell me they won't be able to make your baby really clever or good-looking if you're prepared to pay enough.

One of the most exciting chances genetic engineering can give us is the development of gene therapy. Most of our research is looking at changing the genetic material in the affected cells by adding healthy genes or switching off damaged genes. But we are only treating the disease, not curing it. The affected person can pass on the faulty genes to their children, who in turn will need to be treated.

Gene therapy actually offers a way of curing genetic diseases. It would involve changing the genes of a fertilised egg or very early embryo so that the baby is born with healthy genes in all its cells. This is known as germ line gene therapy. We know it raises some major ethical issues and opinion across the world is strongly divided about it. In fact most countries, including the UK, have so far completely banned germ line gene therapy.

I'm more concerned about GM foods. Who knows what we're all eating nowadays. I don't want strange genes inside me, thank you very much. We've got plenty of fruit and vegetables as it is – why do we need more?

I think GM food is such a good idea. If the scientists can modify crops so they don't go off so quickly, food should get cheaper, and there will be more to go around. And what about these plants that produce pesticides – that'll stop a lot of crop spraying, so that should make our food cleaner and cheaper. It's typical of us in the UK that we moan and panic about it all.

We have some real worries about the GM crops which don't form fertile seeds. It does mean the growers in the countries where we do a lot of work are going to struggle. In the past they just kept seeds from the previous year's crops, so it was cheap and easy. On the other hand, these GM crops don't need spraying very much. They grow well in our dry conditions and they keep well too – so there are some advantages.

ACTIVITY

You are going to produce a 10-minute slot for a daytime television show entitled '**Genetic engineering – a good thing or not?**' You can ask any of the people shown here to come on your show and express their views. Using this and the information about genetic engineering on pages 102 and 103 to help you, plan out the script for your time on air. Remember that you have to inform the public about genetic engineering, entertain them and make them think about the issues involved.

SUMMARY QUESTIONS

1

a) How has the small plant shown in diagram A been produced?

b) What sort of reproduction is this?

c) How were the seeds in B produced?

d) How are the new plants which you would grow from the packet of seeds shown in B different from the new plants shown in A?

2 Tissue culture techniques mean that 50 000 new raspberry plants can be gained from one old one instead of 2 or 3 taking cuttings. Cloning embryos from the best bred cows means that they can be genetically responsible for thirty or more calves each year instead of two or three.

a) How does tissue culture differ from taking cuttings?

b) How can one cow produce thirty or more calves in a year?

c) What are the similarities between cloning plants and cloning animals in this way?

d) What are the differences in the techniques for cloning animals and plants?

e) Why do you think there is so much interest in finding different ways to make the breeding of farm animals and plants increasingly efficient?

3 Human growth is usually controlled by the pituitary gland in your brain. If you don't make enough hormone, you don't grow properly and remain very small. This condition affects 1 in every 5000 children. Until recently the only way to get growth hormone was from the pituitary glands of dead bodies. Genetically engineered bacteria can now make plenty of pure growth hormone.

a) Draw and label a diagram to explain how a healthy human gene for making growth hormone can be taken from a human chromosome and put into a working bacterial cell.

b) What are the advantages of producing substances like growth hormone using genetic engineering?

EXAM-STYLE QUESTIONS

1 Young plants and animals resemble their parents. This is because characteristics are inherited by the young from their parents.

Match words **A**, **B**, **C** and **D** with the spaces **1** to **4** in the sentences.

A Chromosomes   **B** DNA

C Gametes   **D** Genes

Genetic information is passed from parents to offspring in sex cells called …**1**… The sex cells contain thread-like structures known as …**2**… .

The thread-like structures contain thousands of …**3**…. that are the units of inheritance.

These units of inheritance are small sections of a double helix called …**4**… .   (4)

2 The table is about the production of offspring.

Match words **A**, **B**, **C** and **D** with the processes **1** to **4** in the table.

A Sexual reproduction   **B** Asexual reproduction

C Inheritance   **D** Cloning

	Process
1	Joining of male and female sex cells (gametes) to produce young
2	Making young that are identical to both their parents and to each other, especially in agriculture
3	Producing offspring without sex cells (gametes)
4	Passing on characteristics from parents to offspring

(4)

3 Plant tissue culture is a method used to create new plants. One method is described below:

- A small piece of tissue is removed from a plant.
- Under sterile conditions, the tissue is placed in a vessel containing nutrients.
- A mass of identical plant cells develops.
- These cells are placed in a medium containing nutrients and plant growth regulators (hormones).
- Young plants develop that are separated and grown to maturity.

(a) What type of reproduction is involved in plant tissue culture?   (1)

(b) Why is a disease more likely to kill every one of a group of plants produced by tissue culture than a group grown from seeds?   (2)

(c) Suggest one advantage of growing plants from tissue culture over growing them from cuttings or seeds.

(1)

4 Some humans suffer from diabetes. One form of diabetes is caused by the inability to make the hormone insulin. Diabetics can lead normal lives providing they can inject insulin into their bloodstream at regular intervals. The diagram shows how insulin can be made using genetic engineering.

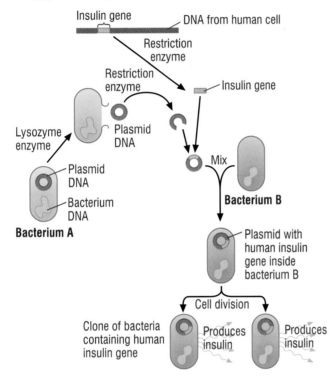

(a) **Using only the information in the diagram** suggest what the following enzymes do:

   (i) lysozyme enzyme (1)

   (ii) restriction enzyme. (1)

(b) **Using only the information in the diagram**, describe how the human insulin gene is transferred into a bacterium which then makes insulin. (6)

(c) Insulin is produced from cloned bacteria that contain the human insulin gene.

   (i) What is a clone? (1)

   (ii) Do you think that cloning is the result of sexual or asexual reproduction? Explain your answer. (1)

   (iii) A bacterium can divide every 20 minutes. Starting with 50 bacteria, how many bacteria would there be after 2 hours? Show your working. (1)

(d) Insulin to treat diabetes used to be extracted from the pancreases of pigs and cattle. Give two advantages of insulin produced by genetic engineering over extracting it from animals. (2)

HOW SCIENCE WORKS QUESTIONS

1 DNA outside the nucleus

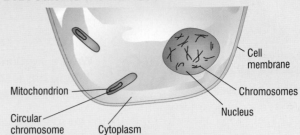

Mitochondria are sometimes referred to as the powerhouse of the cell. They carry out much of the cell's respiration. They contain their own DNA as a circular chromosome. Research has shown that this DNA is passed almost completely from the mother to the child. This is because the mitochondria are to be found in the cytoplasm. Some very rare diseases can be caused by the mitochondrial DNA mutating. These diseases can cause blindness or a disease very much like type 2 diabetes and deafness. Some children who inherit mutated mitochondrial DNA die at a very young age.

Use your scientific knowledge of cell structure and your understanding of the technology of cloning to answer the following questions.

a) Suggest how women who are at risk of passing on this mutated mitochondrial DNA might be helped. (1)

b) What are the ethical issues around this research? (2)

c) Who should be making decisions about whether or not families should be helped in this way? (3)

d) What is the role of the scientist in helping to make these decisions? (1)

2 Genetically modified (GM) crops

The first trials of GM crops were destroyed during the summer of 1999. Protesters, concerned that pollen from the crop could affect other local plants, were very pleased. The government, who started the trials, were not pleased! The government's Food Safety minister said:

'We can't operate food safety policy on a *hunch* – we have to have the science and that's why we need the trials.'

a) Explain the difference between a 'hunch' and 'science'. (2)

b) Do you think the protesters had based their ideas on a 'hunch'? Explain your answer. (1)

c) One of the purposes of the trial was to find out how far pollen from the GM crop could travel. Describe how the trial might be set up. (5)

B1b 7.1

The origins of life on Earth

We are surrounded by an amazing variety of life on planet Earth. Questions like 'Where has it all come from?' and 'When did life on Earth begin?' have puzzled people for generations.

There is no record of the origins of life on Earth – it is a puzzle which can never be completely solved. No-one was there to see it and there is no direct evidence for what happened. We don't even know when life on Earth began. However, most scientists think it was somewhere between 3 billion and 4.5 billion years ago!

There are some interesting ideas and well-respected theories which explain most of what you can see around you. The biggest problem we have is finding the evidence to support the ideas.

a) When do scientists think life on Earth began?

What can we learn from fossils?

We share the Earth with millions of different species of living organisms. But this is tiny compared to the 4 billion species that scientists believe have lived on Earth during its history.

Most of these species have disappeared again in the mists of time. Some of them have gone completely. Others have left living relatives. The fossil record gives us an insight into how much – and how little – organisms have changed since life developed on Earth.

b) How many species of living organisms are thought to have existed on Earth over the years?

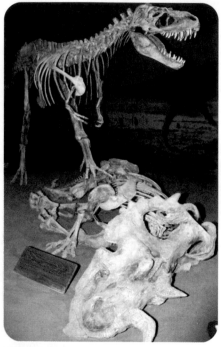

Figure 1 This amazing fossil shows two dinosaurs – prehistoric animals which died out millions of years before we appeared on Earth. Fossils can only give us a brief glimpse into the past. We will never know exactly what snuffed out the life of these spectacular reptiles all those years ago.

Fossils are the remains of plants or animals from many thousands or millions of years ago which are found in rocks. You have probably seen a fossil in a museum or on TV or – if you are really lucky – found one yourself.

The fossil record is not complete because so much rock has been broken down, worn away, buried or melted over the years. In spite of this, it can still give us a 'snapshot' of life millions of years before we were born.

Fossils can be formed in a number of ways:

- Many fossils were formed when harder parts of the animal or plant were replaced by other minerals over long periods of time. These are the most common fossils.
- Another type of fossil was formed when an animal or plant did not decay after it died. Sometimes the temperature was too low for decay to take place and the animals and plants were preserved in ice. However, these fossils are rare.

Some of the fossils we find are not of actual animals or plants, but rather of traces they have left behind. Fossil footprints and droppings all help us to build up a picture of life on Earth long ago.

Often the fossil record is very limited. Small bits of skeletons are found, or little bits of shells. Luckily we have a very complete fossil record for a few animals, including the horse. What's more, fossils show us that not all animals have changed over time. For example, fossil sharks from millions of years ago look very like their modern descendants.

c) Why do ice fossils give us clear evidence of animals that lived in the past?

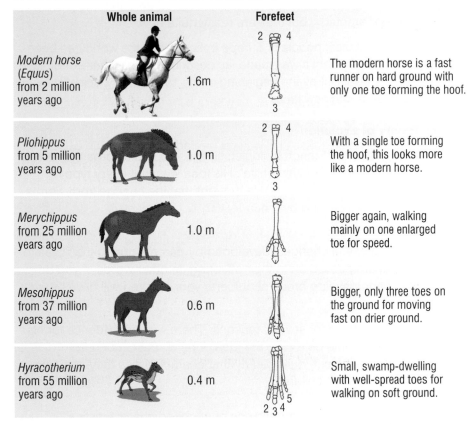

Whole animal		Forefeet	
Modern horse (Equus) from 2 million years ago	1.6m		The modern horse is a fast runner on hard ground with only one toe forming the hoof.
Pliohippus from 5 million years ago	1.0 m		With a single toe forming the hoof, this looks more like a modern horse.
Merychippus from 25 million years ago	1.0 m		Bigger again, walking mainly on one enlarged toe for speed.
Mesohippus from 37 million years ago	0.6 m		Bigger, only three toes on the ground for moving fast on drier ground.
Hyracotherium from 55 million years ago	0.4 m		Small, swamp-dwelling with well-spread toes for walking on soft ground.

Figure 3 The story of the horse. The horse as we know it today has evolved from some very different animals. We know they existed from the very clear record left in the fossils.

Figure 2 This baby mammoth has been preserved in ice for about 40 000 years. Ice fossils are very rare. They can give us an amazing glimpse of what a prehistoric animal looked like, the colour of its skin or fur and even what it had been eating.

FOUL FACTS

Fossils have revealed the world of the dinosaurs, lizards that dominated the Earth millions of years ago. The biggest plant-eater found so far is *Argentinosaurus huinculensis,* which was over 40 metres long and probably weighed about 80–100 tonnes! More scary was the biggest carnivore found. *Giganotosaurus* was about 14 metres long, walked on two legs, had a brain the size of a banana and enormous jaws with 20 cm long serrated teeth.

SUMMARY QUESTIONS

1 Copy and complete using the words below:

animal   decay   evidence   fossils   ice fossils   minerals   plant

One important piece of …… for how life has developed on Earth are …… . The most common type are formed when parts of the …… or …… are replaced by …… as they decay. Some fossils were formed when an organism did not …… after it died. These …… …… are very rare.

2 a) There are several theories as to how life on Earth began. Why is it impossible to know for sure?

b) Make a timeline to show how our ideas about the age of the Earth have changed since the 17th century as more information has become available.

c) Why are fossils such important evidence for the way life has developed?

3 Look at the evolutionary tree of the horse in Figure 3. Explain how the fossil evidence of the legs helps us to understand what the animals were like and how they lived.

KEY POINTS

1 Fossils provide us with evidence of how much – or how little – different organisms have changed since life developed on Earth.

2 It is very difficult for scientists to know exactly how life on Earth began because there is no direct evidence.

B1b 7.2

Theories of evolution

The theory of **evolution** tells us that all the species of living things alive today have evolved from the first simple life forms that existed on Earth. Most of us take these ideas for granted – but they are really quite new.

Up to the 18th century most people in Europe believed that the world had been created by God. They thought it was made, as described in the Christian Bible, a few thousand years ago. But by the beginning of the 19th century, scientists were beginning to come up with new ideas.

Lamarck's theory of evolution

Jean-Baptiste Lamarck, a French biologist, suggested that all organisms were linked by what he called a 'fountain of life'. His idea was that every type of animal evolved from primitive worms. He thought that the change from worms to other organisms was caused by the **inheritance of acquired characteristics**.

The theory was that useful changes, developed by parents during their lives to help them survive, are passed on to their offspring. In other words, if you do lots of swimming and develop broad shoulders, your children will have broad shoulders as well!

Lamarck's theory fell down for several reasons. There was no evidence for his 'fountain of life' and people didn't like the idea of being descended from worms. People could also see quite clearly that characteristics they acquired were not passed on to their children.

a) What is meant by the phrase 'inheritance of acquired characteristics'?

Figure 1 In Lamarck's model of evolution, giraffes have long necks because each generation stretched up to reach the highest leaves. So each new generation had a slightly longer neck!

Charles Darwin and the origin of species

Our modern ideas about evolution began with the work of one of the most famous scientists of all time – **Charles Darwin**. Darwin set out in 1831 as the ship's naturalist on *HMS Beagle.* He was only 22 years old and the voyage to South America and the South Sea Islands would take five years.

Darwin planned to study geology on the trip. But as the voyage went on he became as excited by his collection of animals and plants as by his rock samples.

b) What was the name of the ship that Darwin sailed on?

Figure 2 Darwin was very impressed by the giant tortoises he found on the Galapagos Islands. The tortoises on each island had different shaped shells and a slightly different way of life – and Darwin made careful drawings of them all.

In South America, Darwin discovered a new form of the common rhea, an ostrich-like bird – but not until his party had cooked and eaten one of them! Two different types of the same bird living in slightly different areas set Darwin thinking.

On the Galapagos Islands he was amazed by the variety of species and the way they differed from island to island. Darwin found strong similarities between types of finches, iguanas and tortoises on the different islands. Yet each was different and adapted to make the most of local conditions.

Darwin collected huge numbers of specimens of animals and plants during the explorations of the *HMS Beagle*. He also made detailed drawings and kept written observations. The long voyage home gave him plenty of time to think about what he had seen. Charles Darwin returned home after five years with some new and different ideas forming in his mind.

c) What is the name of the famous islands where Darwin found so many interesting species?

After Darwin returned to England he spent the next 20 years working on his ideas. He knew they would meet a lot of opposition. He realised he would need lots of evidence to support his theories. He used the amazing animals and plants he had seen on his journeys as part of that evidence.

He also built up evidence from breeding experiments on pigeons to support his theory. In 1859, he published his famous book *On the Origin of Species by means of Natural Selection* (often known as *The Origin of Species*).

Darwin's central theory is that all living organisms have evolved from simpler life forms. This evolution has come about by a process of **natural selection**. Reproduction always gives more offspring than the environment can support. Only those which are most suited to their environment – the 'fittest' – will survive. When they breed, they pass on those useful characteristics to their offspring. Darwin suggested that this was how evolution takes place.

d) What was the name of Darwin's famous book?

The evidence for evolution

The fossil record shows us how species have evolved over millions of years, and how different species are related to each other. Studying the similarities and differences between species, can also help us to understand the evolutionary relationships between them. Observing how changes in the genes can be passed from one generation to another gives us more evidence still.

GET IT RIGHT!

Don't get confused between:
- the **theory of evolution** and
- the **process of natural selection**.

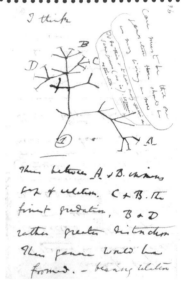

Figure 3 As Darwin studied his specimens, he began to build up a branching picture of evolution, which he tried out in different forms in his notebooks

KEY POINTS

1 The **theory of evolution** states that all the species which are alive today – and many more which are now extinct – evolved from simple life forms which first developed more than three billion years ago.
2 Darwin's theory is that evolution takes place through **natural selection**.
3 Studying the similarities and differences between species helps us to understand how they have evolved and how closely related they are to each other.

SUMMARY QUESTIONS

1 What is meant by the following terms:

   a) evolution?   b) natural selection?
   c) inheritance of acquired characteristics?

2 Why was Darwin's theory of natural selection only accepted very gradually?

3 What was the importance of the following in the development of Darwin's ideas?

   a) South American rheas.
   b) Galapagos tortoises, iguanas and finches.
   c) The long voyage of *HMS Beagle*.
   d) The twenty years from his return to the publication of his book *The Origin of Species*.

4 Suggest reasons why Lamarck and Darwin were convinced by their theories explaining life on Earth.

Natural selection

Figure 1 If all of these dandelions seeds developed to become adults and then breed themselves, there would be problems! In fact, very few of them will make it. The survivors will have a combination of genes that gives them a competitive edge over all the others.

Figure 2 The natural world is often brutal. Only the best adapted predators capture prey – and only the best adapted prey animals escape!

Scientists explain the variety of life today as the result of a process called *natural selection*. The idea was first suggested 150 years ago by Charles Darwin.

The natural world is a harsh place, and as you saw in Chapter 5 animals and plants are always in competition with each other. Sometimes an animal or plant gains an advantage in the competition. This might be against other species or against other members of its own species. That individual is more likely to survive and breed – and this is known as *natural selection*.

a) Who first suggested the idea of natural selection?

Survival of the fittest

Charles Darwin described natural selection as the 'survival of the fittest'. Reproduction is a very wasteful process. Animals and plants always produce more offspring than the environment can support.

Fruit flies can produce 200 offspring every two weeks. The yellow star thistle, an American weed, produces around 150 000 seeds per plant per year! If all those offspring survived we'd be overrun with fruit flies and yellow star thistles!

But the individual organisms in any species show lots of variation. This is because of differences in the genes they inherit. Only the offspring with the genes best suited to their habitat manage to stay alive and breed successfully. This is natural selection at work.

Think about rabbits. The rabbits with the best all-round eyesight, the sharpest hearing and the longest legs will be the ones which are most likely to escape being eaten by a fox. They will be the ones most likely to live long enough to breed. What's more, they will pass those useful genes on to their babies. The slower, less alert rabbits will get eaten and their genes will be digested with the rest of them!

b) Why would a rabbit with sharp hearing be more likely to survive than one with less keen hearing?

The part played by mutation

New forms of genes (new alleles) result from changes in existing genes. These changes are known as **mutations**. They are tiny changes in the long strands of DNA.

Mutations occur quite naturally through mistakes made in copying your DNA when your cells divide. Mutations introduce more variety into the genes of a species. In terms of survival, this is very important.

c) What is a mutation?

Many mutations have no effect on the characteristics of an organism, and some mutations are harmful. However, just occasionally a mutation has a good effect. It produces an adaptation which makes an organism better suited to its environment. This makes it more likely to survive and breed.

Whatever the adaptation, if it helps an organism survive and reproduce it will get passed on to the next generation. The mutant gene will gradually become more common in the population. It will cause the species to evolve.

When new forms of a gene arise from mutation, it may cause a more rapid change in a species. This is particularly true if circumstances change as well.

Natural selection in action

Have you ever eaten oysters? They are an expensive treat! They are collected from special oyster beds under the sea. Malpeque Bay in Canada has some very large oyster beds. In 1915, the oyster fishermen noticed a few oysters which were small and flabby with pus-filled blisters.

By 1922 the oyster beds were almost empty. The oysters had been wiped out by a new and devastating disease (soon known as Malpeque disease).

Fortunately a few of the shellfish carried a mutation which made them resistant to the disease. Not surprisingly, these were the only ones to survive and breed. The oyster beds filled up again and by 1940 they were producing more oysters than ever.

But the new population of oysters had evolved. As a result of natural selection, almost every oyster in Malpeque Bay now carries the allele which makes them resistant to Malpeque disease. So the disease is no longer a problem.

d) What is Malpeque disease?

Figure 3 Oyster yields from Malpeque Bay 1915–40. As you can see, disease devastated the oyster beds. However thanks to the process of natural selection, a healthy population of oysters managed to survive and reproduce again.

SUMMARY QUESTIONS

1 Copy and complete using the words below:

   **adaptation   breed   environment   generation   mutation
   natural selection   organism   survive**

   When a ......has a good effect it produces an ...... which makes an ......better suited to its .... This makes it more likely to ...... and .... The mutation then gets passed on to the next .... This is ....

2 Give three examples from this spread of characteristics which are the result of natural selection, e.g. all-around eyesight in rabbits.

3 Explain how the following characteristics of animals and plants have come about in terms of natural selection.

   a) Male peacocks have large and brightly coloured tails.
   b) Cacti have spines instead of leaves.
   c) Camels can tolerate their body temperature rising far higher than most other mammals.

4 Explain how mutation affects the evolution of a species.

KEY POINTS

1 New forms of genes result from changes (mutations) in existing genes.
2 Different organisms in a species show a wide range of variation because of differences in their genes.
3 The individuals with the characteristics most suited to their environment are most likely to survive and breed successfully.
4 The genes which have produced these successful characteristics are then passed on to the next generation.

B1b 7.4 Extinction

LEARNING OBJECTIVES

1 Why do species become extinct?
2 What killed off the dinosaurs?

Millions of years ago	Plants	Animals
NOW	Flowering plants dominant	Humans
	Conifers and flowering plants dominant	Many mammals
65		
	MAJOR EXTINCTION EVENT Dinosaurs extinct	
136		Bony fish spread
		Dinosaurs dominant
		Modern crustaceans
190	Conifers and ferns dominant	Dinosaur ancestors
		First mammals and birds
225		
	MAJOR EXTINCTION EVENT Many amphibians and invertebrates extinct	
	Conifers appeared	
280	Fern forests	First reptiles
		Many amphibians
345	Plants with veins	Many fish
		First insects
	MAJOR EXTINCTION EVENT 70% of species lost	
		Sharks and amphibians
395	Algae common	Fish with jaws
		First land arthropods
430		
	MAJOR EXTINCTION EVENT	
	First land plants with veins	Jawless fish and molluscs
500		
	FIRST MAJOR EXTINCTION EVENT	
	Algae dominant	Trilobites common
570		Worm-like animals
2500	Origins of life	

Figure 1 A summary of the main events in the evolution of life on Earth

Throughout the history of life on Earth, we think a total of about 4 billion different species have existed. Yet only a few million species of living organisms are alive today. The rest have become ***extinct***.

Extinction is the permanent loss of all the members of a species from the face of the Earth.

As conditions change, new species evolve which are better fitted to survive the new conditions. At the same time older species which cannot cope with the changes, and which do not compete so well for food and other resources, gradually die out. This is how evolution takes place and the balance of species on Earth gradually changes. Extinction is very important.

a) What is extinction?

Environmental changes

Throughout history, the climate and environment of the Earth has been changing. At times the Earth has been very hot. At other times, temperatures have fallen and the Earth has been in the grip of an Ice Age.

Organisms which do well in the heat of a tropical climate won't thrive in an icy landscape. Many of them become extinct through lack of food or being too cold to breed. New species, which cope well in cold climates, evolve and thrive by natural selection.

Changes to the climate or the environment are the main cause of extinction throughout history. For example, most scientists think it was a big climate change that caused the dinosaurs to become extinct millions of years ago. This was possibly caused by a giant meteorite crashing into the Earth, creating drastic changes to the climate.

There have been five occasions during the history of the Earth when big climate changes have led to extinction on an enormous scale. Look at the major extinction events in Figure 1. These are part of the process by which evolution takes place.

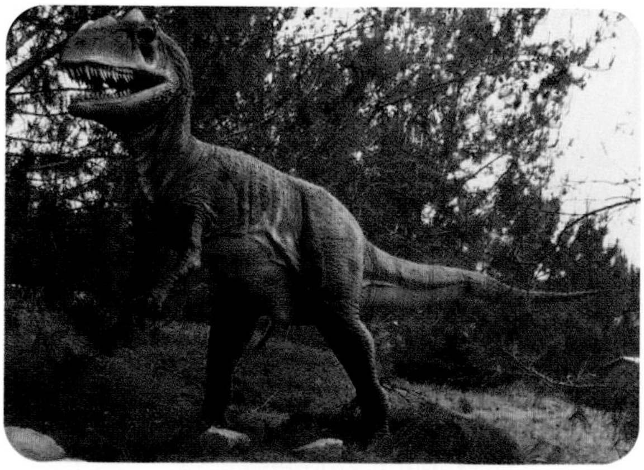

Figure 2 The dinosaurs ruled the Earth for millions of years, but when the whole environment changed, they could not adapt and died out. By the time things began to warm up again, mammals, which could control their own body temperature, were becoming dominant. The age of reptiles was over.

b) How does the climate change during an Ice Age?

Organisms which cause extinction

The other main cause of extinction is other living things. This can take place in several different ways:

- If a new **predator** turns up in an area, it can wipe out unsuspecting prey animals very quickly. That's because the prey do not have adaptations to avoid it.

  A new predator may evolve, or an existing species might simply move into new territory. Sometimes it is our fault. The brown tree snake from Australia was brought to Guam by people following World War II. By the 1960s many bird species on Guam were becoming extinct at a rapid rate – eaten by the snakes which attacked their nests at night!

  The birds had no chance to evolve a defence to this new night-time predator. Because many of the birds of Guam are now extinct, the snakes have started eating lizards instead!

- New *diseases* (caused by microorganisms) can drive a species to the point of extinction. They are most likely to cause extinctions on islands, where the whole population of an animal or plant is close together.

  The Tasmanian devil in New Zealand is one example where this may happen. These rare animals are dying from a new form of cancer, which seems to attack and kill them very quickly.

- Finally, one species can drive another to extinction through successful *competition*. You may see a new mutation which gives one type of organism a real advantage over another, or you may find people have introduced a new species by mistake.

  If the new species is really successful it may take over from the original animal or plant and make it extinct. In Australia the introduction of rabbits has been a nightmare because they eat so much and breed so fast! Other native Australian animals are dying out because they cannot compete.

GET IT RIGHT!

Always mention a **change** when you suggest reasons for extinction.
Watch your timescales – remember humans were not responsible for the extinction of the dinosaurs!

Figure 3 The Scottish island of North Uist has a similar problem to Guam. Someone brought a few hedgehogs onto the island to tackle garden slugs. The hedgehogs bred rapidly and are eating the eggs and chicks of the many rare sea birds which breed on the island. Now people are trying to kill the hedgehogs to save the birds!

SUMMARY QUESTIONS

1 Copy and complete using the words below:

> **climate   competitors   diseases   Earth   environment**
> **Extinction   predators   species**

…… is the permanent loss of all the members of a …… from the …… . It may be caused by changes to the …… or ……, to new ……, new …… or possibly new …… .

2 Explain how we think the dinosaurs became extinct.

3 a) Explain how each of the following situations might cause a species of animal or plant to become extinct.
  i) Mouse Island has a rare species of black-tailed mice. They are preyed on by hawks and owls, but there are no mammals which eat them. A new family bring their pregnant pet cat to the island.
  ii) English primroses have quite small leaves. Several people bring home packets of seeds from a European primrose which has bigger leaves and flowers very early in the spring.
 b) Why is extinction an important part of evolution?

KEY POINT

1 Extinction may be caused by changes to the environment, new predators, new competitors and possibly new diseases.

Evolution – the debate goes on

Science versus religion?

When Charles Darwin published his book *The Origin of Species*, he knew that it would cause trouble between the scientific community and the Church. He wanted to put forward his ideas, but he did not want to unsettle faithful Christians – his beloved wife was one! Of course the book caused an uproar, and the debate still continues in some places today.

Darwin's basic principles are not in dispute among scientists, although they often like to discuss the fine details of evolution. But not everyone agrees. For some people there is no conflict between a deep faith in God and an acceptance of evolution. Others find this a problem. Religion is a system of faith and unquestioned belief. It deals with spiritual things which cannot be explained simply by using scientific methods, based on collecting evidence and data on the natural universe.

One area where the Church and science have clashed is on the age of the Earth itself. Here are some of the stages in the debate!

The age of the Earth

Fossils provide evidence that animals and plants have changed and developed over a very long time. This process is known as **evolution**. The idea of evolution suggests that the Earth itself must also have existed for billions of years. This view of the origins of life on Earth is quite recent.

In the 17th century the story of the creation of the Earth in the Christian Bible was still largely unquestioned. One famous historian, Archbishop Usher, used it to calculate that the Earth was less than 6000 years old!

During the 18th century people began to travel more. They not only discovered new lands, but amazing plants and animals as well. Our ancestors unearthed the fossil remains of massive creatures and strange plants. They began to build up evidence that the Earth had changed dramatically during its history.

By the beginning of the 19th century, the evidence for evolution was building up. Sir Charles Lyell was a British geologist. He showed that the Earth was very ancient, and was shaped by rivers and 'subterranean fires'. He estimated that the Earth was several hundred million years old and published his ideas from 1830 onwards.

Lyell's work was important to Darwin. He came up with a theory that all living organisms have arisen from evolution by natural selection. This would have taken many millions of years. Fossils also helped to confirm Darwin's ideas. They showed some of the stages at which the different animals and plants appeared and how they have changed over time. However, some people still believe that fossils were put into rocks by God to test our faith.

Born in 1890, Arthur Holmes was the first person to use radioactivity to date rocks. He established the age of the Earth as around 3 billion years old, giving plenty of time for evolution to have happened!

In some states of America fundamentalist Christians have a powerful voice. They believe that the Earth and everything in it was created exactly as it is described in the Bible. They would like to prevent the theory of evolution from being written about in school text books or taught in schools. At the very least they would like to see the Creationist view given equal emphasis.

ACTIVITY

You have been asked to give evidence at the School Board meeting in the United States. The panel is trying to decide whether to allow the scientific theory of evolution to be taught in their school.

Put together a short report on why it is important that students know about the theory of evolution. Bring together some evidence of evolution occurring in the world around them. Show that it is important for students to understand the scientific view which is held by the majority of their peers in the developed world. This is true even if they choose not to accept the evidence and to hold their own beliefs.

Making extinction extinct?

Plant species are under threat all around the world – but we're fighting back. We have a plan to protect 24 000 plant species by storing their seeds! It's a massive international project masterminded from the Royal Botanic Gardens at Kew in the UK.

In the UK we have collected seeds from all the wild flowering plants for the Millennium Seed Bank Project (MSBP). Once we've collected them they are put into storage. These UK seeds are not alone in their storage jars – seeds from around the world are sent to join them.

Sixteen countries from Jordan to Madagascar, and from Botswana to Mexico are involved in the Millennium Seed Bank project.

South Western Australia in particular has many, many plants – at least 12 000 species are known! It is one of the world's top 'hot spots' for biodiversity and seeds from all of the threatened species are being collected. Once the seeds are stored, even if the plants become extinct in the wild, we have the chance of introducing them again in the future! These plants are so important – they could be a source of new medicines in the future.

Once the seeds for the MSBP have been collected, they are checked, dried and then stored at about −20°C. Under these conditions they can survive for decades – or even centuries!

So many fascinating animals have become extinct. I m[...] I'd love to see a dodo, wouldn't you? And going further ba[...] what about a woolly mammoth walking the Earth again? These examples are a bit extreme! But there are teams around the world who are using modern technology to save species which are on the brink of extinction, or have just become extinct.

We hope cloning will be the key to the future for some species. Even if we can't clone them successfully now, by saving some of their tissue we may be able to bring the species back in the future!

For example, Banteng, the wild cattle of Java, Burma and Borneo are endangered in the wild. In 2003 a team in the US managed to produce one healthy cloned Banteng calf, and another that was abnormal. Before that, in 2000, we cloned a healthy Guar calf. Guar are very rare cattle indeed so this was quite a triumph. Sadly little Noah died from an infection just two days after he was born – but we can and will try again!

We've got some tissue from the ear of a Pyranean Ibex. The last one died in 2000. It would be fantastic to bring them back soon. And going back to mammoths . . . They are very like our modern elephants in many ways and we've found very well preserved cells in some of the ice fossils. I'm sure that one day someone will clone an extinct animal and bring the species back into existence. What's more, I'm sure prehistoric animals might even be cloned one day – and I hope I'm here to see it!

New species are being cloned all the time. In future we may be able to clone species of animals and plants that are threatened with extinction (like this banteng) and keep them going. We may even be able to clone species which became extinct some time ago using ice fossils or material found on dried specimens in museums.

ACTIVITIES

a) Extinction!

Your task is to make a poster titled 'Extinction!' for display in the school science department. You can choose what to highlight in your poster. You can use information from this chapter and you might like to use other resources as well. The following might give you some ideas:

- How extinction comes about and why it is important.
- Comparing extinction rates in the past and now – why is there so much concern?
- How can extinction be prevented/undone?

b) Interfering with nature?

Write a letter to *The Times*:

Either: express your enthusiasm for using new technologies to prevent extinction, keep species going and bring back extinct species.

Or: express your disgust at using new technologies to prevent extinction, keep species going and bring back extinct species.

QUESTIONS

ck fossils formed?

ce fossils formed?

dence do fossils give us about how life
on Earth has developed?

d) Why is the fossil record not complete?

2 a) Summarise the similarities and differences
between Darwin's theory of evolution and
Lamarck's.

b) Why do you think Lamarck's theory was so
important to the way Darwin's theory was
subsequently received?

3

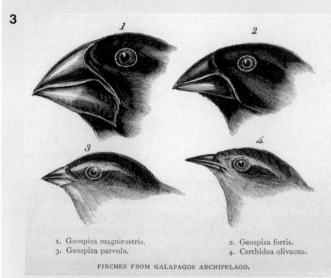

1. Geospiza magnirostris.   2. Geospiza fortis.
3. Geospiza parvula.   4. Certhidea olivacea.
FINCHES FROM GALAPAGOS ARCHIPELAGO.

Darwin's finches – more evidence for evolution

Look at the birds in the picture. They are known as
Darwin's finches. They live on the Galapagos
Islands. Each one has a slightly different beak and
eats a different type of food.
Explain carefully how natural selection can result in
so many different beak shapes from one original
type of founder finch.

4 a) Why is extinction important to the success of
evolution?

b) Why are scientists so worried about the rate at
which extinction is occurring now?

c) Find out and write about one animal or plant
which has become extinct in the last twenty years.

d) Find out and write about one animal or plant
which is close to extinction now. Explain what, if
anything, is being done to help prevent it from
becoming extinct.

e) Many groups are keen to prevent animals and
plants becoming extinct at all costs. Why is it not
necessarily a good idea to prevent extinction?

EXAM-STYLE QUESTIONS

1 The list contains factors that have played a part in the
development of the species we see on Earth today:

A Extinction   B Evolution

C Natural selection   D Mutation

Match words A, B, C and D with the processes 1, 2, 3
and 4 in the table.

	Process
1	Change and development of organisms over a long period of time
2	Change to the amount or arrangement of genetic material within a cell
3	Permanent loss of all members of a species
4	Passing of genes to offspring by the organisms most suited to their environment

(4)

2 Which of the following does **not** play a part in evolution
by natural selection?

A Inheritance of acquired characteristics.

B Mutation of existing genes.

C Variation of individuals within a species.

D Production of offspring by individuals best suited to
their environment.   (1)

3 Which of the following would **not** normally cause the
extinction of a species?

A Environmental change

B Fewer competitors

C New diseases

D More predators   (1)

4 Not all scientists agree on the exact evolutionary
relationship between different primates. The diagram
shows a timeline for one version of this relationship.

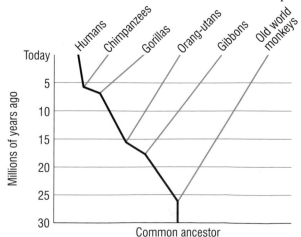

(a) Which group is the closest relative to humans?

   A Old world monkeys

   B Orang-utans

   C Gorillas

   D Chimpanzees

(b) How many million years after the old world monkeys evolved from the common ancestor did the gorillas evolve?

   A 27   **B** 20   **C** 10   **D** 7

(c) How many of the primate groups shown in the diagram were on Earth 20 million years ago?

   A 6   **B** 5   **C** 3   **D** 1

(d) Many of the ancestors of the present-day primates are now extinct. How do we know these ancestors once lived?

   A By studying DNA samples.

   B By studying fossil records.

   C By studying blood samples.

   D By studying cell structure.   (4)

5 The Galapagos Islands are a group of islands in the Pacific Ocean. The nearest country on the mainland is Ecuador, 1000 km away. By some means, a few seed-eating finches were the first birds to reach the islands. This single ancestral species has since evolved into many different species. Charles Darwin visited the islands and noted that each species had a beak adapted to the type of food it ate.

Using the theory of natural selection, explain how the ancestral species might have evolved into birds with different-shaped beaks.   (6)

HOW SCIENCE WORKS QUESTIONS

It is difficult to gather data that illustrates evolution. It is possible to gather data to show natural selection, but this usually takes a long time. Simulations are useful because, while they are not factually correct, they do show how natural selection might work.

Darwin used evidence from his visit to the Galapagos Islands to show how natural selection might have worked. He used this as evidence for evolution, by natural selection. Some of the evidence he gathered was about the size and shape of the beaks of the finches on the different islands.

A class decided to simulate natural selection, by seeing if they could use different tools to pick up seeds. This is what they did.

Four students each chose a particular tool to pick up seeds. The teacher then scattered hundreds of seeds onto a patch of grass outside the lab. The four students were then given five minutes to pick up as many seeds as they could.

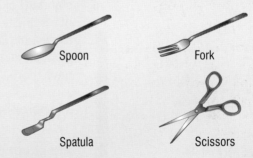

Spoon   Fork

Spatula   Scissors

James, who was using a spoon, picked up 23 seeds, whilst Farzana, using a fork, could only pick up two. Claire managed seven seeds with the spatula, but Jenny struggled to pick up her two seeds with a pair of scissors.

a) Put the essential data into a table.   (3)

b) How would this data be best presented?   (1)

c) Was this a fair test? Explain your answer.   (1)

d) What conclusion can you draw from this simulation?   (1)

The effects of the population explosion

LEARNING OBJECTIVE

1   What effect is the growth in the number of people having on the Earth and its resources?

Figure 1 The Earth from space. As the human population of the Earth grows, our impact on the planet gets bigger every day.

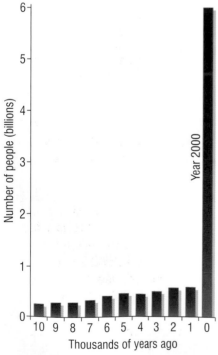

Figure 2 This record of human population growth shows the massive increase during the last few hundred years

We have only been around on the surface of the Earth for a relatively short time – less than a million years. Yet our activity has changed the balance of nature on the planet enormously. Some of the changes we have made seem to be driving many other species to extinction. Some people worry that we may even threaten our own survival.

Human population growth

For many thousands of years people lived on the Earth in quite small numbers. There were only a few hundred millions of us! We were scattered all over the world, and the effects of our activity were usually small and local. Any changes could easily be absorbed by the environment where we lived.

But in the last 200 years or so, the human population has grown very quickly. At the end of the 20th century the human population was over 6 billion, and it is still growing

If the **population** of any other species of animal or plant had suddenly increased in this way, natural predators, lack of food, build up of waste products or diseases would have reduced it again. But we have discovered how to grow more food than we could ever gather from the wild. We can cure or prevent many killer diseases. We have no natural predators. This helps to explain why the human population has grown so fast.

Not only have our numbers grown hugely, but in large parts of the world our standard of living has also improved enormously. In the UK we use vast amounts of electricity and fuel to heat and light our homes and places of work. We use fossil fuels like oil to produce this electricity. We also use it to move about in cars, planes, trains and boats at high speed and to make materials like plastics. We have more than enough to eat and if we are ill we can often be made better.

a) Approximately how many people are living on the Earth today?

The effect on land and resources

The increase in the numbers of people has had a big effect on our environment. All these billions of people need land to live on. More and more land is used for the building of houses, shops, industrial sites and roads. Some of these building projects destroy the habitats of rare species of other living organisms.

We use billions of acres of land around the world for farming, to grow food and other crops for human use. Wherever people farm, the natural animal and plant population is destroyed.

In quarrying we dig up great areas of land for the resources it holds such as gravel, metal ores and diamonds. This also reduces the land available for other organisms.

b) How do people reduce the amount of land available for other animals and plants?

Figure 3 In the UK alone hundreds of thousands of new houses are being built, and miles of new road systems. Every time we clear land like this, the homes of countless animals and plants are destroyed.

The huge human population is an enormous drain on the resources of the Earth. Raw materials are rapidly being used up. This includes non-renewable energy resources such as oil and natural gas and metal ores which cannot be replaced.

Pollution

The growing human population also means vastly increased amounts of waste. This is both human bodily waste and the rubbish from packaging, uneaten food and disposable goods. The dumping of this waste makes large areas of land unavailable for any other life except scavengers.

There has also been an enormous increase in manufacturing and industry to produce the goods we want. This in turn has led to industrial waste.

The waste we produce presents us with some very difficult problems. If it is not handled properly it can cause serious **pollution**. Our water may be polluted by sewage, by fertilisers from farms and by toxic chemicals from industry. The air we breathe may be polluted with smoke and poisonous gases such as sulfur dioxide. (See page 122.)

The land itself can be polluted with toxic chemicals from farming such as pesticides and herbicides, and with industrial waste such as heavy metals. These chemicals in turn can be washed from the land into the water. If the ecology of the Earth is affected by our population explosion, our use of resources and our waste, everyone will pay the price.

c) What substances commonly pollute
   i) water, ii) air, and iii) land?

GET IT RIGHT!

Make sure you know exactly which pollutants affect air, land and water!

SCIENCE @ WORK

Scientists working for groups as diverse as the United Nations and Greenpeace are involved both in monitoring the world population growth and in measuring and controlling levels of pollution.

Figure 4 In countries like ours, we have a very high standard of living. But a kitchen like this uses lots of resources – wood, metals, plastics and energy. This all results in pollution and the removal of resources which can never be replaced.

KEY POINTS

1 The human population is growing rapidly and the standard of living is increasing.
2 More waste is being produced. If it is not handled properly it can pollute the water, the air and the land.
3 Humans reduce the amount of land available for other animals and plants.
4 Raw materials, including non-renewable resources, are being used up rapidly.

SUMMARY QUESTIONS

1 Copy and complete using the words below:

   diseases   farming   food   increase   population
   predators   treat   two hundred

   The human ...... has increased dramatically in the last ...... years. Better ...... methods mean we have more .... We can ...... and prevent many .... We have no natural .... All this has allowed the numbers to ....

2 a) How has the standard of living increased over the last hundred years?
   b) Give three ways in which people have used up different resources.

3 Write a clear paragraph explaining how the ever-increasing human population causes pollution in a number of different ways.

B1b 8.2

Acid rain

1 How is acid rain formed?
2 What are the effects of acid rain on living organisms?

Figure 1 Air pollution is usually invisible. Just occasionally the level is so high it can actually be seen. The brown haze you can see over this city is caused by high levels of nitrogen oxides produced from car exhausts.

DID YOU KNOW?

Acid rain has been measured with a pH of 2.0 – more acidic than vinegar!

Figure 2 These trees should be covered in leaves and full of insects, birds and other animal life. Instead they are dead and bare, killed by the action of acid rain.

Human activities can have far-reaching effects on the environment and all the other living things which share the Earth. One of the biggest problems is the way we produce pollution.

Everybody needs air – so when the air we breathe is polluted, no-one escapes the effects. One of the major sources of air pollution is the burning of fossil fuels. We are using more and more oil, coal and natural gas. We also burn huge amounts of the fuels made from them, such as petrol, diesel and aviation fuel for planes. Fossil fuel is a non-renewable resource – there is a limited amount of it on Earth and eventually it will all be used up.

a) Name three fossil fuels.

The formation of acid rain

When fossil fuels are burned, carbon dioxide is released into the atmosphere as a waste product. However, carbon dioxide is not the only waste gas produced. Fossil fuels often contain sulfur impurities. When these burn they react with oxygen to form sulfur dioxide gas. At high temperatures, for example in car engines, nitrogen oxides are also released into the atmosphere.

These gases pollute the air and can cause serious breathing problems for people if the concentration gets too high. They are also involved in the formation of acid rain. This pollutes land and water over a wide area.

The sulfur dioxide and nitrogen oxides dissolve in the rain and react with oxygen in the air to form dilute sulfuric acid and nitric acid. This makes the rain more acidic – it is known as **acid rain**.

b) What are the main gases involved in the formation of acid rain?

The effects of acid rain

Not surprisingly, acid rain has a damaging effect on the environment. If it falls onto trees, the acid rain can cause direct damage. It may kill the leaves and, as it soaks into the soil, even the roots of the trees may be destroyed. In some parts of Europe and America, huge areas of woodland are dying as a result of acid rain.

Acid rain has an indirect effect on our environment, as well as its very direct effect on plants. As acid rain falls into lakes, rivers and streams the water in them becomes acidic. If the concentration of acid gets too high, plants and animals can no longer survive. Many lakes and streams have become dead, no longer able to support life.

c) How does acid rain kill trees?

Acid rain is a difficult form of air pollution to pin down and control. It is formed by pollution from factories. It also comes from the cars and other vehicles we use every day. The source of the gases is pretty widespread. Many Western countries have worked hard to stop their factories and power stations from producing these acidic gases. Unfortunately there are still many places in the world where these gases are not controlled.

The worst effects of acid rain are often not felt by the country which produced the pollution in the first place. The sulfur and nitrogen oxides are carried high in the air by the prevailing winds. As a result, it is often relatively 'clean' countries which get the pollution and the acid rain from their dirtier neighbours. Their own clean air goes on to benefit someone else!

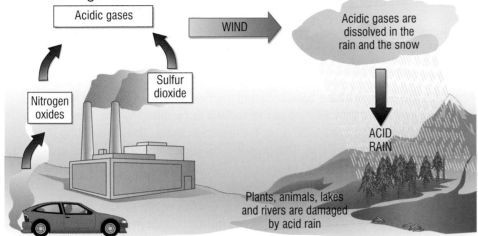

Figure 3 Air pollution in one place can cause acid rain – and serious pollution problems – somewhere else entirely. Depending on the prevailing winds, it can even be in another country!

People have become more aware of the problems caused by acid rain. The UK and other countries have introduced measures to reduce the levels of sulfur dioxide and nitrogen oxides in the air. More and more cars are fitted with catalytic converters. Once hot, these remove the acidic gases before they are released into the air. There are strict rules about the levels of sulfur dioxide and nitrogen oxides in the exhaust fumes of new cars.

Power stations are one of the main sources of acidic gases. In the UK we have introduced cleaner, low-sulfur fuels and started generating more electricity from gas and nuclear power. We have also put systems in power station chimneys to clean the flue gases before they are released into the atmosphere. As a result, the levels of sulfur dioxide in the air, and of acid rain, have fallen steadily over the last 40 years. (See Figure 4.)

SUMMARY QUESTIONS

1 Copy and complete using the words below:

   **acid rain   carbon dioxide   fossil   nitric   nitrogen oxides
   sulfur   sulfuric**

   When ...... fuels are burned the pollutant gases ...., ... dioxide and .... are released into the atmosphere. The sulfur dioxide and nitrogen oxides dissolve in the rain, reacting with water and oxygen from the air, to form dilute ...... acid and ...... acid. This forms ....

2 Explain how pollution from cars and factories burning fossil fuels pollute:
   a) the air   b) the water   c) the land.

3 To get rid of acid rain it is important that all the countries in an area control their production of sulfur and nitrogen oxides. If only one or two clean up their factories and cars it will not be effective. Explain why this is.

4 Use Figure 4 to help you answer this question.
   a) Produce a bar chart to show the approximate levels of sulfur dioxide in the air in the UK at five-year intervals from 1980 to 2000.
   b) Explain the trend you can see on your chart.

GET IT RIGHT!

Be clear about the effects of the different combustion gases produced – sulphur dioxide does **not** affect global warming but it **does** cause acid rain!

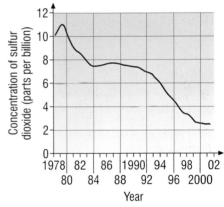

Figure 4 Graph to show levels of sulfur dioxide concentrations in the air in the UK, 1978–2002

KEY POINTS

1 When we burn fossil fuels, carbon dioxide is released into the atmosphere.
2 Sulfur dioxide and nitrogen oxides can be released when fossil fuels are burnt. These gases dissolve in the rain and make it more acidic.
3 Acid rain may damage trees directly. It can make lakes and rivers too acidic so plants and animals cannot live in them.

B1b 8.3 Global warming

LEARNING OBJECTIVES

1 How does combustion affect the atmosphere?
2 How do living things affect the atmosphere?
3 What will be the consequences of global warming?

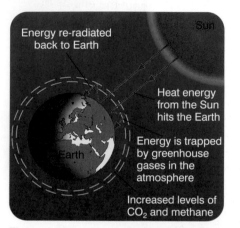

Figure 1 Many scientists believe that this simple warming effect could, if it is not controlled, change life on Earth, as we know it

GET IT RIGHT!

Respiration and combustion **produce** carbon dioxide, photosynthesis **removes** it from the atmosphere. Greenhouse gases in the atmosphere **re-radiate** heat energy back to the surface of the Earth.

Many scientists are very worried that the climate of the Earth is getting warmer. This is often called **global warming**.

The greenhouse effect

Normally the Earth radiates back much of the heat energy it absorbs from the Sun. This keeps the temperature at the surface acceptable for life. Now carbon dioxide and methane are building up in the atmosphere. They are acting like a greenhouse around the Earth. The greenhouse gases absorb much of the energy which is radiated away. It can't escape out into space. As a result, the Earth and its surrounding atmosphere are warmer than they should otherwise be.

The effect is to raise the temperature of the Earth's surface. The change is very small, about 0.06°C every ten years at the moment. Not much – but an increase of only a few degrees Celsius could cause quite large changes in the Earth's climate.

Many scientists think that an increase in severe and unpredictable weather will be one of the changes we see due to global warming. Some people think the very high winds and extensive flooding seen around the world in the 21st century are early examples of the effects of global warming.

If the Earth warms up, the ice caps at the north and south poles will melt. This will cause sea levels to rise. There is evidence that this is already happening. It will mean more flooding for low-lying shores of all countries all over the world. Eventually parts of countries, or even whole countries, will disappear beneath the seas.

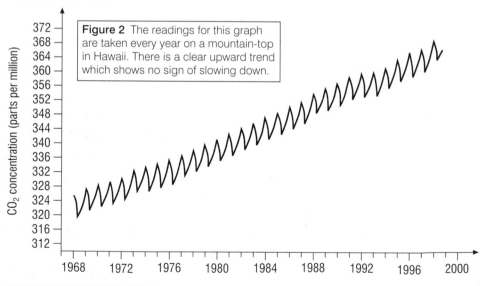

Figure 2 The readings for this graph are taken every year on a mountain-top in Hawaii. There is a clear upward trend which shows no sign of slowing down.

a) Name two greenhouse gases.

The effects of combustion

Carbon dioxide is made when we burn fossil fuels in cars, in our homes and in power stations. The number of cars and power stations around the world is steadily increasing. And respiration by all the living organisms on Earth produces carbon dioxide as well! So carbon dioxide levels are rising.

The effects of deforestation

All around the world large-scale deforestation is taking place. We are cutting down trees over vast areas of land for timber and to clear the land for farming. The trees are felled and burned in what is known as 'slash-and-burn' farming. The land produced is only fertile for a short time, after which more forest is destroyed. No trees are planted to replace those cut down.

Deforestation increases the amount of carbon dioxide released into the atmosphere. Burning the trees leads to an increase in carbon dioxide levels from combustion. The dead vegetation left behind decays. It is attacked by decomposing microorganisms, which releases more carbon dioxide.

Normally trees and other plants use carbon dioxide in photosynthesis. They take it from the air and it gets locked up in plant material like wood for years. So when we destroy trees we lose a vital carbon dioxide 'sink'. Dead trees don't take carbon dioxide out of the atmosphere.

For millions of years the levels of carbon dioxide released by living things into the atmosphere have been matched by the plants taking it out and the gas dissolving in the seas. As a result the levels in the air stayed about the same from year to year. But now the amount of carbon dioxide produced is increasing fast as a result of human activities. This speed means that the natural sinks cannot cope, and so the levels of carbon dioxide are building up.

b) What is deforestation?

Cows, rice and methane

Methane levels are rising too. It has two major sources. As rice grows in swampy conditions, known as paddy fields, methane is released. As the population of the world has grown so has the farming of rice, the staple diet of many countries.

The other source of methane is cattle. Cows produce methane during their digestive processes and release it at regular intervals.

In recent years the number of cattle raised to produce cheap meat for fast food, such as burgers, has grown enormously. So the levels of methane are rising. Many of these cattle are raised on farms produced by deforestation.

c) Where does the methane that is building up in the atmosphere come from?

Figure 3 Tropical rainforests are being destroyed at an alarming rate to supply the developed world with goods like mahogany toilet seats and cheap burgers

FOUL FACTS

When we lose forests, we lose biodiversity. Lots of plants and animals die out. We may well be destroying a source of new medicines or food for the future!

KEY POINTS

1 There is large-scale deforestation in tropical areas.
2 Large-scale deforestation has led to an increase of carbon dioxide into the atmosphere (from burning and the actions of microorganisms). It has also reduced the rate at which carbon dioxide is removed by plants.
3 More rice fields and cattle have led to increased levels of methane in the atmosphere.
4 Increased levels of the greenhouse gases carbon dioxide and methane may be causing global warming as a result of the greenhouse effect.

SUMMARY QUESTIONS

1 Define the following terms:
   global warming; greenhouse gases; deforestation; a carbon sink.

2 a) Why are the numbers of
      i) rice fields, and   ii) cattle
      in the world increasing?

   b) Why is this a cause for concern?

3 Give three reasons why deforestation increases the amount of greenhouse gases in the atmosphere.

4 a) Use the data in Figure 2 to produce a bar chart showing the maximum recorded level of carbon dioxide in the atmosphere every tenth year from 1970 to the year 2000.
   b) Explain the trend you can see on your chart.
   c) Explain the greenhouse effect. How might it affect the conditions on Earth?

B1b 8.4

Sustainable development

GET IT RIGHT!

Be clear about the meaning of the term 'sustainable'.
Be able to give examples of how families can help conserve natural resources.

As our world gets more and more crowded, we are becoming increasingly aware of the need for **sustainable development**. This combines human progress and environmental stability. It improves the quality of our lives without risking the future of generations to come.

Sustainable development

Sustainable development means looking after the environment. We need to conserve natural resources. So, for example, farmers need to look after the land. They can plough the remains of crop plants into the soil, and use animal waste instead of chemical fertilisers. They can also replant hedgerows to prevent soil erosion and avoid deforestation. These will help to make sure that growing crops will be possible for years to come.

We can see another example of sustainable development in our woodlands. We use an enormous amount of wood and paper, but we have fewer trees than almost all of our European neighbours.

However, over the last 80 years or so the Forestry Commission has developed sustainable commercial woodlands. Felling can only take place as long as replanting replaces the felled trees. Now farmed woodlands not only provide a sustainable resource but also a rich environment for a wide variety of species.

Sustainable development has to be a global idea – it is no good if it only happens in the UK. Everywhere that deforestation occurs, replanting needs to take place.

Figure 1 Sustainable woodlands have become an important and attractive part of sustainable development in the UK

Another example of the importance of sustainable development is in the management of our fishing stocks. In many areas we have taken so many fish from the sea that the populations can no longer replace themselves. The numbers of fish like cod are dropping fast.

To save the fish stocks, the numbers of fish caught *must* be reduced. But it needs agreement by fishermen everywhere for this sustainable development of the sea to become a reality.

a) What is sustainable development?

Conserving resources

An important part of sustainable development is using natural resources wisely. We must use only what we need, and conserve natural resources as much as possible. You and your family can help to do this in lots of different ways.

We live in a throw-away society. We use something – and then put it in the bin. This uses up resources and means land is wasted under rubbish tips. But if we recycle our waste, we save resources, use land wisely and use less energy. You can recycle your old newspapers (saving trees), glass bottles (saving energy) and aluminium cans (saving aluminium ore and energy!).

Figure 2 Our throw-away society causes problems in lots of ways. Not only do we waste resources, but a tip like this uses lots of land, and pollutes the area all around it.

Another way we can help is to make our homes more energy-efficient. Energy is one of our most important resources. Unfortunately, using electricity, gas or oil in our homes uses up some of these valuable resources. It adds to the carbon dioxide in the atmosphere as well, so the less we use the better.

We use huge amounts of energy heating our homes – and then lose it through the windows, roof and gaps around the doors. Making our homes energy-efficient helps save resources and prevent global warming. By insulating your roof spaces and the walls of your homes, and having double glazing, you will save a lot of energy. Energy-efficient boilers make the best use of your fuel. Switching things off when you have finished using them helps as well!

Finally we can look at our transport. Use your car less, and walk or cycle to places nearby. This will help you save petrol (a non-renewable resource) and also avoid adding more carbon dioxide to the atmosphere. Public transport can help as well, carrying lots of people at the same time. It may not be as convenient as using your own car, but it is certainly better for the environment.

Figure 3 Just changing to energy-efficient light bulbs like these will help to conserve resources and reduce carbon dioxide levels

---

SUMMARY QUESTIONS

1 Copy and complete using the words below:

   cars conserving energy efficient recycle resources sustainable

   An important part of …… development is using natural …… wisely. This means using only what we need, and …… natural resources as much as possible. We can …… waste, make our homes more …… …… and make less use of our …… .

2 List as many ways as possible in which you and your family could help to conserve natural resources.

3 Choose one aspect of sustainable development and produce a report on how it works and why it is so important.

KEY POINTS

1 Improving your quality of life without compromising future generations is known as sustainable development.

2 Sustainable development involves using natural resources wisely.

3 We can help by recycling, making our homes energy-efficient and avoiding using our cars when possible.

B1b 8.5 Planning for the future

PRACTICAL

Pollution indicators

Have a look at your local area and see how many different types of lichens you can find. How clean do you think your air is?

- Is the information we gain from lichens reliable and valid evidence of pollution?
- Compare them to using data-logging equipment to monitor pollution.

Figure 1 Lichens grow well where the air is clean. In a polluted area there would be far fewer species of lichen growing. This is why they are useful bioindicators.

Figure 2 This water looks clean and inviting – a look at some pollution indicator species would tell us if it is as clean as it looks!

Most of the population of the world lives in the developing countries. Their way of life has not changed much over the centuries. There are many people, but each one uses few resources.

There are relatively few people in the developed world. However, those people are surrounded by technology, almost all of which uses a great deal of energy. Planning is needed at local, regional and global levels to make sure that the resources of the world are used in a fair and sustainable way.

a) Why does the developed world use so many more resources than the developing world?

Making local planning decisions

All around the UK people want to build homes, shopping centres, roads and factories. Before any building can take place, you have to apply for planning permission. For sustainable development to work, we need to consider the environment every time a planning decision is made.

To help us make the right decision we need information from field studies, carried out by scientists. They can give us information about the environment at the site of a proposed development. They can also show us the impact of a similar development elsewhere. One important tool we have in these field studies is to use living organisms as indicators of pollution.

Lichens grow on places like rocks, roofs and the bark of trees. They are very sensitive to air pollution. When the air is clean, scientists will find many different types of lichen growing. The more polluted the air, the fewer lichen species there will be. So a field check on lichen levels will give us the information we need about possible air pollution from a factory or road.

In the same way we can use invertebrate animals as water pollution indicators. The cleaner the water, the more species you will find. There are some species which are only found in the cleanest water. Others can be found even in very polluted waters.

By using living organisms as pollution indicators in this way we can reach sensible decisions about the impact a new development might have on the environment. This helps us make planning decisions which allow for the sustainable development of an area.

b) How can we use living organisms as indicators of pollution?

Brown field or green field?

As the UK population grows we need more houses. We can build them on 'green field sites' or 'brown field sites'.

Green field sites are countryside which has not been built on before. Brown field sites are usually within towns and cities, and have already been used. They are often the sites of old industrial buildings, factories, petrol stations or even rubbish tips.

Using brown field sites has many advantages for building. Water, sewage, electricity, gas and transport systems are usually already in place or easily available. No farmland or countryside is spoilt or lost.

However, brown field sites are often contaminated with chemicals and can suffer from subsidence. It is often expensive to use brown field sites because they have to be cleaned up (decontaminated) before they can be used. What's more, they can be a valuable environment in their own right.

Although it is a good idea to build on brown field sites when it is possible, the decision is not always an easy one!

Protecting SSSIs

Around the country there are many areas which are *SSSIs* (*Sites of Special Scientific Interest*). These sites may have a particularly interesting or unique landscape. They may be home to rare species of plants or animals. They may be important in bird migration or as a breeding ground.

Environmentalists believe it is very important that these SSSIs are not disturbed and – most importantly – not built on. When planning decisions are made, the environmental importance of an area needs to be considered carefully.

The idea of protected areas is common now in Europe, and is spreading around the world. In terms of global planning, it is important that governments everywhere take their environmental responsibilities seriously. That is the only way we can continue to develop while sustaining the variety of life around us – and protecting the Earth from the worst effects of pollution and waste.

Figure 3 People will soon be living in these homes – yet only a short time ago this brown field site was an old bus station

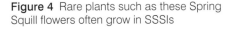

Figure 4 Rare plants such as these Spring Squill flowers often grow in SSSIs

DID YOU KNOW?

Protecting the environment all over the world is vital if we are to look after the variety of living things – the biodiversity – we now enjoy. We just don't know which ones will be useful in the future!

SUMMARY QUESTIONS

1 Define the following terms: pollution indicator; brown field site; green field site; SSSI.

2 Think about your local area. Choose a brown field site which you think could be used for building. Write a proposal which puts forward your plans, explaining why you have chosen the site rather than a local farmer's field.

3 a) Here is some information about four different species of living things. Explain how you would use each one as an indicator to help make planning decisions about sustainable development.
   i) Salmon are fish which are only found in clean water with lots of oxygen in it.
   ii) Lichens are very sensitive to air pollution.
   iii) Lizards are badly affected by pesticides in their insect prey.
   iv) Blood worms can survive in very polluted water.
   b) Evaluate different methods you could use to collect environmental data. Comment on their reliability and usefulness as sources of evidence.

KEY POINTS

1 We can use information gained from field studies to help us make planning decisions about sustainable development. One type of information is to use living organisms as indicators of pollution levels.

2 Brown field sites are thought to be more suitable than green field sites for new buildings.

3 Environmentalists believe it is important not to build on Sites of Special Scientific Interest (SSSIs).

B1b 8.6 · Environmental issues

How can we be sure?

The build-up of greenhouse gases cannot be denied, because there is hard evidence for it. However, there is still debate among scientists about whether these really cause the problems which are blamed on the 'greenhouse effect'. The majority of scientists now believe that global warming is at least partly linked to human activities such as burning of fossil fuels and deforestation. But not everyone agrees.

Some extreme weather patterns have certainly been recorded in recent years. But go back in history and it is possible to find other equally violent periods of weather long before fossil fuels were used so heavily and deforestation was happening.

The temperature of the Earth has varied greatly over millions of years. We have had both Ice Ages and times when almost the entire Earth was covered in tropical vegetation, all long before humans evolved. So some scientists argue that what we are seeing is the result of natural changes rather than a direct result of human activities.

ACTIVITY

You are going to present a 3–4 minute slot on global warming for a news programme for young people. You need to explain the different scientific views and why it is so hard to be certain what is happening. You also need to present some of the scientific evidence and plan some informative and attention-grabbing graphics. Make sure you present a balanced picture, with plenty of facts – but make it interesting too or they'll all change channels!

Scientists suggest that flooding is due to global warming

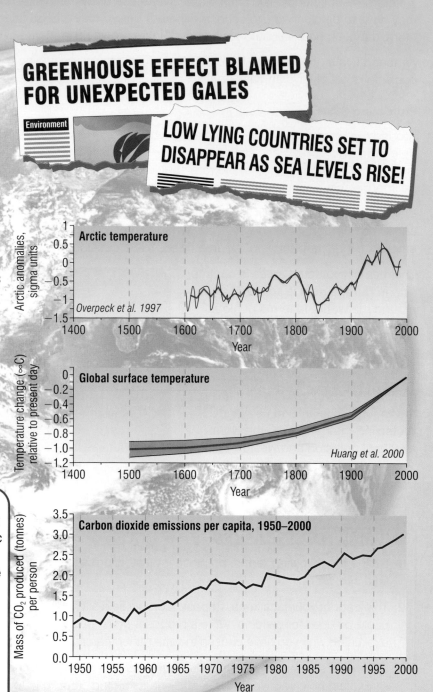

GREENHOUSE EFFECT BLAMED FOR UNEXPECTED GALES

Environment

LOW LYING COUNTRIES SET TO DISAPPEAR AS SEA LEVELS RISE!

Arctic temperature
Arctic anomalies, sigma units
Overpeck et al. 1997
Year

Global surface temperature
Temperature change ($^\circ$C) relative to present day
Huang et al. 2000
Year

Carbon dioxide emissions per capita, 1950–2000
Mass of CO_2 produced (tonnes) per person
Year

What can I do?

What people do in their lives really does affect our planet. More and more evidence shows that we cannot go on using the resources of the Earth in the way we do. But there are lots of different ways of looking at the problems. Here are just a few of them.

Many people are now trying to control and reduce the amount of greenhouse gases we produce. The problem is that the whole world has to agree because the Earth only has one atmosphere and this is affected by things which happen everywhere in the world. There are enormous problems with this.

One easy way to stop deforestation is to stop buying what they produce. They only cut the forests down to provide timber and beef. Don't buy mahogany loo seats, and don't eat cheap burgers! If we all stopped buying, they'd soon stop cutting down the rain forests.

It's the wealthy developed nations who have to take responsibility for most of the carbon dioxide emissions. We need to persuade our citizens to cut back a lot on using their cars. They won't do it because they're afraid it would affect their jobs and industry. I mean, America wouldn't even sign up to the Kyoto agreement, and they produce more pollution than anyone else!

I just want to earn a living to support my family. We have no proper schools, hospitals and roads. I must work at the logging camp – it's the only work for miles around and I want my children to have an education.

I've got so little money, I need to buy things as cheaply as I can. I don't really care where the beef comes from – I can feed my kids really cheaply and give them meat nearly every day if we eat burgers and things. We're never going to see a rainforest, are we?

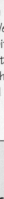

It looks a bit different from where I live. If we can't expand our industries, what chance do we have of getting richer? We need to make more money to improve the life and health of our people. Why should we have to sacrifice our development to help solve a problem caused by other, much richer nations?

There are a lot of us who are trying to monitor and support a more responsible use of the forests. The trouble is, it's very difficult to enforce guidelines, and there are always people ready to make money fast and illegally.

ACTIVITY

Problems like greenhouse emissions, global warming and the use of non-renewable resources can seem overwhelming. But each one of us can make a difference in the choices we make in our everyday lives.

You are going to develop some web pages to be used by your school to help everyone recognise some of the problems there are. More importantly, you are going to suggest ways in which they can help to conserve resources and change attitudes.

Design a flyer to be handed out at registration time that will give everyone the web address of your site and make them want to visit it.

Think of some targets you can set for your school, or ways in which students can set targets for their own families. Make a web site – and make a difference!

SUMMARY QUESTIONS

1 a) List the main ways in which humans reduce the amount of land available for other living things.

   b) Explain why each of these land uses is necessary.

   c) Suggest ways in which two of these different types of land use might be reduced.

2 a) Draw a flow diagram showing acid rain formation.

   b) Look at Figure 4, on page 123.
   i) What was the level of sulfur dioxide in the air in the UK in 1980?
   ii) What was the approximate level of sulfur dioxide in the air in the UK in the year that you were born? (Make sure you give your birth year.)
   iii) What was the level of sulfur dioxide in the air in 2001?

   c) Explain how the levels of sulfur dioxide have been reduced in the UK since 1978.

3 In Figure 2, on page 124 you can see clearly annual variations in the levels of carbon dioxide recorded each year. These fluctuations are thought to be due to seasonal changes in the way plants are growing and photosynthesising through the year.

   a) Explain how changes in plant growth and rate of photosynthesis might affect carbon dioxide levels.

   b) Explain how you could use this as evidence to try and prevent the loss of plant life by deforestation.

   c) How is the ever-increasing human population affecting the build-up of greenhouse gases – and what effect are they having on world climate?

4 Explain carefully how the following measures can help to support sustainable development.

   a) Encouraging farmers to plough the remains of crop plants into the soil, to use animal waste as well as chemical fertilisers and to replant hedgerows.

   b) Recycling all the aluminium drinks cans you use.

   c) Monitoring the types of invertebrates found in a river both above and below the sewage outfall.

   d) Setting up a 'walking bus' for a primary school.

   e) Enforcing standards for the building trade. These include the thickness of the insulation, the materials which can be used and the fittings (e.g. boilers, light bulbs) to be put in new houses.

EXAM-STYLE QUESTIONS

1 The diagram shows a town and some of its surroundings.

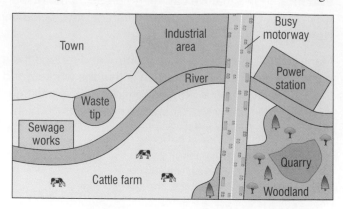

(a) A site that reduces the level of carbon dioxide in the atmosphere is:

   A Quarry   **B** Waste tip

   C Woodland   **D** Town

(b) Methane is another greenhouse gas. The site that produces the most methane is:

   A Industrial area

   B Cattle farm

   C Power station

   D Woodland

(c) Three sites produce gases that contribute to acid rain. These are:

   A Town, industrial area, quarry

   B Industrial area, busy motorway, power station

   C Busy motorway, power station, waste tip

   D Town, quarry, waste tip

(d) The two sites most likely to be damaged by acid rain falling on them are:

   A Cattle farm and quarry

   B Quarry and woodland

   C Woodland and river

   D River and waste tip

(e) Air pollution in the town is monitored continuously. This is most likely to be done by:

   A A large team of council technicians working shifts.

   B A small team of elite scientists on call 24 hours a day.

   C Local people using biological indicators such as lichen.

   D Electronic sensors attached to data logging equipment.   (5)

HOW SCIENCE WORKS QUESTIONS

2 In 1997 the World Climate Summit took place in Kyoto. Agreement was reached to control global warming by cutting emissions of greenhouse gases.

   (a) List two possible consequences of global warming. (2)

   (b) One of the most important greenhouse gases is methane.
     (i) What are the two main sources of methane? (2)
     (ii) Why has the amount of methane produced increased steadily over the past 200 years? (2)

   (c) Another group of greenhouse gases are the oxides of nitrogen. State the main source of these gases. (1)

   (d) State two ways in which deforestation may also contribute to global warming. (2)

3 Lichens are plants that do not grow well where there is air pollution. The number of lichen species growing along a 15 km line from the centre of a UK city was recorded. The results were plotted on a graph.

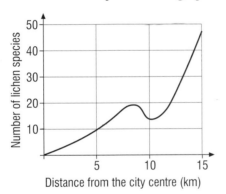

Distance from the city centre (km)

   (a) (i) Name the independent variable shown above. (1)
     (ii) Is this variable categoric, ordered, discrete or continuous? (1)
     (iii) Is the dependent variable categoric, ordered, discrete or continuous? (1)

   (b) Draw the headings for the table used to record the data for the graph. (1)

   (c) How many species of lichen are found at a distance of 5 km from the city centre? (1)

   (d) At what distance from the city centre is the least polluted air found? (1)

   (e) What is the relationship between the number of lichen species and the distance from the city centre to a point 8 km from the centre? (1)

   (f) Give a possible reason for the fall in the number of lichen species at a distance of 10 km from the city. (2)

Kim needs some help to experiment on the effect of acid rain on some radish plants. She is not too certain of exactly what to do. She has the idea that if she set up some dishes with some soil and planted some radish seeds, she could then put different amounts of vinegar onto them to see what happens.

a) What advice would you give Kim as to how many different dishes she should use? (2)

b) How would you decide how many seeds to plant in each dish? (2)

c) Kim said that she wanted to keep the type of soil the same in each dish. Is this a good idea? Explain your answer. (1)

d) Kim was interested in the effects of acid rain. Do you think that vinegar was a good choice to simulate the effects of acid rain? Explain your answer. (2)

e) You could use universal indicator papers or a pH meter to measure the pH of the soil and the acid. Which would you choose . . . and why? (2)

f) Name the independent variable in Kim's investigation. (1)

g) Suggest to Kim what might be a suitable dependent variable. (1)

h) Where would you suggest Kim keeps the dishes? (2)

i) It's always a good idea to prepare a table before you start the investigation. Prepare a table for Kim to collect her results. (3)

EXAMINATION-STYLE QUESTIONS

Science A
Questions **1** and **2**
In these questions match the letters with the numbers.
Use each answer only once.

1 To survive, animals and plants must be adapted to live in their environments.
Four different adaptations **A**, **B**, **C** and **D** are listed below.
Match words **A**, **B**, **C** and **D** with the spaces **1** to **4** in the sentences.

See pages 82–5

   A Water storage tissue

   B Reduced surface area

   C Thick coat

   D Camouflage

   Polar bears are insulated by a ......**1**......
   The white coat of an arctic fox acts as ......**2**......
   The stem of a cactus is used as ......**3**......
   To reduce water loss some desert plants have a ......**4**......   *(4 marks)*

2 Animals and plants can now be produced with features that we choose.
Match **A**, **B**, **C** and **D** with the spaces **1** to **4** in the sentences.

See pages 98–103

   A Tissue culture

   B Genetic engineering

   C Embryo transplantation

   D Cloning

   Producing genetically identical individuals from a single parent is called
   ......**1**...... Transferring genes from one organism to another is known as
   ......**2**......
   Using small groups of cells to grow new organisms is called ......**3**......
   Splitting apart cells from an organism before they specialise and placing them in
   a host mother is known as ......**4**......   *(4 marks)*

3 A scientist wished to find out which plants are best to grow on soil near the sea.
He performs an experiment to measure the mass of crop produced for different
plants at a range of salt concentrations.

See page 7

   What would be the dependent variable in this experiment?

   A Varieties of plant used.

   B Number of plants used.

   C Salt concentration.

   D Mass of crop produced.   *(1 mark)*

Science B

1 The rock shown in the diagram contains a fossil of a plant which lived millions of years ago.

See pages 108–11

(a) Describe **one** way in which this fossil might have been formed. *(2 marks)*

(b) (i) Evidence from fossils supports the theory of evolution. What is the theory of evolution? *(2 marks)*

(ii) How does fossil evidence support this theory? *(2 marks)*

2 The concentration of gases in glacier ice gives a measure of the composition of the air at the time the ice was formed. Because air moves rapidly around the Earth, this measure also applied to areas far away from where the measurements were taken. The graph shows the concentration, in parts per million, of carbon dioxide which has dissolved in glacier ice over the past 250 years.

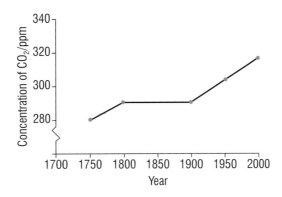

(a) Using the graph, describe the changes in concentration of carbon dioxide in the glacier ice from 1750 to 2000. *(3 marks)*

(b) Give one piece of evidence that suggests the instruments used in this investigation had to be very sensitive. *(1 mark)*

(c) List two sources of the carbon dioxide found in the atmosphere. *(2 marks)*

(d) Explain why there has been an increase in carbon dioxide in the atmosphere since 1900. *(5 marks)*

(e) Thirty-eight nations signed the Kyoto Treaty and agreed to cut their carbon dioxide emissions by just over 5% by 2012. Suggest three different actions that governments could take to help achieve their targets. *(3 marks)*

(f) If the concentration of carbon dioxide in the atmosphere increases further, how might this affect the temperature of the air at the Earth's surface? *(1 mark)*

(g) Explain **how** carbon dioxide in the atmosphere might cause this temperature change. *(2 marks)*

GET IT RIGHT!

'**Using the graph**' is how question 2 part (a) begins. Another expression used by examiners is 'using only information in the diagram'. In all these cases you should only use the information provided. Do not panic if you have not met the information before. Simply work logically through the facts provided. In this case start at year 1750 and describe (say what is happening) as you work along the plotted line to the year 2000.

For example, 'from 1750 to 1800 there is an increase in CO_2 . . . etc.' No explanations are required – not in this stage at least.

C1a | Products from rocks

Rocks, such as limestone from this quarry, provide us with many useful materials

What you already know

Here is a quick reminder of previous work that you will find useful in this unit:

- The chemical elements consist of atoms, which we represent by symbols.

- We can arrange the chemical elements in the periodic table.

- The substances in a mixture are not joined together chemically, so we can separate them again.

- We can represent chemicals using chemical formulae.

- We can summarise what happens in a chemical reaction using a word equation.

- The mass of the products formed in a chemical reaction is the same as the mass of the reactants they were formed from.

- When we burn fossil fuels we may produce new substances that can affect the environment.

- We can place the metals in a reactivity series. Metals higher up the series can displace metals lower down the series.

RECAP QUESTIONS

1 How many different types of atom are there in a jar of pure sulfur?

2 What are the symbols for the following elements: iron, oxygen, copper, sodium, chlorine, aluminium, calcium?

3 How could you separate a mixture of sand and salt?

4 Iron and chlorine react together to make iron chloride. Write a word equation for this reaction.

5 a) Describe in words what happens in this reaction:

magnesium + oxygen → magnesium oxide

b) In the reaction between magnesium and oxygen, the mass of magnesium and oxygen was 2.5 grams. How much magnesium oxide was formed?

6 When fossil fuels burn in plenty of air, what new substances are produced?

7 Magnesium ribbon is put into blue copper sulphate solution. The solution becomes colourless and a pinkish metal is produced.

Then copper metal is put into a solution of silver nitrate. The solution turns blue and silver metal is produced.

Arrange the three metals, copper, silver and magnesium, in order of reactivity. Put the most reactive first.

Making connections

My job takes me all round the world! I enjoy all the travelling – I always have. When I'm not flying for work I fly for pleasure – I've got a little light aircraft that I keep near my home.

This new airport building really will be a fantastic place for international passengers to arrive. It couldn't have been designed ten years ago – we just didn't have the materials that could have been used to build it then.

We have to keep all the runways in excellent condition for safety reasons. The runways are made of concrete, and with aeroplanes getting heavier all the time we have to be on the lookout for cracks. There's lots of research going on to work out the best way to make runways last as long as possible.

I have to drive about forty miles to work each day. The drive doesn't worry me, but the cost does. So I bought a diesel car which uses much less fuel. It still costs a lot, but it's a price I'm prepared to pay for living somewhere nice.

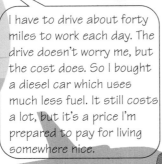

We have to make sure that the aircraft have all the fuel they need – it is no good if they can't take off because we haven't finished our job! A big airliner needs thousands of kilograms of fuel on board – they burn about six tonnes an hour, you know!

We've only flown once before, and we're very nervous. Our son and his wife emigrated to Australia last year, so this is the only way we can get to see them. Flying is exciting though, isn't it?!

It's our job to keep the aircraft well-maintained. Modern materials – especially alloys – make aeroplanes lighter, stronger and safer than ever before.

ACTIVITY

An airport shows how important science has become for the world we live in. Building materials, engineering materials for modern aircraft, and fuels – each of these relies on up-to-date scientific knowledge.

Work in groups to plan an exhibition to go on display for the passengers and travellers passing through an airport. You should design your exhibition to show how air transport relies on materials made by scientists. It should show the importance of these materials from the time people leave home to the time the aircraft gets off the ground.

You'll need to think about the environmental impact of airports and flying too. How will you get the message across that too much flying is bad for the planet without upsetting your airport sponsors? Use pictures, words, computer presentations – anything to make your message clear.

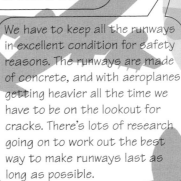

Flying is supposed to be so glamorous, but I fly so often it's just a bore. I mean, one first class lounge is just like another, isn't it?

Chapters in this unit

Rocks and building   Rocks and metals   Crude oil

Atoms, elements and compounds

LEARNING OBJECTIVES

1 What are elements made of?
2 How do we represent atoms and elements?
3 What are atoms and how do they join together?

Figure 1 An element contains only *one* type of atom – in this case bromine

Look at the things around you and the substances that they are made from – wood, metal, plastic, glass . . . the list is almost endless. Look further and the number of different substances is mind-boggling.

All substances are made of **atoms**. We think that there are about 100 different types of atom. These can combine in a huge variety of ways to give us all those different substances.

Some substances are made up of only one type of atom. We call these substances **elements**. As there are only about 100 different types of atom, there are only about 100 different elements.

a) How many different types of atom are there?
b) Why can you make millions of different substances from these different types of atom?

Elements can have very different properties. For example, elements such as silver, copper and gold are shiny solids. Other elements such as oxygen, nitrogen and chlorine are gases.

Atoms have their own symbols

The name that an element has depends on the language. For example, sulfur is called *Schwefel* in German and *azufre* in Spanish! Because a lot of scientific work is international, it is important that we have symbols for elements that everyone understands. You can see these symbols in the periodic table in Figure 2.

Group numbers

1	**2**												**3**	**4**	**5**	**6**	**7**	**0**

H 1 Hydrogen																		He 2 Helium
Li 3 Lithium	Be 4 Beryllium												B 5 Boron	C 6 Carbon	N 7 Nitrogen	O 8 Oxygen	F 9 Fluorine	Ne 10 Neon
Na 11 Sodium	Mg 12 Magnesium												Al 13 Aluminium	Si 14 Silicon	P 15 Phosphorus	S 16 Sulfur	Cl 17 Chlorine	Ar 18 Argon
K 19 Potassium	Ca 20 Calcium	Sc 21 Scandium	Ti 22 Titanium	V 23 Vanadium	Cr 24 Chromium	Mn 25 Manganese	Fe 26 Iron	Co 27 Cobalt	Ni 28 Nickel	Cu 29 Copper	Zn 30 Zinc	Ga 31 Gallium	Ge 32 Germanium	As 33 Arsenic	Se 34 Selenium	Br 35 Bromine	Kr 36 Krypton
Rb 37 Rubidium	Sr 38 Strontium	Y 39 Yttrium	Zr 40 Zirconium	Nb 41 Niobium	Mo 42 Molybdenum	Tc 43 Technetium	Ru 44 Ruthenium	Rh 45 Rhodium	Pd 46 Palladium	Ag 47 Silver	Cd 48 Cadmium	In 49 Indium	Sn 50 Tin	Sb 51 Antimony	Te 52 Tellurium	I 53 Iodine	Xe 54 Xenon
Cs 55 Caesium	Ba 56 Barium	Lanthanum see below	Hf 72 Hafnium	Ta 73 Tantalum	W 74 Tungsten	Re 75 Rhenium	Os 76 Osmium	Ir 77 Iridium	Pt 78 Platinum	Au 79 Gold	Hg 80 Mercury	Tl 81 Thallium	Pb 82 Lead	Bi 83 Bismuth	Po 84 Polonium	At 85 Astatine	Rn 86 Radon
Fr 87 Francium	Ra 88 Radium	Actinium see below															

The transition metals

The alkali metals
The alkaline earth metals

The halogens   The noble gases

Lanthanides

La 57 Lanthanum	Ce 58 Cerium	Pr 59 Praseodymium	Nd 60 Neodymium	Pm 61 Promethium	Sm 62 Samarium	Eu 63 Europium	Gd 64 Gadolinium	Tb 65 Terbium	Dy 66 Dysprosium	Ho 67 Holmium	Er 68 Erbium	Tm 69 Thulium	Yb 70 Ytterbium	Lu 71 Lutetium

Actinides

Ac 89 Actinium	Th 90 Thorium	Pa 91 Protactinium	U 92 Uranium	Np 93 Neptunium	Pu 94 Plutonium	Am 95 Americium	Cm 96 Curium	Bk 97 Berkelium	Cf 98 Californium	Es 99 Einsteinium	Fm 100 Fermium	Md 101 Mendelevium	No 102 Nobelium	Lr 103 Lawrencium

Figure 2 The periodic table shows the symbols for the elements and a lot of other information too

The symbols in the periodic table represent atoms. For example, O represents an atom of oxygen while Na represents an atom of sodium. The elements in the table are arranged in vertical columns, called *groups*. Each group contains elements with similar chemical properties.

c) Why do we need symbols to represent atoms of different elements?
d) A particular element in the periodic table is a reactive metal. What properties are the elements in the same column as this element likely to have?

Atoms, elements and compounds

Almost all of the substances around us are not pure elements. Most substances are made up of different atoms joined together to form **compounds**. Some compounds are made from just two types of atom joined together (e.g. water, made from hydrogen and oxygen). Other compounds consist of many different types of atom.

Atoms are made up of a tiny central *nucleus* with *electrons* around it.

Look at the diagram in Figure 3.

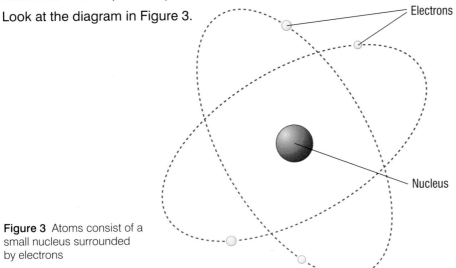

Electrons

Nucleus

Figure 3 Atoms consist of a small nucleus surrounded by electrons

Sometimes atoms react together by transferring electrons to form chemical bonds. Other atoms may share electrons to form chemical bonds. No matter how they are formed, chemical bonds hold atoms tightly to each other.

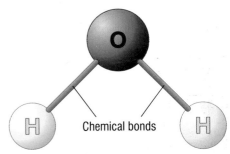

Chemical bonds

Figure 4 Two or more atoms bonded together are called a **molecule**. Look at the chemical bonds holding the hydrogen and oxygen atoms together in the water molecule.

SUMMARY QUESTIONS

1 Copy and complete using the words below:

atoms   **bonds**   **molecule**   **sharing**

All elements are made up of .... When two or more atoms join together a ...... is formed. The atoms in elements and compounds are held together by giving, taking or ...... electrons. We say that they are held tightly together by chemical ....

2 Explain the following statement: 'When two elements are mixed together they can often be separated quite easily, but when two elements are combined in a compound they can be very difficult to separate.'

3 a) Draw diagrams to help explain the difference between an element and a compound, using actual examples.
   b) Draw a labelled diagram to show the structure of an atom.

KEY POINTS

1 All substances are made up of atoms.
2 Elements contain only one type of atom.
3 Different atoms can bond together by giving, taking or sharing electrons, to form compounds.

C1a 1.2

Limestone and its uses

LEARNING OBJECTIVES

1 What are the uses of limestone?
2 What happens when we heat limestone?

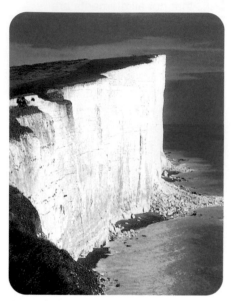

Figure 1 These white cliffs are made of chalk. This is one type of limestone, formed from the shells of tiny sea plants.

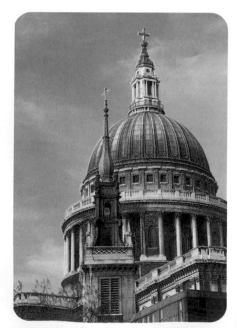

Figure 2 St Paul's Cathedral in London is built from limestone blocks

Uses of limestone

Limestone is a rock that is made mainly of ***calcium carbonate***. Some types of limestone were formed from the remains of tiny animals and plants that lived in the sea millions of years ago. We dig limestone out of the ground in quarries all around the world. It has many important uses, including use as a building material.

Many important buildings around the world are made of limestone. We can cut and shape the stone taken from the ground into blocks. These can be placed one on top of the other, like bricks in a wall. We have used limestone in this way to make buildings for hundreds of years.

We can also process limestone to make other building materials. Powdered limestone can be heated to high temperatures with a mixture of sand and sodium carbonate (soda) to make **glass**.

Powdered limestone can also be heated with powdered clay to make **cement**. When we mix cement powder with water, sand and crushed rock, a slow chemical reaction takes place. The reaction produces a hard, stone-like building material called **concrete**.

a) What is limestone made of?
b) How do we use limestone to make buildings?

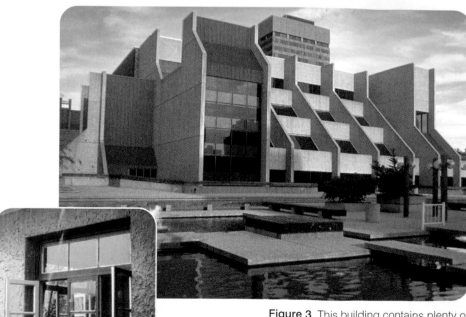

Figure 3 This building contains plenty of concrete which is made from limestone

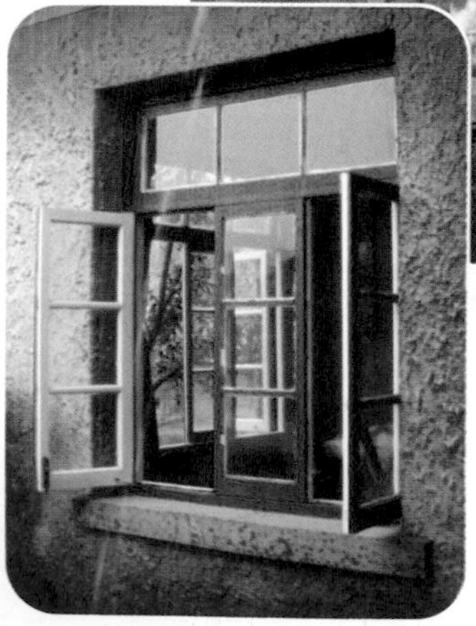

Figure 4 Glass gives us buildings that let in natural light and which protect us from the weather

Heating limestone

The chemical formula for calcium carbonate is $CaCO_3$. This tells us that for every calcium atom (Ca) there is one carbon atom (C) and three oxygen atoms (O).

When we heat limestone strongly it breaks down to form a substance called **quicklime** (calcium oxide). Carbon dioxide is also produced in this reaction. Breaking down a chemical by heating is called **thermal decomposition**.

To make lots of quicklime this reaction is done in a furnace called a *lime kiln*. We fill the kiln with crushed limestone and heat it strongly using a supply of hot air. Quicklime comes out of the bottom of the kiln, while waste gases leave the kiln at the top.

We can show the thermal decomposition reaction by a word equation:

$$\text{calcium carbonate} \xrightarrow{\text{heat}} \text{calcium oxide} + \text{carbon dioxide}$$
$$\text{(quicklime)}$$

A rotary lime kiln

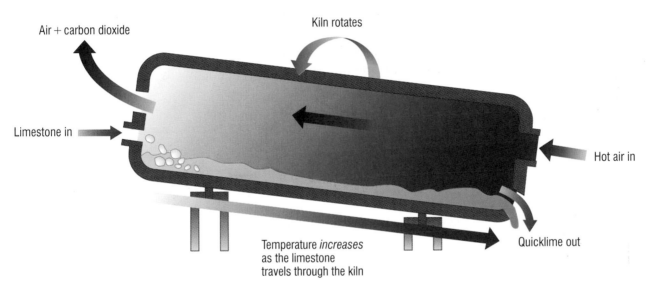

Figure 5 Quicklime is produced in a lime kiln. It is often produced in a **rotary kiln**, where the limestone is heated in a rotating drum. This makes sure that the limestone is thoroughly mixed with the stream of hot air so that it decomposes completely.

SUMMARY QUESTIONS

1 Copy and complete using the words below:

    $CaCO_3$ calcium   cement   concrete   glass

    Limestone is mostly made of …… carbonate (whose chemical formula is ……). As well as being used to produce blocks of building material, limestone can be used to produce …… , …… and …… that can also be used in building.

2 Produce a poster to show how limestone is used in building.

3 The stone roof of a building is supported by columns made of limestone. Why might this be unsafe after a fire in the building? Explain the chemical process involved in weakening the structure.

KEY POINTS

1 Limestone is made mainly of calcium carbonate.
2 Limestone is widely used in building.
3 Limestone breaks down when we heat it strongly (thermal decomposition) to make quicklime and carbon dioxide.

Decomposing carbonates

C1a 1.3

LEARNING OBJECTIVES

1 Do other carbonates behave in the same way as limestone?
2 What happens to the atoms in a chemical reaction?
3 Is the mass of the products in a chemical reaction related to the mass of the reactants?

PRACTICAL

Investigating carbonates
You can investigate the thermal decomposition of carbonates by heating samples in a Bunsen flame. Imagine that you have been given samples of the carbonates listed below. What kinds of changes might tell you if a sample decomposes when you heat it?

Figure 1 Investigating the thermal decomposition of a solid

Carbonate samples:
sodium carbonate,
potassium carbonate,
magnesium carbonate,
zinc carbonate,
copper carbonate.

In the last spread we saw that limestone consists mainly of calcium carbonate. This decomposes when we heat it, producing quicklime (calcium oxide) and carbon dioxide. Calcium is an element in Group 2 of the periodic table (see periodic table on page 138). As we have already seen, the elements in a group tend to behave in the same way – so does magnesium carbonate also decompose when you heat it? And what about other carbonates too?

a) Why might you expect magnesium carbonate to behave in a similar way to calcium carbonate?

In an investigation into the behaviour of carbonates, a student draws the following conclusions when he heats samples of carbonates:

Calcium carbonate	Sodium carbonate	Potassium carbonate	Magnesium carbonate	Zinc carbonate	Copper carbonate
✓	✗	✗	✓	✓	✓

(✓ = decomposes, ✗ = does not decompose)

b) To which group in the periodic table do sodium and potassium belong? (See page 138.)
c) To which group in the periodic table do magnesium and calcium belong?
d) What do these conclusions suggest about the behaviour of the carbonates of elements in Group 1 and Group 2?
e) Can you be certain about your answer to question d)? Give reasons.

Investigations like this show that when many carbonates are heated in a Bunsen flame they decompose. They form the metal oxide and carbon dioxide – just as calcium carbonate does. Sodium and potassium carbonate do not decompose at the temperature in the Bunsen flame.

Balancing equations

We can represent the decomposition of carbonates using chemical equations. These use chemical formulae for elements and compounds, and help us to see how much of each chemical is reacting. Representing reactions in this way is better than using word equations, for three reasons:

- word equations are only useful if everyone who needs to read them speaks the same language,
- word equations do not tell us how much of each substance is involved in the reaction,
- word equations can get very complicated when lots of chemicals are involved.

When calcium carbonate decomposes, we can show the chemical reaction like this:

$$CaCO_3 \rightarrow CaO + CO_2$$

This equation is *balanced* – there is the same number of each type of atom on both sides of the equation. This is very important, because atoms cannot be created or destroyed in a chemical reaction. This also means that the mass of the products formed in the reaction is equal to the mass of the reactants.

We can check if an equation is balanced by counting the number of each type of atom on either side of the equation. If the numbers are equal, then the equation is balanced.

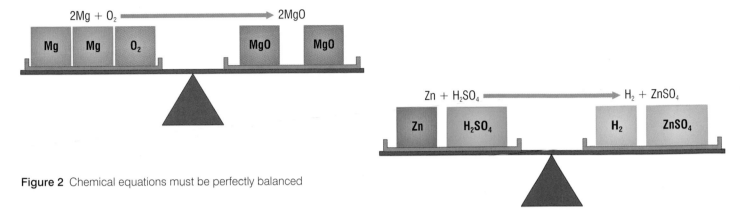

Figure 2 Chemical equations must be perfectly balanced

Magnesium carbonate decomposes like this:

$$MgCO_3 \rightarrow MgO + CO_2$$

Notice how the symbol for magnesium (Mg) has simply replaced the symbol for calcium (Ca). Everything else has stayed the same.

f) Zinc carbonate has a similar formula to calcium carbonate and decomposes in the same way. Write the balanced equation for this reaction.

In some cases it is not so easy to balance an equation. Here is an example of a reaction which shows how it must be written to produce a balanced equation.

When magnesium reacts with hydrochloric acid, hydrogen gas is given off. Magnesium chloride is also formed. Magnesium chloride has the formula $MgCl_2$ and hydrogen gas is H_2 so the equation starts off as:

$$Mg + HCl \rightarrow MgCl_2 + H_2$$

But this means that there are not enough H and Cl atoms on the left-hand side. By adding another HCl molecule we can balance the equation, which now reads:

$$Mg + 2HCl \rightarrow MgCl_2 + H_2$$

To check that this equation is balanced, we add up the different atoms on either side. There are two magnesium atoms, two hydrogen atoms and two chlorine atoms on each side of the equation – so it is balanced.

SUMMARY QUESTIONS

1 Give the general word equation for the thermal decomposition of a metal carbonate.

2 a) A mass of 2.5 g of barium carbonate ($BaCO_3$) completely decomposes when it is heated. What is the total mass of products formed in this reaction?

 b) Write a balanced equation to show the reaction in a).

KEY POINTS

1 Some carbonates decompose when we heat them in a Bunsen flame.

2 There is the same number of each type of atom on each side of a balanced chemical equation.

3 The mass of reactants is the same as the mass of the products in a chemical reaction.

Quicklime and slaked lime

LEARNING OBJECTIVES

1 What is slaked lime?
2 How is it produced?
3 How is lime mortar made?

Limestone is used very widely as a building material. We can also process it to make other materials used for building.

As we saw on page 141, quicklime is produced when we heat limestone strongly. The calcium carbonate in the limestone undergoes thermal decomposition.

When we add water to quicklime it reacts to produce **slaked lime**. Slaked lime's chemical name is **_calcium hydroxide_**. This reaction gives out a lot of heat.

$$calcium\ oxide\ +\ water\ \rightarrow\ calcium\ hydroxide$$
$$CaO\ \ \ \ \ +\ H_2O\ \rightarrow\ \ \ \ \ \ Ca(OH)_2$$

Although it is not very soluble, we can dissolve a little calcium hydroxide in water. After filtering, this produces a colourless solution called **_lime water_**. We can use lime water to test for carbon dioxide. When carbon dioxide is bubbled through clear lime water the solution turns cloudy. This is because calcium carbonate is formed, which is insoluble in water.

$$calcium\ hydroxide\ +\ carbon\ dioxide\ \rightarrow\ calcium\ carbonate\ +\ water$$
$$Ca(OH)_2\ \ \ \ \ +\ \ \ \ \ CO_2\ \ \ \ \ \rightarrow\ \ \ \ \ \ CaCO_3\ \ \ \ \ +\ H_2O$$

a) What substance do we get when quicklime reacts with water?
b) Describe how we can make lime water from this substance.

You can explore the reactions of limestone using some very simple apparatus.

GET IT RIGHT!

Make sure that you know the limestone story and the word equations for the reactions.

PRACTICAL

Investigating the chemical reactions of limestone

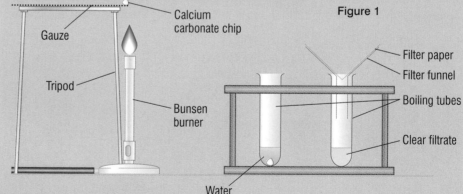

Heat the calcium carbonate chip very strongly, making it glow. This produces calcium oxide (quicklime). Let the calcium oxide cool down.
Then you react the calcium oxide with a few drops of water to produce calcium hydroxide (slaked lime).
When you dissolve this in water and filter, it produces lime water.
Carbon dioxide bubbled through the lime water produces calcium carbonate and the solution goes cloudy.

● The reaction between quicklime and water gives out a lot of energy. What do you observe during the reaction?
● Why does bubbling carbon dioxide through lime water make the solution go cloudy?

DID YOU KNOW?

The saying 'to be in the limelight' originated from the theatre, well before the days of electricity. Stages were lit up by heated limestone before electric lamps were invented.

Mortar

About 6000 years ago the Egyptians heated limestone strongly in a fire and then combined it with water. This produced a material that hardened with age. They used this material to plaster the pyramids. Nearly 4000 years later, the Romans mixed slaked lime with sand and water to produce **mortar**.

Mortar holds other building materials together – for example, stone blocks or bricks. It works because the lime in the mortar reacts with carbon dioxide in the air, producing calcium carbonate again. This means that the bricks or stone blocks are effectively held together by stone, which makes the construction very strong.

$$\text{slaked lime} + \text{carbon dioxide} \rightarrow \text{calcium carbonate} + \text{water}$$
$$Ca(OH)_2 + CO_2 \rightarrow CaCO_3 + H_2O$$

The amount of sand in the mixture is very important – too little sand and the mortar shrinks as it dries. Too much sand makes it too weak.

Even today, mortar is still used widely as a building material. However, modern mortars can be used in a much wider range of ways than in ancient Egyptian and Roman times.

Figure 2 The original lime mortar has flaked away from the surface of the Sphinx in Egypt, and many of the stones are now missing

Figure 3 Lime mortar should be used to repair old buildings

NEXT TIME YOU...

... look at a building made of bricks, think how long people have been using mortar to join blocks of stone and brick like this.

SUMMARY QUESTIONS

1 Copy and complete using the words below:

> **calcium hydroxide   carbonate   carbon dioxide   mortar**
> **quicklime   thousands**

Limestone has been used in building for ...... of years. When it is heated, limestone produces ......, also called calcium oxide. If calcium oxide is reacted with water .... is produced. This can be combined with sand to produce ......, used to hold building materials like bricks together. Mortar relies on .... in the air to produce a chemical reaction in which calcium ...... is formed.

2 a) When quicklime reacts with water, slaked lime is produced. Write a balanced equation to show the reaction.
   b) The slaked lime in mortar reacts with carbon dioxide in the air to produce calcium carbonate and water. Write a balanced equation for this reaction.

KEY POINTS

1 When water is added to quicklime it produces slaked lime.
2 Lime mortar is made by mixing slaked lime with sand and adding water.

C1a 1.5

Cement, concrete and glass

LEARNING OBJECTIVES

1 How do we make cement?
2 What is concrete?
3 How is glass made?

Figure 1 Lime mortar is not suitable for building pools since it will not harden when in contact with water

Cement

Although lime mortar holds bricks and stone together very strongly it does not work in all situations. In particular, lime mortar does not harden very quickly. It will not set at all where water prevents it from reacting with carbon dioxide.

The Romans realised that they needed to add something to lime mortar to make it set in wet conditions. They found that adding brick dust or volcanic ash to the mortar mixture enabled the mortar to harden even under water. This method remained in use until the 18th century.

Then people found that heating limestone with clay in a kiln produced **cement**. Much experimenting led to the invention of *Portland cement*. We make this from a mixture of limestone, clay and other minerals which are heated and then ground into a fine powder.

This type of cement is still in use today. The mortar used to build a modern house consists of a mixture of Portland cement and sand. This sets when it is mixed thoroughly with water and left for a few days.

Figure 2 Portland cement was invented nearly 200 years ago. It is still in use all around the world today.

a) What does lime mortar need in order to set hard?
b) Why will lime mortar not set under water?

Concrete

Sometimes builders add stones or crushed rocks to the mixture of water, cement and sand. When this sets, it forms a hard, stone-like building material called **concrete**.

This material is very strong. It is especially good at resisting forces which tend to squeeze or crush it. We can make concrete even stronger by pouring it around steel rods or bars and then allowing it to set. This makes *reinforced concrete*, which is also good at resisting forces that tend to pull it apart.

Figure 3 Badly designed reinforcement caused the concrete roof of this terminal building at Charles de Gaulle Airport in France to collapse in May 2004

PRACTICAL

Which mixture makes the strongest concrete?

Try mixing different proportions of cement, gravel and sand, then adding water, to find out how to make the strongest concrete.

- How did you test the concrete's strength?
- How could you improve the reliability of your data?

Figure 4 Glass can produce some spectacular buildings

Glass

We can also use limestone to make a very different kind of building material. When powdered limestone is mixed with sand and sodium carbonate and then heated strongly it produces **glass**.

Glass is very important in buildings since it allows us to make them both weatherproof and light. Hundreds of years ago only very rich people could afford glass for their windows. So ordinary people's buildings must have been very cold and very dark during the winter!

Modern chemists have developed glass with many different properties. This makes it possible to design buildings that would have been impossible to build even 50 years ago.

GET IT RIGHT!

You don't need to know the details of the industrial processes for making glass – just how it is made.

SUMMARY QUESTIONS

1 Write down three ways that limestone is used in building other than as limestone blocks.

2 List the different ways in which limestone has been used to build your home or school.

3 Glass has been advertised as 'insulation you can see through!' Explain what the advertiser meant by this.

4 Concrete and glass are commonly used building materials. Evaluate the use of:
   a) concrete to make a path rather than using bricks
   b) glass to make a window pane rather than using perspex.

KEY POINTS

1 Cement is made by heating limestone with clay in a kiln.
2 Concrete is made by mixing crushed rocks, cement and sand with water.
3 Glass is made by heating powdered limestone, sand and sodium carbonate together very strongly.

Building materials – from old to new

Out of the past . . .

Since ancient times people have always needed somewhere to shelter. In hot countries people need to find somewhere cool during the day. In cold countries they need somewhere warm – and in wet, windy Britain we often need somewhere to get out of the rain!

In the past, people often had to move around since they did not work in one place. People who looked after animals or who worked in different places at different times of the year built shelters. They used wood and any other natural materials that they could find at that place. They built very simple homes that could be put up quickly. They did not use large amounts of material that had to be carried over long distances.

Once people started to settle down and live in one place, it was worth building a house which was bigger and which took longer to build. People began to use materials like stone, and to develop new materials like bricks and concrete. It was more difficult to build well with these new materials. This meant it was often necessary for people with special skills to be involved in building.

Figure 1 A charcoal burner's hut. The charcoal burner had to watch the kiln constantly, so it was necessary to live right next to it. Charcoal was sold as a fuel. It burned with an intense heat, much hotter than wood.

QUESTIONS

1 Why did ancient people use local materials for building rather than materials that had to be brought from a long distance away?

2 Why did people need special skills to work with materials like bricks and mortar?

. . . and into the future

As new building materials were developed, new ways of using them were found. Buildings could be made bigger and taller, and new designs were possible. People could live in big houses, grand houses, small houses – and even flats and bungalows.

As our understanding of the chemistry of materials grew, artificial materials like plastics and metals began to replace natural materials like wood. Because the properties of these new materials can be controlled, they can be made to order. They are produced to have exactly the right strength for the job they have to do. As materials scientists continue to produce exciting new materials, who knows what homes in the future may look like?!

ACTIVITY

a) Find some pictures of old houses. Find out what they are made of and think about why people chose these materials.

b) Design a house of the future. Clearly label the materials you use and their properties. Be creative – you can use materials that haven't been invented yet!

lifestyle

Ancient or Modern?

Anna and Simon are buying their second house. It's a modern one, on a big estate. They're hoping to make it look a bit different, to reflect their personalities and lifestyle.

We'd really like an old house, but there's nothing like that around here. So we want to use old building materials to make the house look older than it really is. We think that we might use natural limestone to cover up the bricks, and we're going to put an old wooden door in to replace the PVC front door. We might even take out the modern windows too. We want it to look more like a cottage than a house on a big estate.

Roger and Su moved into their house a while ago. It's old, and needs some work done on it. Some of the windows leak, and there's a lot of painting to do! They want to try to make the house easy to maintain.

This house is very old, and it really needs updating. With our children growing up, we really don't have time to look after the house. If we put modern doors and windows in, we won't have to worry about painting them. The windows will be double-glazed too, so the house will be much warmer in winter. And that old metal gutter round the roof – we'll replace that with white PVC plastic. So much more modern!

ACTIVITY

Imagine that you are the host of a TV makeover show. Both couples in the magazine are guests on the show. They need your advice about how they should makeover their houses. Write and present a short script for the programme, presenting your advice to each couple. Remember that you'll not only need to tell them what they should do but why they should do it.

ACTIVITY

Decisions, decisions

There's a huge choice of building materials available for people who want to update their homes – modern scientists have seen to that!

Discuss the issues below:
- Should people just 'do want they fancy' when they decide to update a house?
- Are Anna and Simon right to think about making their modern house look old? Why?
- Should Roger and Su try to make their house as easy as possible to maintain? What advice would you give them?

SUMMARY QUESTIONS

1 a) Jim has a sample of a pure element. How many different types of atom are there in it?

b) What do we call the vertical columns of elements in the periodic table?

c) Three elements are in the same column of the periodic table. What does this tell us about the reactions of these three elements?

d) The atoms of two different elements react together to form a compound. Why is it difficult to separate the two elements when this has happened?

2 The diagram shows a design for a handwarmer which uses quicklime and water. To activate the handwarmer you squeeze the container. This breaks the capsule containing the water so that it mixes with the quicklime.

a) Using complete sentences, describe how the handwarmer works.

b) Write down a balanced chemical equation for the reaction in the handwarmer, using the correct symbols.

c) Is the handwarmer re-usable or disposable? Give reasons for your answer.

3 a) A set of instructions for making concrete reads:

'To make good, strong concrete, thoroughly mix together 4 buckets of gravel, 3 buckets of sand and one bucket of cement. When you have done this, add half a bucket of water.'

Copy and complete the table showing the percentage of each ingredient in the concrete mixture. Give your answers to the nearest whole number.

Ingredient	gravel	sand	cement	water
Number of buckets				
Percentage				

b) Describe an investigation you could try to see which particular mixture of gravel, sand and cement makes the strongest concrete. What would you vary, what would you keep the same and how would you test the 'strength' of the concrete?

EXAM-STYLE QUESTIONS

1 The diagram shows a molecule of ammonia, NH_3.

H—N—H
|
H

Match the words **A**, **B**, **C** and **D** with spaces **1** to **4** in the sentences.

A bonds

B electrons

C elements

D symbols

Ammonia is a compound made from two ....**1**......... .
The atoms in the molecule are represented by ....**2**......... .
The atoms in ammonia are held together by chemical ....**3**.......... .
Each atom has a nucleus surrounded by ....**4**.......... .

(4)

2 The diagram shows stages in making cement and concrete.

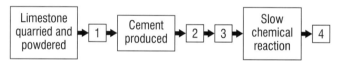

Match statements **A**, **B**, **C** and **D** with the numbers **1** to **4** to describe what happens in this process.

A cement mixed with sand and crushed rock

B concrete produced

C limestone heated in a kiln with clay

D water added to mixture

(4)

3 (a) Slaked lime is made by reacting quicklime with:

A carbon dioxide

B oxygen

C sulfuric acid

D water

(1)

(b) The chemical name for slaked lime is:

A calcium chloride

B calcium hydroxide

C calcium oxide

D calcium sulfate

(1)

(c) Slaked lime can be used to make:

A bricks

B clay

C mortar

D quicklime

(1)

(d) Lime water goes cloudy when reacted with carbon dioxide. Which substance is produced?

   A   calcium carbonate

   B   calcium chloride

   C   calcium oxide

   D   calcium sulfate   (1)

4   Glass is used in almost all buildings.

   (a) Suggest **two** properties of glass that make it useful in buildings.   (2)

   (b) Suggest and explain one disadvantage of using glass in buildings.   (2)

5   One of the largest limestone quarries in the United Kingdom is near the town of Buxton. It is in the Peak District National Park, an area popular with tourists.

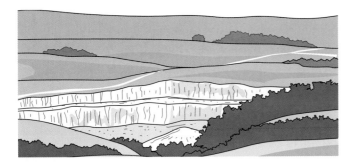

Suggest **three** social or environmental issues involved in quarrying limestone in the Peak District.   (3)

6   Mortars used in most modern buildings are made using cement.
A student tested the strength of a ready-mixed mortar. He did this by dropping a mass onto a small mortar beam from increasing heights until the beam broke in half. He tested 4 beams made from the mortar. His results were 20 cm, 50 cm, 65 cm and 15 cm.

   (a) What was the range of the student's results?   (2)

   (b) Work out the mean of his results.   (1)

   (c) Comment on the precision of his results.   (1)

   (d) (i)   Besides cement, what was the other solid in the ready-mixed mortar?   (1)

       (ii) What other solid is needed to make concrete instead of mortar?   (1)

HOW SCIENCE WORKS QUESTIONS

Look at the standards for the testing of cement shown below and answer the questions that follow.

Standards for the testing of cement

Cement must be tested to the following standards:

- After 7 days use compressive test standard equipment to test the strength of the cement.

- Three batches of already tested cement must also be treated in the same way.

- The bowl must be wiped clean and 400 g of test cement added during 30 seconds. Add 400 g of the sand and mix for 120 seconds. Add 200 g of water and mix for 240 seconds.

- The sand being used must be washed, heated in a kiln at 110°C and then passed through a sieve with holes of diameter 1 mm.

- The mixing bowl used must be between 20 cm wide at the top, narrowing in a curve to 8 cm at the bottom and made of stainless steel.

- The apparatus used must conform to that described in CTM19.

a) What was the dependent variable in this testing? (1)

b) What was the 'control group' in this testing? (1)

c) Why is it important to test cement that has already been tested? (1)

d) Why is it important to test the cement? (1)

e) Why is it important to give so much detail of how to test the cement? (1)

f) Who should NOT carry out these tests? Explain your answer. (2)

g) Who should carry out these tests? Explain your answer. (2)

C1a 2.1

Extracting metals

1 Where do metals come from?
2 How do we extract metals from the Earth?

Figure 1 The Angel of the North stands 20 metres tall, and is made of steel which contains a small amount of copper

Metals have been important to people for thousands of years. You can follow the course of history by the materials people used – from the Stone Age to the Bronze Age and then on to the Iron Age.

Where do metals come from?

Metals are found in the Earth's crust. We find most metals combined with other chemical elements, often with oxygen. This means that the metal must be chemically separated from its compounds before you can use it.

In some places there is enough of a metal or metal compound in a rock to make it worth extracting the metal. Then we call the rock a metal **ore**.

Whether it is worth extracting a particular metal depends on:

● how easy it is to extract it from its ore,
● how much metal the ore contains.

A few metals, such as gold and silver, are so unreactive that they are found in the Earth as the metals (elements) themselves. We say that they exist in their *native* state.

Sometimes a nugget of gold is so large it can simply be picked up. At other times tiny flakes have to be physically separated from sand and rocks by panning.

a) Where do we find metals in nature?
b) If there is enough metal in a rock to make it economic to extract it, what do we call the rock?
c) Why are silver and gold found as metals rather than combined with other elements?

Figure 2 Panning for gold. Mud and stones are washed away while the dense gold remains in the pan.

DID YOU KNOW...

. . . that gold in Wales is found in seams, just like coal – although not as thick, unfortunately! Gold jewellery was worn by early Welsh princes as a badge of rank. Welsh gold has been used in modern times to make the wedding rings of Royal brides.

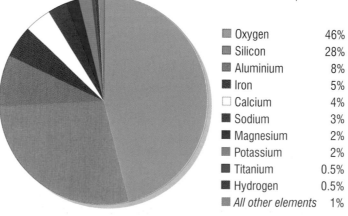

■ Oxygen	46%
■ Silicon	28%
■ Aluminium	8%
■ Iron	5%
□ Calcium	4%
■ Sodium	3%
■ Magnesium	2%
■ Potassium	2%
■ Titanium	0.5%
■ Hydrogen	0.5%
■ *All other elements*	1%

Figure 3 There are many different elements that go to make up the Earth's crust

How do we extract metals?

The way in which we extract a metal depends on its place in the *reactivity series*. The reactivity series lists the metals in order of their reactivity. The most reactive are placed at the top and the least reactive at the bottom.

A more reactive metal will displace a less reactive metal from its compounds. Carbon (a non-metal) will also displace less reactive metals from their oxides. We use carbon to extract metals from their ores commercially.

> d) A metal cannot be extracted from its ore using carbon. Where is this metal in the reactivity series?

You can find many metals, such as copper, lead, iron and zinc, combined with oxygen. The compounds are called **metal oxides**. Because carbon is more reactive than each of these metals, you can use it to extract them from their ores.

When you heat the metal oxide with carbon, the carbon removes the oxygen from the metal oxide to form carbon dioxide. The reaction leaves the metal, as the element, behind:

$$\text{metal oxide} + \text{carbon} \rightarrow \text{metal} + \text{carbon dioxide}$$

For example:

$$\text{lead oxide} + \text{carbon} \rightarrow \text{lead} + \text{carbon dioxide}$$
$$2PbO + C \rightarrow 2Pb + CO_2$$

We call the removal of oxygen in this way a **reduction reaction**.

> e) What do chemists mean by a reduction reaction?

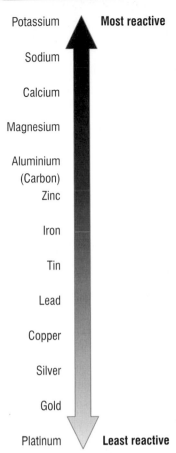

Products from rocks

Most reactive — Potassium, Sodium, Calcium, Magnesium, Aluminium, (Carbon), Zinc, Iron, Tin, Lead, Copper, Silver, Gold, Platinum — Least reactive

Figure 4 This reactivity series shows how reactive each element is compared to the other elements

---

PRACTICAL

Reduction by carbon

Heat some copper oxide with carbon powder strongly in a test tube.

Empty the contents into an evaporating dish.

You can repeat the experiment with lead oxide and carbon if you have a fume cupboard to work in.

- Explain your observations. Include a word equation or balanced symbol equation.

SUMMARY QUESTIONS

1 Copy and complete using the words below:

crust   extracted   native   reduction

Metals come from the Earth's …… . Some metals are very unreactive and are found in their …… state. Metals, such as zinc and iron, are found combined with oxygen and can be …… using chemical reactions. These are known as …… reactions, as oxygen is removed from the oxide.

2 Make a list of all the metal objects found in your classroom or at home. Try to name the metal(s) used to make each object.

3 Platinum is never found combined with oxygen. What does this tell you about its reactivity? Give a use of platinum that depends on this property.

4 Zinc oxide (ZnO) can be reduced to zinc by heating it in a furnace with carbon. Carbon monoxide (CO) is given off in the reaction. Write a word equation and a balanced equation for the reduction of zinc oxide.

KEY POINTS

1 The Earth's crust contains many different elements.

2 A metal ore contains enough of the metal to make it economically worth extracting the metal.

3 We can find gold and other unreactive metals in their native state.

4 The reactivity series is useful in deciding the best way to extract a metal from its ore.

5 Metals more reactive than carbon cannot be extracted from their ores using carbon.

Extracting iron

LEARNING OBJECTIVES

1 What are the raw materials for making iron?
2 How is iron ore reduced?

Figure 1 Making iron is a hot, dirty process – and it can be quite spectacular too!

Extracting iron from its ore is a huge industry. Iron is the second most common metal in the Earth's crust. Iron ore contains iron combined with oxygen. Iron is less reactive than carbon. So we can extract iron by using carbon to remove oxygen from the iron oxide in the ore.

We extract iron using a **blast furnace**. This is a large container made of steel. It is lined with fireproof bricks to withstand the high temperatures inside. There are solid raw materials which we use in the blast furnace, as well as lots of air. They are:

● **Haematite** – this is the most common iron ore. It contains mainly iron(III) oxide (Fe_2O_3) and sand.
● *Coke* – this is made from coal and is almost pure carbon. It will provide the reducing agent to remove the oxygen from iron(III) oxide.

We also add limestone to remove impurities.

a) List the three solid raw materials for making iron.

Hot air is blown into the blast furnace. This makes the coke burn, which heats the furnace and forms carbon dioxide gas.

$$C + O_2 \rightarrow CO_2$$

At the high temperatures in the blast furnace, this carbon dioxide reacts with more coke to form carbon monoxide gas.

$$CO_2 + C \rightarrow 2CO$$

The carbon monoxide reacts with the iron oxide, removing its oxygen, and reducing it to molten iron. This flows to the bottom of the blast furnace.

$$Fe_2O_3 + 3CO \rightarrow 2Fe + 3CO_2$$

DID YOU KNOW?

The oldest iron smelting plants that we know about were built in China over 2000 years ago. Iron goods around 500 years older than this have been found in China, so it seems likely that iron-making in China goes back at least 2500 years. The first iron smelter in Europe was not built until about 1200 CE.

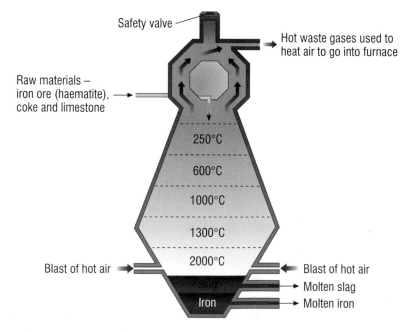

Figure 2 A blast furnace produces molten iron. Impurities from the iron ore are removed as slag.

Some of the molten iron is left to solidify in moulds – we call this **cast iron**. It contains about 96% iron. Most of the iron is kept molten to be turned into **steel**.

We make steel by removing more of the impurities from the iron. Then we mix it with other elements to change its properties.

b) What is the name of the substance that reduces the iron oxide to iron?

Iron – an important metal

Iron played a vital part in the Industrial Revolution, which happened in Britain during the 1700s and 1800s. Three generations of the same family – all called Abraham Darby – improved and developed new ways of making iron.

Up until this time charcoal had been used as the source of carbon for reducing the iron oxide. The Darby family replaced this with coke. They developed the blast furnace as a way of making iron continuously rather than in batches.

Most of the world's steel is now made in Asia. That's because the cost of making it is much lower than in Europe and North America. Iron and steel-making used to employ many thousands of people in the UK but now employs far fewer. Other jobs for these people have had to be made in other industries.

Figure 3 The youngest Darby built the world's first cast-iron bridge in 1779. It still spans the River Severn at Coalbrookdale in Shropshire. Coalbrookdale is now much better known as Ironbridge Gorge.

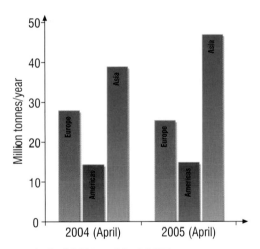

Figure 4 World steel output in April 2004 and April 2005

SUMMARY QUESTIONS

1 Copy and complete using the words below:

carbon   **coke**   **haematite**   **limestone**

We can extract iron from iron ore using .... The most common type of iron ore is .... Other raw materials for making iron are ...... and ....

2 How is steel different from cast iron?

3 a) Some of the iron(III) oxide (Fe_2O_3) in the blast furnace is reduced by carbon, giving off carbon dioxide. Write a word equation for this reaction.

   b) Write a balanced symbol equation for the reaction described in part a).

KEY POINTS

1 We extract iron from iron ore by reducing it with carbon in a blast furnace.

2 The solid raw materials used to make iron are iron ore (haematite), coke and limestone.

3 Molten iron is tapped off from the bottom of the blast furnace.

Properties of iron and steels

LEARNING OBJECTIVES

1 Why is iron from a blast furnace not very useful?
2 How is iron changed to make it more useful?

The iron produced by a blast furnace, called pig iron, is not very useful. It is about 96% iron and contains impurities, mainly of carbon. This makes pig iron very brittle. However, we can treat the iron from the blast furnace to remove some of the carbon.

Figure 1 The iron which has just come out of a blast furnace contains about 96% iron. The main impurity is carbon.

If we remove all of the carbon and other impurities from pig iron, we get pure iron. This is very soft and easily-shaped, but it is too soft for most uses. If we want to make iron really useful we have to make sure that it contains tiny amounts of other elements, including carbon and certain metals.

We call a metal that contains other elements an **alloy**.

Iron that has been alloyed with other elements is called **steel**. By adding elements in carefully controlled amounts, we can change the properties of the steel.

a) Why is iron from a blast furnace very brittle?
b) Why is pure iron not very useful?
c) How do we control the properties of steel?

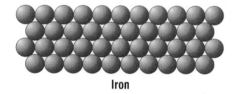

Iron

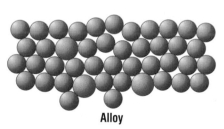

Alloy

Figure 2 The atoms in pure iron are arranged in layers which can easily slide over one another. In alloys the layers cannot slide so easily because atoms of other elements change the regular structure.

Look at Figure 2:
The atoms in pure iron are arranged in layers. Because of this regular arrangement, the atoms can slide over one another very easily. This is why pure iron is so soft and easily shaped.

Steels

Steel is not a single substance. There are lots of different types of steel. All of them are alloys of iron with carbon and/or other elements.

The simplest steels are the *carbon steels*. We make these by alloying iron with small amounts of carbon (from 0.03% to 1.5%). These are the cheapest steels to make. We use them in many products, like the bodies of cars, knives, machinery, ships, containers and structural steel for buildings.

Often these carbon steels have small amounts of other elements in them as well. High carbon steel, with a relatively high carbon content, is very strong but brittle. On the other hand, low carbon steel is soft and easily shaped. It is not as strong, but it is much less likely to shatter.

Mild steel is one type of low carbon steel. It contains less than 0.1% carbon. It is very easily pressed into shape. This makes it particularly useful in mass production, for example, making car bodies.

Low-alloy steels are more expensive than carbon steels because they contain between 1% and 5% of other metals. Examples of the metals used include nickel, chromium, manganese, vanadium, titanium and tungsten. Each of these metals gives a steel that is well-suited for a particular use. We use nickel–steel alloys to build long-span bridges, bicycle chains and military armour-plating. That's because they are very resistant to stretching forces. Tungsten steel operates well under very hot conditions so it is used to make high-speed tools.

Even more expensive are the *high-alloy steels*. These contain a much higher percentage of other metals. For example, chromium steels have between 12% and 15% chromium mixed with the iron, and often some nickel is mixed in too – this provides strength and chemical stability.

These chromium–nickel steels are more commonly known as *stainless steels*. We use them to make cooking utensils and cutlery. They are also used to make chemical reaction vessels because they combine hardness and strength with great resistance to corrosion. Unlike most other steels, they do not rust!

Figure 3 The properties of steel alloys make them ideal for use in suspension bridges

GET IT RIGHT!

Know how the hardness of steels is related to their carbon content.

SUMMARY QUESTIONS

1 Complete the following sentences using the terms below:

> **carbon steel**   **pure iron**   **steel**

If all of the carbon and other impurities are removed from pig iron we get ....

Iron that has been alloyed with other elements is called ....

Iron that has been alloyed with a little carbon is called ....

2 a) What is the difference between high-alloy and low-alloy steels?
   b) Why are surgical instruments made from steel containing chromium and nickel?
   c) Make a table to summarise the composition and properties of different types of steel.
   d) Using diagrams, explain how alloying a metal with atoms of another element changes its properties.

KEY POINTS

1 Pure iron is too soft for it to be very useful.
2 Carefully controlled quantities of elements are added to iron to make alloys of steel with different properties.

Alloys in everyday use

Clara

LEARNING OBJECTIVES

1 Why are alloys more useful than pure metals?
2 What are smart alloys?

As we saw on page 156, pure iron is not very useful because it is too soft. In order to use the iron, we must turn it into an alloy by adding other elements. Other metals must also be turned into alloys to make them as useful as possible.

Copper

Copper is a soft, reddish coloured metal. It conducts heat and electricity very well. We have used copper for thousands of years. In fact some people think that it was the first metal to be used by humans. Copper articles have been found which are well over ten thousand years old.

Just like pure iron, pure copper is very soft.

Figure 1 The Statue of Liberty in New York contains over 80 tonnes of copper

Bronze was probably the first alloy made by humans, about 5500 years ago. It is usually made by mixing copper with tin, but small amounts of other elements can be added as well. For example, we can add phosphorus. This gives the alloy properties which make it ideal to use for bearings where we want very low friction.

We make *brass* by alloying copper with zinc. Brass is much harder than copper and it is workable.

It can be hammered into sheets and bent into different shapes. This property is used to make musical instruments.

a) What are the names of two common alloys of copper?
b) Why are copper alloys more useful than pure copper?

DID YOU KNOW?

Bronze makes an excellent material for the barrels of cannons. The low friction between the bronze barrel and the iron cannon ball makes it unlikely that the ball will stick in the barrel and cause an explosion.

Gold and aluminium

Just like copper and iron, we can make gold and aluminium harder by adding other elements. We usually alloy gold with copper and silver when we want to use it in jewellery. By varying the proportions of the two metals added we can get differently coloured 'gold' objects. They can vary from yellow to red, and even a shade of green!

Aluminium is a metal with a low density. It can be alloyed with a wide range of other elements – there are over 300 alloys of aluminium! These alloys have very different properties. We can use some to build aircraft while others can be used as amour plating on tanks and other military vehicles.

c) Apart from making gold harder, what else can alloying change?

d) What property of aluminium makes it useful for making alloys in the aircraft industry?

Smart alloys

Some alloys have a very special property. If we bend (or **deform**) them into a different shape and then heat them, they return to their original shape all by themselves. These alloys are sometimes called **smart alloys**. Their technical name is **shape memory alloys** (SMAs), which describes the way they behave. They seem to 'remember' their original shape!

We can use the clever properties of shape memory alloys in many ways. Some of the most interesting uses of SMAs have been in medicine. Doctors treating a badly broken bone can use smart alloys to hold the bones in place while they heal. They cool the alloy before it is wrapped around the broken bone.

When it heats up again the alloy goes back to its original shape. This pulls the bones together and holds them while they heal. Dentists have made braces to push teeth into the right position using this technique.

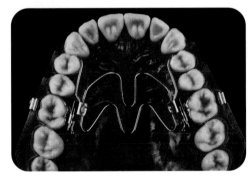

Figure 2 This dental brace pulls the teeth into the right position as it warms up. That's smart!

SUMMARY QUESTIONS

1 Copy and complete using the words below:

   aluminium   brass   bronze   smart   soft   thousands

   Copper has been used by people for ...... of years. Like pure iron, pure copper is too ...... to be very useful. Copper can be alloyed with tin to make ......, and with zinc to make .... There are over 300 alloys of .... Some alloys can 'remember' their shape when they are heated after they have been bent – they are called ...... alloys.

2 Why can aluminium alloys be used in so many different ways?

3 a) Explain the advantages of a dental brace made of 'smart' alloy over one made of a conventional alloy.

   b) Do some research to find some other uses of smart alloys, explaining why they are used.

KEY POINTS

1 Copper, gold and aluminium are all alloyed with other metals to make them more useful.

2 We can control the properties of alloys by adding different amounts of different elements.

3 Smart alloys are also called shape memory alloys. When deformed they return to their original shape on heating.

4 Shape memory alloys can be used in medicine and dentistry.

Transition metals

C1a 2.5

LEARNING OBJECTIVES

1 What are transition metals and why are they useful?
2 Why do we use so much copper?
3 How can we produce enough copper?

In the centre of the periodic table there is a large block of metallic elements. Here we find the elements called the **transition metals** or *transition elements*. Many of them have similar properties.

Like all metals, the transition metals are very good conductors of electricity and heat. They are also generally hard, tough and strong. Yet we can easily bend or hammer them into useful shapes. We say they are **malleable**. With the exception of mercury, which is a liquid at room temperature, the transition metals have very high melting points.

a) In which part of the periodic table do we find the transition metals?
b) Name *three* properties of these elements.

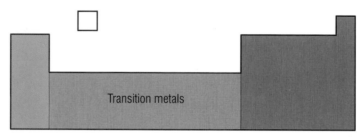

Figure 1 The transition metals

The properties of the transition metals mean that we can use them in many different ways. You will find them in buildings and in cars, trains and other types of transport. Their strength makes them useful as *construction materials*. We use them in heating systems and for electrical wiring because heat and electricity pass through them easily.

One of the transition metals is copper. Although copper is not particularly strong, we can bend and shape it easily. It also conducts electricity and heat very well and it does not react with water. So it is ideal where we need pipes that will carry water or wires that will conduct electricity.

c) Why do we use transition metals so much?
d) What makes copper so useful?

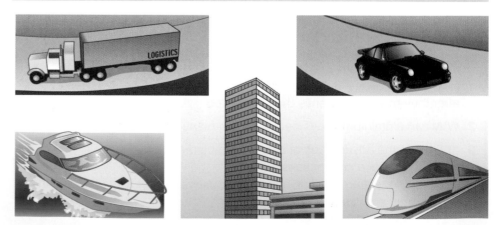

Figure 2 Transition metals are used in many different ways because of their properties. Copper is particularly useful because it is such a good conductor of heat and electricity.

We extract copper from **copper ore**. There are two main methods used to remove the copper from the ore. In one method we use sulfuric acid to produce copper sulfate solution, before extracting the copper.

The other process is called **smelting**. We heat copper ore very strongly in air to produce crude copper. Then we use the impure copper as anodes in electrolysis cells to make pure copper. 85% of copper is still produced by smelting.

e) What chemical do we use to treat copper ore in order to form copper sulfate?

Processing copper ore uses huge amounts of electricity, and costs a lot of money. If we have to smelt an ore, the heating also requires a lot of energy.

New ways to extract copper

Instead of extracting copper using chemicals, heat and electricity, scientists are developing new ways to do the job. They can now use bacteria, fungi and even plants to help extract copper.

If we could extract metals like this on a large scale, it could be a lot cheaper than the way we do it now. It could be a lot 'greener' too.

We would also be able to extract copper from ores which contain very little copper. At the moment it is too expensive to process these 'low grade' ores using conventional methods.

> **NEXT TIME YOU...**
>
> . . . take a walk in your garden or go onto the school field, think about the transition metals under your feet. Although they might be too expensive to extract now, new ways of extracting metals could mean that you are literally standing on a goldmine!

Figure 4 In Australia Dr Jason Plumb looks for bacteria that can extract metals from ores. His search takes him to some exciting places – including volcanoes!

> **KEY POINTS**
>
> 1 The transition metals are found in the middle block of elements in the periodic table.
> 2 Transition metals have properties that make them useful for building and making things.
> 3 Copper is a very useful transition metal because of its high conductivity.
> 4 Scientists are looking for new ways to extract copper that will use less energy.

> **SUMMARY QUESTIONS**
>
> 1 Write a few words describing the following: a) transition metals, b) properties of copper, c) smelting.
>
> 2 Silver and gold are transition metals that conduct electricity even better than copper. Why do we use copper to make electric cables instead of either of these metals?
>
> 3 a) Explain briefly two ways of extracting copper metal.
>   b) Explain the advantages of extracting copper using bacteria or fungi rather than the way most is extracted now.

Aluminium and titanium

LEARNING OBJECTIVES

1 Why are aluminium and titanium so useful?
2 Why does it cost so much to extract aluminium and titanium?
3 Why should we recycle aluminium?

Although they are very strong, many metals are also very dense. This means that we cannot use them if we want to make something that has to be both strong and light. Examples are alloys for making an aeroplane or the frame of a racing bicycle.

Where we need metals which are both strong and light, **aluminium** and **titanium** fit the bill. They are also metals which do not **corrode**.

Aluminium is a silvery, shiny metal which is surprisingly light – it has a relatively low density for a metal. It is an excellent conductor of heat and electricity. We can also shape it and draw it into wires very easily.

Although aluminium is a relatively reactive metal, it does not corrode easily. This is because the aluminium atoms at its surface immediately react with the oxygen in air to form a thin layer of tough aluminium oxide. This layer stops any further corrosion taking place.

Aluminium is not a particularly strong metal, but we can use it to form alloys. These alloys are harder, more rigid and stronger than pure aluminium.

Figure 1 We use aluminium alloys to make aircraft because of their combination of lightness and strength

As a result of these properties, we use aluminium to make a whole range of goods. These range from cans, cooking foil and saucepans through to high-voltage electricity cables, aeroplanes and space vehicles.

a) Why does aluminium resist corrosion?
b) How do we make aluminium stronger?

Titanium is a silvery-white metal, which is very strong and very resistant to corrosion. Like aluminium it has an oxide layer that protects it. Although it is denser than aluminium, it is less dense than most transition metals.

Titanium has a very high melting point – about 1660°C – so we can use it at very high temperatures.

We use it instead of steel and aluminium in the bodies of high-performance aircraft and racing bikes. Here its combination of relative low density and strength is important. We also use titanium to make parts of jet engines because it keeps its strength at high temperatures.

The strength of titanium at high temperatures makes it very useful in nuclear reactors. In reactors we use it to make the pipes and other parts that must stand up to high temperatures. Titanium performs well under these conditions, and its strong oxide layer means that it resists corrosion.

Another use of titanium is also based on its strength and resistance to corrosion – replacement hip joints.

c) Why does titanium resist corrosion?

d) What properties make titanium ideal to use in jet engines and nuclear reactors?

We extract aluminium using electrolysis. Because it is a reactive metal we cannot use carbon to displace it from its ore. Instead we extract aluminium by passing an electric current through molten aluminium oxide at high temperatures.

Titanium is not particularly reactive, so we could produce it by displacing it from its ore with carbon. But carbon reacts with the titanium making it very brittle. So we have to use a more reactive metal. We use sodium or magnesium to do this. However, we have to produce both sodium and magnesium by electrolysis in the first place.

The problem with using electrolysis to extract these metals is that it is very expensive. That's because we need to use high temperatures and a great deal of electricity.

In the UK each person uses around 8 kg of aluminium every year. This is why it is important to *recycle* aluminium. It saves energy, since recycling aluminium does not involve electrolysis.

e) Why do we need electricity to make aluminium and titanium?

f) Why does recycling aluminium save electricity?

Figure 2 We can use titanium inside the body as well as outside. Because it has a low density, is strong and does not corrode, we can use titanium to make alloys that are excellent for artificial joints. These are artificial hip joints, used to replace a natural joint damaged by disease or wear and tear.

SUMMARY QUESTIONS

1 Copy and complete using the words below:

corrode   electrolysis   expensive   light   oxide   reactive   strong

Aluminium and titanium alloys are useful as they are ...... and .... Although aluminium is reactive, it does not ...... because its surface is coated with a thin layer of aluminium .... Titanium does not corrode because it is not very ...... and also has its oxide layer to protect it. ...... is used in the extraction of both metals from their ores which makes them ....

2 Why is titanium used to make artificial hip joints?

3 Each person in the UK uses about 8 kg of aluminium each year.

a) Recycling 1 kg of aluminium saves about enough energy to run a small electric fire for 14 hours. If you recycle 50% of the aluminium you use in one year, how long could you run a small electric fire on the energy you have saved?

b) Explain the benefits of recycling aluminium.

KEY POINTS

1 Aluminium and titanium are useful because they resist corrosion.

2 Aluminium and titanium are expensive because extracting them from their ores involves many stages in the processes and requires large amounts of energy.

3 Recycling aluminium is important because we need to use much less energy to produce 1 kg of recycled aluminium than we use to extract 1 kg of aluminium from its ore.

163

C1a 2.7

Using metals

Metals and society

The way we use metals has literally changed the world. Since ancient times metals have enabled us to do things we could never have done without them. These range from making tools to generating electricity; from creating jewellery to flying in aeroplanes.

The timeline below shows some of the main points in the history of the use of metals, up to the beginning of the 20th century.

5000–4000 BCE
People mine copper ores
and smelt them

3500–3000 BCE
People use bronze, an alloy
of copper and tin, to make
tools and weapons

1000 BCE
People have now developed ways of making
iron. Iron is much better than bronze for making
tools and weapons because it is harder. This
means that blades stay sharp for longer

1903 CE
The Wright brothers fly their first aeroplane
at Kitty Hawk USA. It has an engine
made using cast aluminium to make it
as light as possible

1825 CE
Pure aluminium and titanium
are produced for the first time

1400 CE
In Europe people can now produce
temperatures high enough to melt iron.
This means that it can be cast in moulds
to make more complicated shapes

QUESTIONS

1 Why was the development of iron so important?

2 Find out about the development of metals in the last 100 years and extend the timeline using this information.

Mining with plants

As they grow, plants absorb dissolved chemicals from the ground in the water that they take into their roots. Some plants absorb a lot of one certain chemical, and scientists have found that we can use these plants to help us extract metals from soil.

The technology that we need to do this is quite simple. So we can use plants to mine metals where it would be far too expensive to do it in any other way. Extracting metals like this is called *phytomining*.

QUESTIONS

3 Why can we use phytomining when it would be too expensive to extract metals in other ways?

4 There is a long time delay between planting the crop and extracting the metal. What might happen if the price of the metal falls before the crop can be harvested?

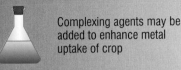

Complexing agents may be added to enhance metal uptake of crop

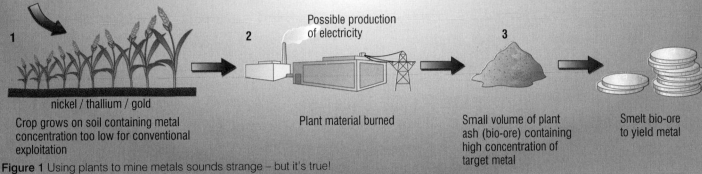

1 nickel / thallium / gold
Crop grows on soil containing metal concentration too low for conventional exploitation

Possible production of electricity

2 Plant material burned

3 Small volume of plant ash (bio-ore) containing high concentration of target metal

Smelt bio-ore to yield metal

Figure 1 Using plants to mine metals sounds strange – but it's true!

Recycling fridges

It is nearly always cheaper for us to recycle metals than to extract new metals from their ores. This is especially true for metals like aluminium, where we need to use large amounts of energy to extract them. But sometimes it is not easy to recycle metals because they are combined with other materials. As an example, look at the problems of recycling fridges.

ACTIVITY

Many people do not realise how important it is to recycle old fridges and other household equipment.
Design a poster to be used in a campaign to persuade people to recycle their old fridge, washing machine, and so on, rather than simply dumping them.

First, we need to remove the chemicals in the cooling system of the fridge

Pliers used to puncture the cooling circuit and extract the liquid coolant

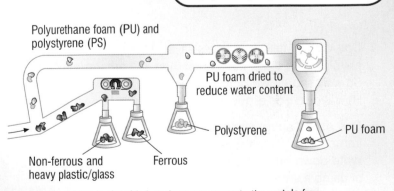

Polyurethane foam (PU) and polystyrene (PS)

PU foam dried to reduce water content

Polystyrene

PU foam

Non-ferrous and heavy plastic/glass

Ferrous

Then we can take out the parts of the fridge like the shelves

After this, the fridge is shredded, and we can separate the metals from other materials like plastics and insulating foam. Some of the metals are **ferrous** metals (they contain iron) while others are **non-ferrous**. Magnets can be used to separate the ferrous metals from the non-ferrous metals

Figure 2 Recycling fridges is difficult but necessary

SUMMARY QUESTIONS

1 Write simple definitions for the following terms:
   a) metal ore
   b) native state
   c) reduction reaction.

2 Gold is a very dense, unreactive metal. Old-fashioned gold prospectors used to 'pan' for gold in streams by scooping up small stones from the stream bed and washing them in water, allowing them slowly to get washed out of the pan.
   Using words and diagrams, explain how they could find gold by using this technique.

3 We can change the properties of metals by alloying them with other elements.
   a) Write down **three** ways that a metal alloy may be different from the pure metal.
   b) Choose **one** of these properties. Use the two diagrams below to help you to explain why a metal alloy behaves differently to the pure metal.

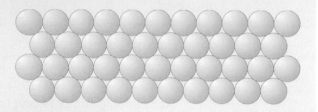

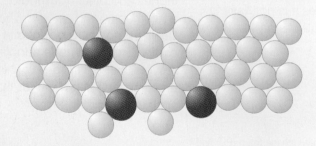

4 One of your fellow students says: 'There is more aluminium in the Earth's crust than any other metal. So why should we bother recycling it?
   How would you argue against this point of view?

5 One use of smart alloys is to make spectacle frames.
   Write down **one** advantage and **one** disadvantage of using a smart alloy like this.

EXAM-STYLE QUESTIONS

1 This question is about the uses of these metals:
   A aluminium   B copper
   C gold   D iron
   Which of these metals is used
   (a) as the main metal in alloys to build aircraft?
   (b) in alloys to make jewellery?
   (c) to make all steels?
   (d) to make water pipes and electrical wiring?   (4)

2 Choose a metal from the list A to D to match each description.
   A aluminium   B chromium
   C gold   D titanium
   (a) A metal that is strong at high temperatures and resists corrosion.
   (b) An unreactive metal found native in the Earth.
   (c) This metal has a low density and is extracted by electrolysis.
   (d) This metal is mixed with iron to make high alloy steels.   (4)

3 Use words from the list A to D to complete the word equations.
   A copper   B iron
   C sodium   D water
   (a) copper oxide + sulfuric acid   copper sulfate + ...
   (b) copper sulfate + iron   ...... + iron sulfate
   (c) iron oxide + carbon   ...... + carbon dioxide
   (d) titanium + ...   titanium + sodium
        chloride   chloride   (4)

4 A student tested the flexibility of four different alloy rods. She suspended a mass from the end of the rods which were fixed at the other end to the edge of a bench. She measured how far each rod bent.
   Which words describe the 'distance the rod bent'?
   A a categoric, independent variable.
   B a continuous, independent variable.
   C a categoric, dependent variable.
   D a continuous, dependent variable.   (1)

5 Name the types of substance described in each part of this question.

(a) These elements are hard, tough and strong, conduct heat and electricity well and are found in the middle of the periodic table.

(b) These rocks contain enough metal to make it worth extracting.

(c) This is a metal that contains other elements to give it specific properties.

(d) These materials are smart because they can return to their original shape when heated and are used by surgeons to hold broken bones while healing. (4)

6 Iron is extracted from iron oxide by removing oxygen.

(a) What name is given to a reaction in which oxygen is removed from a compound? (1)

(b) Name an element that could be used to remove oxygen from iron oxide. (1)

(c) Write a word equation for the reaction that would take place in (b). (2)

7 Titanium is used to make replacement hip joints. One reason why titanium can be used in this way is that it resists corrosion.

(a) How is titanium protected from corrosion? (1)

(b) Suggest **two** other properties of titanium that make it suitable for this use. (2)

8 Most of the world's steel is now made in Asia.

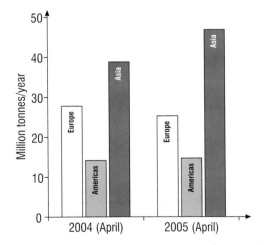

Suggest **two** reasons why it costs less to make steel in Asia than in Europe. (2)

9 New methods using bacteria, fungi and plants are being developed to extract copper. Suggest **three** reasons why these new methods have been developed. (3)

HOW SCIENCE WORKS QUESTIONS

How hard is gold?

The following was overheard in a jeweller's shop:

'I would like to buy a 24 carat gold ring for my husband.'

'Well madam, we would advise that you buy one which is lower carat gold. It looks much the same but the more gold there is, the less hard it is.'

Is this actually the case? Let's have a look scientifically at the data.

Pure gold is 24 carat. A carat is a twenty-fourth, so $24 \times 1/24 = 1$ or pure gold. So a 9 carat gold ring will have 9/24ths gold and 15/24ths of another metal, probably copper or silver. Most 'gold' sold in shops is therefore an alloy. How hard the 'gold' is will depend on the amount of gold and on the type of metal used to make the alloy.

Here is some data on the alloys and the maximum hardness of 'gold'.

Gold alloy (carat)	Maximum hardness (BHN)
9	170
14	180
18	230
22	90
24	70

a) The shop assistant said that 'the more gold there is, the less hard it is.' Was this based on science or was it hearsay? Explain your answer. (2)

b) In this investigation which is the independent variable? (1)

c) Which of the following best describes the hardness of the alloy?
   i)   continuous
   ii) discrete
   iii) categoric
   iv) ordered. (1)

d) Plot a graph of the results. (3)

e) What is the pattern in your results? (2)

f) You might have expected that the 9 carat gold was much harder than the 14 or the 18 carat gold, but it isn't. Can you offer an explanation for this? (1)
(Clue – is there an uncontrolled variable lurking around here?)

Fuels from crude oil

C1a 3.1

LEARNING OBJECTIVES

1 What is in crude oil?
2 What are alkanes?
3 How do we represent alkanes?

Some of the 21st century's most important chemistry involves chemicals that are made from crude oil. These chemicals play a major part in our lives. We use them as fuels in our cars, to warm our homes and to make electricity.

Fuels are important because they keep us warm and on the move. So when oil prices rise, it affects us all. Countries that produce crude oil can have an affect on the whole world economy by the price they charge for their oil.

Figure 1 The price of nearly everything we buy is affected by oil because the cost of moving goods to the shops affects the price we pay for them

a) Why is oil so important?

NEXT TIME YOU...

... go anywhere by car, bus, train or even aeroplane, remember that the energy transporting you comes from the sunlight trapped by plants millions of years ago.

Crude oil

Crude oil is a dark, smelly liquid, which is a *mixture* of lots of different chemical compounds. A mixture contains two or more elements or compounds that are not chemically combined together.

Crude oil straight out of the ground is not much use. There are too many substances in it, all with different boiling points. Before we can use crude oil, we have to separate it into its different substances. Because the properties of substances do not change when they are mixed, we can separate mixtures of substances in crude oil by using *distillation*. Distillation separates liquids with different boiling points.

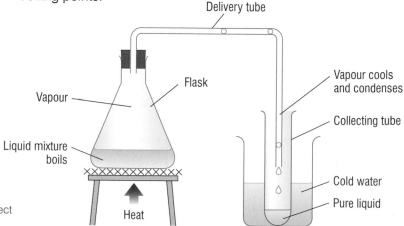

Figure 2 Mixtures of liquids can be separated using distillation. We heat the mixture so that it boils, and collect the vapour that forms by cooling and condensing it.

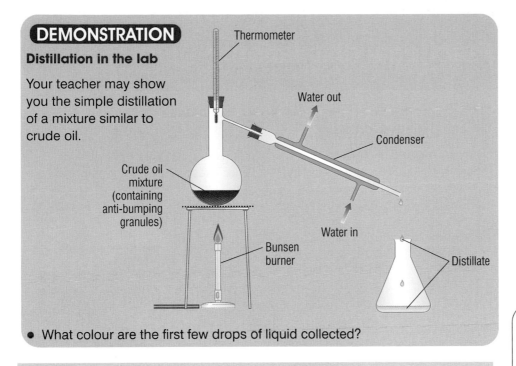

Distillation in the lab

Your teacher may show you the simple distillation of a mixture similar to crude oil.

Thermometer

Water out

Condenser

Crude oil mixture (containing anti-bumping granules)

Water in

Bunsen burner

Distillate

● What colour are the first few drops of liquid collected?

b) What is crude oil?
c) Why can we separate crude oil using distillation?

Nearly all of the compounds in crude oil are made from atoms of just two chemical elements – hydrogen and carbon. We call these compounds **hydrocarbons**. Most of the hydrocarbons in crude oil are **alkanes**. You can see some examples of alkane molecules in Figure 3.

Another way of writing these alkane molecules is like this:

CH_4 (methane); C_2H_6 (ethane); C_3H_8 (propane); C_4H_{10} (butane); C_5H_{12} (pentane).

Can you see a pattern in the formulae of the alkanes? We can write the general formula for alkane molecules like this:

$$C_nH_{2n+2}$$

which means that 'for every n carbon atoms there are $(2n + 2)$ hydrogen atoms'. For example, if an alkane contains 25 carbon atoms its formula will be $C_{25}H_{52}$.

We describe alkanes as **saturated** hydrocarbons. This means that they contain as many hydrogen atoms as possible in each molecule. We cannot add any more.

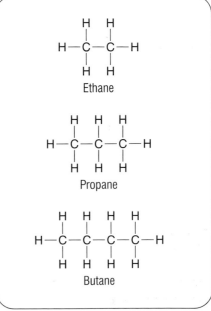

Figure 3 We can represent alkanes like this, showing all of the atoms in the molecule. The line between two atoms in the molecule is the chemical bond holding them together.

SUMMARY QUESTIONS

1 Copy and complete using the words below:

   carbon   distillation   hydrocarbons   hydrogen   mixture

   Crude oil is a …… of compounds. Many of these contain only atoms of …… and …… . They are called …… . The compounds in crude oil can be separated using …… .

2 Why is crude oil not very useful before we have processed it?

3 a) Write down the formula of the alkanes which have 6, 7 and 8 carbon atoms. Then find out their names.
   b) How many carbon atoms are there in an alkane which has 30 hydrogen atoms?

KEY POINTS

1 Crude oil is a mixture of many different compounds.
2 Many of the compounds in crude oil are hydrocarbons – they contain only carbon and hydrogen.
3 Alkanes are saturated hydrocarbons. They contain as much hydrogen as possible in their molecules.

CRUDE OIL

Fractional distillation

The compounds in crude oil

Hydrocarbon molecules can be very different. Some are quite small, with relatively few carbon atoms and short chains. These short-chain molecules are the hydrocarbons that tend to be most useful. Other hydrocarbons have lots of carbon atoms, and may have branches or side chains.

The boiling point of a hydrocarbon depends on the size of its molecules. We can use the differences in boiling points to separate the hydrocarbons in crude oil.

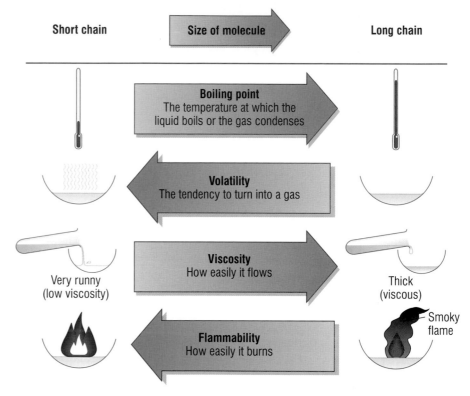

Figure 1 The properties of hydrocarbons depend on the chain length of their molecules.

a) How does the length of the hydrocarbon chain affect:
   (i) the boiling point, and
   (ii) the viscosity (thickness) of a hydrocarbon?
b) A hydrocarbon catches fire very easily. Is it likely to have molecules with long hydrocarbon chains or short ones?

Refining crude oil

We separate out crude oil into hydrocarbons with similar properties, called *fractions*. We call this process **fractional distillation**. Each hydrocarbon fraction contains molecules with similar numbers of carbon atoms. Each of these fractions boils at different temperatures. That's because of the different numbers of carbon atoms in their molecules.

Crude oil is fed in near the bottom of a tall tower (a fractionating column) as hot vapour. The tower is kept very hot at the bottom and much cooler at the top, so the temperature decreases going up the column. The gases in the column condense when they reach their boiling points, and the different fractions are collected at different levels.

Hydrocarbons with the smallest molecules have the lowest boiling points, so they are collected at the cool top of the tower. At the bottom of the tower the fractions have high boiling points. They cool to form very thick liquids or solids at room temperature.

PRACTICAL

Comparing fractions

Your teacher might compare some fractions (mixtures of hydrocarbons with similar boiling points) that we get from crude oil.

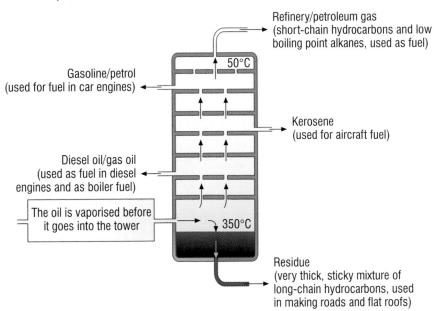

Refinery/petroleum gas (short-chain hydrocarbons and low boiling point alkanes, used as fuel)

Gasoline/petrol (used for fuel in car engines)

50°C

Kerosene (used for aircraft fuel)

Diesel oil/gas oil (used as fuel in diesel engines and as boiler fuel)

The oil is vaporised before it goes into the tower

350°C

Residue (very thick, sticky mixture of long-chain hydrocarbons, used in making roads and flat roofs)

Figure 2 We use fractional distillation to turn the mixture of hydrocarbons in crude oil into fractions, each containing compounds with similar boiling points

Once we have collected them, the fractions need more refining before they can be used.

There are many different types of crude oil. For example, crude oil from Venezuela contains many long-chain hydrocarbons. It is very dark and thick and we call it 'heavy' crude. Other countries, such as Nigeria and Saudi Arabia, produce crude oil which is much paler in colour and more runny. This is 'light' crude.

Light crude oil contains many more of the smaller molecules we can use as fuels and for making other chemicals. So light crude oils cost more to buy than the heavier crude oils which contain a higher percentage of the larger hydrocarbons.

c) Why is 'light' crude oil likely to cost more than 'heavy' crude?

SUMMARY QUESTIONS

1 Copy and complete using the words below:

easily   fractional distillation   fractions   high
mixture   viscosity

Crude oil is a …… of many different hydrocarbons. We can separate crude oil into different …… using …… …… . Hydrocarbon molecules with many carbon atoms tend to have …… boiling points and …… . Hydrocarbon molecules with few carbon atoms catch fire …… .

2 a) Explain the steps involved in the fractional distillation of crude oil.
  b) Make a table to summarise how the properties of hydrocarbons depend on the size of their molecules.

KEY POINTS

1 We separate crude oil into fractions using fractional distillation.
2 Properties of each fraction depend on the size of the hydrocarbon molecules.
3 Lighter fractions make better fuels.

Burning fuels

LEARNING OBJECTIVES

1 What do we produce when we burn fuels?
2 When we burn fuels, how do changes in conditions affect what is produced?
3 What pollutants are produced when we burn fuels?

Figure 1 On a cold day we can often see the water produced when fossil fuels burn

As we saw on page 171, the lighter fractions produced from crude oil are very useful as fuels. When hydrocarbons burn in plenty of air they produce two new substances – carbon dioxide and water.

For example, when propane burns we can write:

$$propane + oxygen \rightarrow carbon\ dioxide + water$$

or

$$C_3H_8 + 5O_2 \rightarrow 3CO_2 + 4H_2O$$

Notice how we need five molecules of oxygen for the propane to burn. This makes three molecules of carbon dioxide and four molecules of water. The equation is *balanced*!

PRACTICAL

Products of combustion

We can test the products given off when hydrocarbon burns as shown in Figure 2.

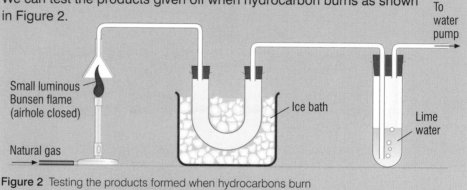

Figure 2 Testing the products formed when hydrocarbons burn

● What happens to the lime water?   ● What collects in the U-tube?

a) What are the names of the two substances produced when hydrocarbons burn in plenty of air?
b) Write a balanced equation for methane (CH$_4$) burning in plenty of air.

All fossil fuels – oil, coal and natural gas – produce carbon dioxide and water when they burn in plenty of air. But as well as hydrocarbons, these fuels also contain other substances. These produce different compounds when we burn the fuel, and this can cause problems for us.

Impurities of sulfur cause us major problems. All fossil fuels contain at least some of this element, which reacts with oxygen when we burn the fuel. It forms a gas called **sulfur dioxide**. This gas is poisonous. It is also acidic. This is bad for the environment, as it is a cause of acid rain.

Sulfur dioxide can also cause engine corrosion.

c) When hydrocarbons burn, what element present in the impurities in a fossil fuel may produce sulfur dioxide?
d) Why is it bad if sulfur dioxide is produced?

When we burn hydrocarbons in a car engine, even more substances can be produced. When there is not enough oxygen inside the cylinders of the engine, we get *incomplete combustion*. Instead of all the carbon in the fuel turning into carbon dioxide, we also get *carbon monoxide* (CO).

Carbon monoxide is a poisonous gas. Your red blood cells pick up this gas and carry it around in your blood instead of oxygen. So even quite small amounts of carbon monoxide gas are very bad for you.

The high temperature inside an engine also allows the nitrogen and oxygen in the air to react together. This makes *nitrogen oxides*, which are poisonous and which can trigger some people's asthma. They also make acid rain.

Diesel engines burn hydrocarbons with much bigger molecules than petrol engines. When these big molecules react with oxygen in an engine they do not always burn completely. Tiny particles containing carbon and unburnt hydrocarbons are produced. These *particulates* get carried into the air of our towns and cities. We do not understand fully what particulates may do when we breathe them in. However, scientists think that they may damage the cells in our lungs and perhaps even cause cancer.

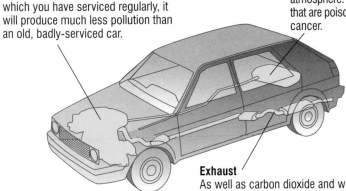

Engine
If you have a car with a modern engine which you have serviced regularly, it will produce much less pollution than an old, badly-serviced car.

Fuel tank
When we fill up our cars some hydrocarbons escape into the atmosphere. Fuels contain hydrocarbons that are poisonous and which may cause cancer.

Exhaust
As well as carbon dioxide and water vapour, exhaust gases may contain carbon monoxide, nitrogen oxides, sulfur dioxide and tiny particles of unburnt carbon and hydrocarbons.

Figure 3 The effect of many cars in a small area. Under the right weather conditions smog can be formed. This is a mixture of SMoke and fOG. Smog formed from car pollution contains many different types of chemicals which can be harmful to us.

SUMMARY QUESTIONS

1 Copy and complete using the words below:

carbon   carbon dioxide   nitrogen   particulates   sulfur   water

When we burn hydrocarbons in plenty of air .... and ...... are made. As well as these compounds other substances like ...... dioxide may be made. Other pollutants that may be formed include ...... oxides, ... monoxide and ....

2 a) Natural gas is mainly methane (CH_4). Write a balanced equation for the complete combustion of methane.
   b) When natural gas burns in a faulty gas heater it can produce carbon monoxide (and water).
      Write a balanced equation to show this reaction.
   c) Explain how i) sulfur dioxide ii) nitrogen oxides and iii) particulates are produced when fuels burn in vehicles.

173

Cleaner fuels

When we burn hydrocarbons, as well as producing carbon dioxide and water, we produce other compounds. Many of these are not good for the environment, and can affect our health.

Pollution from our cars does not stay in one place but spreads through the whole of the Earth's atmosphere. For a long time the Earth's atmosphere seemed to cope with all this pollution. But the increase in the number of cars in the last 50 years means that pollution is a real concern now.

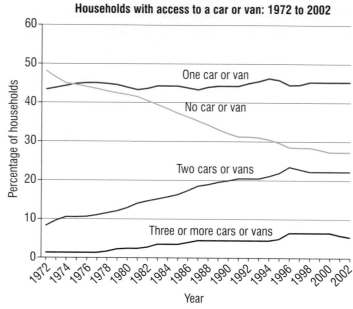

Figure 1 In 1972, about 50% of people had a car they could use. Now around 75% of people have access to a car.

a) Why is there more pollution from cars now than there was 50 years ago?
b) Why is pollution from cars in Moscow as important as pollution from cars in London?

What kinds of pollution?

When we burn any fuel it makes carbon dioxide. Carbon dioxide is a **greenhouse gas**. It collects in the atmosphere and reduces the amount of heat lost by radiation from the surface of the Earth. Most scientists think that this is causing **global warming**, which changes temperatures around the world.

Burning fuels in engines also produce other substances. One group of pollutants is called the *particulates*. These are tiny particles made up of unburnt hydrocarbons, which scientists think may be especially bad for young children. Particulates may also be bad for the environment too – they travel into the upper atmosphere, reflecting sunlight back into space, causing **global dimming**.

Other pollutants produced by burning fuels include carbon monoxide, sulfur dioxide and nitrogen oxides. Carbon monoxide is formed when there is not enough oxygen present for the fuel to react with oxygen to form carbon dioxide.

Carbon monoxide is a serious pollutant because it affects the amount of oxygen that our blood is able to carry. This is particularly serious for people who have problems with their hearts.

Sulfur dioxide and nitrogen oxides damage us and our environment. In Britain scientists think that the number of people who suffer from asthma and hayfever have increased because of air pollution.

Sulfur dioxide and nitrogen oxides also form acid rain. These gases dissolve in water droplets in the atmosphere and form sulfuric and nitric acids. The rain with a low pH can damage plant and animals.

c) Name four harmful pollutants that may be produced when fossil fuels burn.

Cleaning up our act

We can reduce the effect of burning fossil fuels in several ways. The most obvious way is to remove the pollutants from the gases that are produced when we burn fuels. For some time the exhaust systems of cars have been fitted with *catalytic converters*.

The exhaust gases from the engine travel through the catalytic converter where they pass over transition metals. These are arranged so that they have a very large surface area. This causes the carbon monoxide and nitrogen oxides in the exhaust gases to react. They produce carbon dioxide and nitrogen:

carbon monoxide + nitrogen oxides → carbon dioxide + nitrogen

In power stations, sulfur dioxide is removed from the flue gases by reacting it with quicklime. This is called *flue gas desulfurisation* or FGD for short.

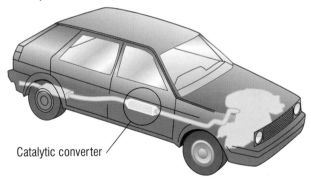

Catalytic converter

Figure 2 A catalytic converter greatly reduces the carbon monoxide and nitrogen oxides produced by a car engine

The methods described here reduce pollution by tackling it after it has been produced. The next page shows how we can also reduce the pollution that fuels produce if we start using alternative fuels.

SUMMARY QUESTIONS

1 Write definitions for the following terms:

a) greenhouse gas, b) global warming, c) global dimming, d) acid rain

2 A molecule of carbon monoxide requires another atom of oxygen if it is to become a molecule of carbon dioxide. How does a catalytic converter supply this?

3 a) Explain how acid rain is formed and how we are reducing the problem.
  b) Compare the effects of global warming and global dimming.

KEY POINTS

1 Burning fuels releases substances that spread throughout the atmosphere.

2 Some of these substances dissolve in droplets of water in the air, which then fall as acid rain.

3 Carbon dioxide produced from burning fuels is a greenhouse gas. It reduces the rate at which energy is lost from the surface of the Earth by radiation.

4 The pollution produced by burning fuels may be reduced by treating the products of combustion. This can remove substances like nitrogen oxides, sulfur dioxide and carbon monoxide.

Alternative fuels

Fuel from plants . . .

The fossil fuels that we use to produce electricity and to drive our cars have some big disadvantages. They produce carbon dioxide, which is a greenhouse gas, and they produce other pollutants too. What's more, once they have all been used there will be no other similar fuels that we can use to replace them – unless we think about the problem and do something about it.

Plants may be one answer to the problem of fuels. For thousands of years people have burned wood to keep themselves warm. Obviously we cannot use wood as a fuel for cars, but there are two ways that plants may be able to keep us on the road. These are explained in Figures 1 and 2.

Figure 1 We can use plants that make sugar to produce ethanol by fermenting the sugar using yeast. We can then add the ethanol to petrol, making **gasohol**. Not only does this reduce the amount of oil needed, it also produces less pollution because gasohol burns more cleanly than pure petrol.

Figure 2 Another new fuel is **biodiesel**. Some plants, like this oilseed rape, produce oils which can be used in diesel engines. We hardly need to make any changes to the engine to do this, and the biodiesel burns very cleanly, like gasohol. These bio-fuels also help tackle global warming. That's because the plants take in carbon dioxide gas during photosynthesis. They still give off carbon dioxide when we burn the bio-fuel – but overall they make little contribution to the greenhouse effect compared with burning fossil fuels.

ACTIVITY

One problem with switching to a new fuel for cars is that people may not trust it, preferring to stick with what they know. One way of convincing them to switch may be to produce advertising material like stickers. People who have switched fuels can then use these on their cars to show other people that the new fuel works just as well as the old fuel.

Design a set of stickers which can be used to make other people think about switching fuels.

... fuel from rubbish

Another way that we could reduce the amount of fuels we use is to replace them with something else – and rubbish seems a good answer! By burning rubbish we could produce some of the energy we need to heat our homes, and we would get rid of a big problem too.

Figure 3 Getting rid of all our rubbish usually means burying it in holes in the ground. This is not a good solution since it is messy, smelly and produces pollution.

Figure 4 We can burn rubbish in an incinerator like this. We can use the energy to heat water which can then heat our homes – or the energy can be used to make electricity.

But producing energy from rubbish is not straightforward. Unless the incinerator is run very carefully, dangerous chemicals called *dioxins* may be produced when the rubbish burns. Although no-one is exactly certain what dioxins do, many people think that they may cause cancer, and that they may damage us in other ways too. So there are arguments on both sides about the benefits of building incinerators.

RECYCLE **NOT** INCINERATE!

BURNING WASTE CAN REPLACE OIL!!

INCINERATOR WILL **POISON** OUR CHILDREN!!

INCINERATOR GOOD FOR ENVIRONMENT AND FOR LOCAL JOBS!!

ACTIVITY

You are going to take part in a planning enquiry which will decide whether or not an incinerator should be built. Choose one of the following rôles:

Director of incinerator company
You believe that incinerating waste is the best option – better for the environment (getting rid of waste and producing energy), and better for the local economy. The incinerator would bring real benefits.

Environmental campaigner
You argue that an incinerator is not good. Although the energy produced would replace fossil fuels, you believe that the pollutants the incinerator would produce will be harmful to local people, especially children.

Parent of young children
Although you like the idea of cheap heating and electricity you are concerned that the pollutants, which may be given off by the incinerator, may harm your children.

Local resident
You have lived and worked in the area for years. You feel that the incinerator will bring many benefits to the local area, and that the cheap energy will be a real bonus for local people.

Chair of enquiry
It is your job to manage the enquiry. You will have to give everyone a chance to speak, and you must ensure that everyone gets a fair hearing.

SUMMARY QUESTIONS

1 The following questions are about using hydrocarbons from crude oil as fuels.

a) When hydrocarbon fuels burn in plenty of air, what are the **two** main products?

b) One of these products is particularly bad for the environment – which one, and why?

c) What other substance may be produced when hydrocarbon fuels burn in plenty of air?

d) When hydrocarbon fuels burn inside an engine, what **two** other substances may be produced?

e) What effect do these two substances have on the environment?

f) Diesel engines produce **particulates**. What are they and what effect may they have on the environment?

2 Look at the graph of crude oil prices:

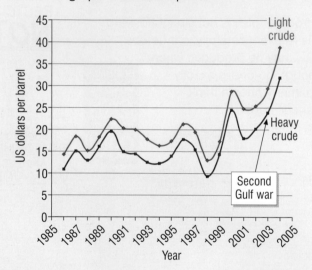

Use the graph to answer the following questions.

a) How do the prices of light crude and heavy crude differ?

b) Why is there this price difference?

c) Why did the price of crude oil rise sharply in 2003?

d) Light crude oil is particularly expensive during summer in the Northern Hemisphere. Suggest **one** possible reason for this.

EXAM-STYLE QUESTIONS

1 The following compounds are found in crude oil:

A C_3H_8 B C_8H_{18}

C $C_{12}H_{26}$ D $C_{16}H_{34}$

You can use A, B, C or D once, more than once or not at all when answering the questions below. Which of these compounds

(a) has the highest boiling point?

(b) catches fire most easily?

(c) is collected at the top of the fractionating column when crude oil is distilled?

(d) is the thickest liquid?   (4)

2 Crude oil is a mixture of many different hydrocarbons. Match the words A, B, C and D with spaces 1 to 4 in the sentences.

A alkanes   B compounds

C fractions   D molecules

(a) Crude oil is separated by distillation into ...... containing hydrocarbons with similar boiling points.

(b) Hydrocarbons with the smallest ...... have the lowest boiling points.

(c) Hydrocarbons are ...... of hydrogen and carbon only.

(d) Crude oil contains mostly saturated hydrocarbons called ....   (4)

3 The table shows the number of carbon atoms in the molecules of four fuels obtained from crude oil.

Fuel	Number of carbon atoms in molecules
petroleum gases	2–4
petrol	4–10
kerosene	10–15
diesel oil	14–19

(a) The fuel with the highest boiling point is . . .

   A petroleum gases   B petrol

   C kerosene   D diesel oil   (1)

(b) Petrol . . .

   A has a higher boiling point than diesel oil.

   B is a thinner liquid than diesel oil.

   C ignites less easily than kerosene.

   D has larger molecules than kerosene.   (1)

(c) The molecule C_4H_{10} could be in . . .

  A petrol only.

  B petrol and kerosene.

  C petrol and petroleum gases.

  D petroleum gases only. (1)

(d) Which one of the following is a saturated hydrocarbon that could be in diesel oil?

  A $C_{12}H_{26}$

  B $C_{16}H_{32}$

  C $C_{17}H_{36}$

  D $C_{18}H_{38}O$ (1)

4 Pentane, C_5H_{12}, is a hydrocarbon fuel. It burns completely in plenty of air.

  (a) Name the gas in air that pentane reacts with when it burns. (1)

  (b) Write a word equation for the combustion of pentane in plenty of air. (2)

  (c) Write a balanced symbol equation for this reaction. (2)

  (d) When the air supply is limited a poisonous gas is produced. Name this gas. (1)

  (e) Write a balanced symbol equation for the combustion of pentane in a limited supply of air. (2)

5 (a) Suggest two fuels that could be used in place of fossil fuels. Give one advantage and one disadvantage for each of the fuels you have named. (6)

6 Oil companies promote the use of low sulfur fuels.

  (a) Explain why it is better to use low sulfur fuels. (3)

  (b) Suggest one other reason why oil companies advertise that their fuels are low in sulfur. (1)

7 Crude oil is separated by fractional distillation. In oil refineries this is done in tall towers called fractionating columns.

Give the main steps in this process and explain how the different fractions are separated in a fractionating column. (4)

→ petroleum gases
→ petrol
→ kerosene
→ diesel oil
→ residue
40°
350°
heated crude oil

HOW SCIENCE WORKS QUESTIONS

Calculating energy from different fuels

This apparatus can be used to determine the heat given out when different fuels are burned.

The burner is weighed before and after to determine the amount of fuel burned. The temperature of the water is taken before and after, so as to calculate the temperature rise. The investigation was repeated. From this the amount of heat produced by burning a known amount of fuel can be calculated.

a) Construct a table that could be used to collect the data from this experiment. (3)

A processed table of results is given below.

Fuel	Mass burned (g)	Temperature rise (°C)	
Ethanol	4.9	48	47
Propanol	5.1	56	56
Butanol	5.2	68	70
Pentanol	5.1	75	76

b) List three variables that need to be controlled. (3)

c) Describe how you would take the temperature of the water to get the most accurate measurement possible. (2)

d) Do these results show precision? Explain your answer? (2)

e) Would you describe these results as accurate? Explain your answer. (Clue – look at the way the investigation was carried out.) (2)

f) How might you present these results? (1)

179

EXAMINATION-STYLE QUESTIONS

Science A

1   Crude oil can be separated by fractional distillation.
    Match the fractions **A**, **B**, **C** and **D** to the outlets **1** to **4** on the diagram.

See page 171

A   diesel oil

B   kerosene

C   petrol

D   petroleum gases

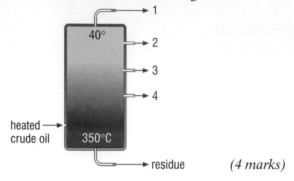

(4 marks)

2   Match the words **A**, **B**, **C** and **D** with the numbers **1** to **4** in the table.

See page 144

A   limestone

B   limewater

C   quicklime

D   slaked lime

1	Goes cloudy when reacted with carbon dioxide.
2	Made by thermal decomposition in a kiln
3	Mainly calcium carbonate
4	Solid calcium hydroxide

(4 marks)

3   A student investigated three unknown metal carbonates to compare how easily each powder decomposed. She wanted to put the powders in order. She bubbled the carbon dioxide gas given off through lime water. She timed how long it took for the lime water to look milky. She repeated each test 3 times.
    Here are her results:

See pages 5, 9, 15

Carbonate	Time for lime water to turn milky (s)
A	186, 275, 157
B	90, 163, 142
C	106, 152, 136

Which statement is true?

A   Her results were accurate because she repeated them 3 times.

B   Her conclusion would be reliable and valid.

C   She collected precise data by repeating her tests.

D   She could not draw a firm conclusion from the data collected.   *(1 mark)*

GET IT RIGHT!

In multiple choice questions that ask you to *match* the letters and numbers, each letter is used only once and so if you know three of the answers, you can answer the fourth one! In questions with parts (a) (b) (c) etc. the letters can be used once, more than once or not at all in each question. It is not worth looking for patterns in letters for answers because they are used randomly, so there is no pattern!

Science B

1 Lead can be extracted from lead oxide using carbon.

See page 153

   (a) Write a word equation for this reaction. *(2 marks)*

   (b) Explain why this is called a reduction. *(1 mark)*

2 Most of the iron produced in a blast furnace is converted into steels.

See page 156

   (a) Why is iron from the blast furnace not very useful? *(1 mark)*

   (b) What are steels? *(2 marks)*

   (c) Explain why steels are harder than pure iron. *(3 marks)*

   (d) Describe how you could compare the strength of an iron wire with a steel wire. You should describe any measurements you would take and how to make it a fair test. *(5 marks)*

See pages 7, 8

3 Complete the equations to show the reactions of limestone and its products.

See pages 142–5

   (a) $CaCO_3 \xrightarrow{\text{heat}} CaO + \ldots\ldots$

   (b) $CaO + \ldots\ldots \longrightarrow Ca(OH)_2$

   (c) $Ca(OH)_2 + \ldots\ldots \longrightarrow CaCO_3 + \ldots\ldots$ *(4 marks)*

4 Read the information in the passage and use it to help you answer these questions.

See page 165

   A new method of mining nickel uses plants to extract nickel compounds from the soil. The plants are grown in fields in parts of Canada where there is a higher than usual amount of nickel, but not enough to make it economical to mine normally. The plants are harvested and burnt to produce energy. The ash that is left after burning the crop contains nickel compounds from which the nickel can be extracted. The yield is up to 400 kg of nickel per hectare and farmers are hoping to receive $2000 per hectare for the nickel in the ash from their crops. The current price of nickel is $14 per kg.

   (a) Why is the nickel in the soil not extracted by normal mining? *(1 mark)*

   (b) Explain how the nickel is concentrated in this process so that it can be extracted. *(3 marks)*

   (c) Suggest a method that could be used to produce nickel metal from the ash. *(2 marks)*

   (d) How much are the farmers hoping to receive for each kg of nickel in the ash from their crops? *(1 mark)*

   (e) Why is the amount the farmers hope to receive less than the current price of nickel? *(1 mark)*

GET IT RIGHT!

Always read any information given in a question very carefully. All of the information is there for a reason – you are expected to use it in your answer. Read the questions carefully and refer back to the information before you write your answers. It may help if you highlight or underline key words on your question paper in exams. If you cannot write in this book, make notes on paper as you do the questions. Ask your teacher how best to do this.

C1b | Oils, Earth and atmosphere

The face of the Earth has changed over millions of years

What you already know

Here is a quick reminder of previous work that you will find useful in this unit:

- We can characterise materials by melting point, boiling point and density.
- Mixtures are made up of different substances that are not chemically combined to each other.
- We can separate a mixture of liquids into its different parts using distillation.
- Changes of state involve energy transfers.
- Rocks are formed by processes that take place over different timescales.
- Burning fossil fuels affects our environment.
- We can use indicators to classify solutions as acidic, neutral or alkaline.

RECAP QUESTIONS

1 Where does the water in a puddle go as the puddle dries out when the Sun shines?

2 A car engine will not work properly if there is dirt in the petrol. The dirty petrol is passed through a funnel containing filter paper. How does this clean the petrol?

3 The tea-leaves in a tea bag contain colours and flavours that dissolve in water. Why is it difficult to make iced tea using a tea bag dipped in very cold water?

4 When you put a tray of water in the freezer its temperature drops. At 0°C the water starts to freeze. The temperature then stays the same until all the water is frozen. What is happening?

5 Two fossils are found in two different layers of rocks, one above the other. Which fossil is likely to be the older one – the one found in the upper layer of rock, or the one found in the lower layer? Explain your answer.

6 When we burn fossil fuels, how can it affect the environment?

7 A solution turns universal indicator red – what does this tell you about the solution?

Making connections

Millions of years ago, small plants and animals lived in the oceans of the world. These tiny organisms were supported by energy from the Sun. When they died, the tiny organisms fell to the bottom of the sea. Then they were covered by mud, sand and rock. Over millions of years they became oil . . .

It was only about 100 years ago that people started driving cars! At first only the very rich could afford to buy them. But slowly, as more and more cars were produced, they became less expensive so that more people could own them.

As more and more people bought more and more cars, there were problems with pollution. As well as carbon dioxide causing global warming, car engines produced other pollutants which were bad for people and for other living things too – including plants.

Biodiesel promises an environmentally friendly fuel which may one day replace fossil fuels. Made from plant oils, biodiesel has much less effect on the environment than ordinary diesel fuel.

ACTIVITY

Imagine that you are going to present a short radio programme about the effect that the motor car has had on our lives. Using the information here, write and present this programme to other people in your class.

Chapters in this unit

Products from oil ——— Plant oils ——— The changing world

Cracking hydrocarbons

C1b 4.1

Many of the fractions that we get by distilling crude oil are not very useful. They contain hydrocarbon molecules which are too long for us to use them as fuels that are in high demand. The hydrocarbons that contain very big molecules are thick liquids or solids with high boiling points. They are difficult to vaporise and do not burn easily – so they are no good as fuels! Yet the main demand from crude oil is for fuels.

Figure 1 Huge crackers like this are used to split large hydrocarbon molecules into smaller ones

Luckily we can break down large hydrocarbon molecules in a process we call **cracking**. The best way of breaking them up uses heat and a catalyst, so we call this *catalytic cracking*. The process takes place in a *cat cracker*.

In the cracker a heavy fraction produced from crude oil is heated strongly to turn the hydrocarbons into a gas. This is passed over a hot catalyst where thermal decomposition reactions take place. The large molecules split apart to form smaller, more useful ones.

a) Why is cracking so important?
b) How are large hydrocarbon molecules cracked?

Example of cracking

Decane is a medium sized molecule with ten carbon atoms. When we heat it to 800°C with a catalyst it breaks down. One of the molecules produced is pentane which is used in petrol. We also get propene and ethene which we can use to produce other chemicals.

$$C_{10}H_{22} \xrightarrow{\text{800°C + catalyst}} C_5H_{12} + C_3H_6 + C_2H_4$$

decane   pentane   propene   ethene

This reaction is an example of **thermal decomposition**.

Notice how this cracking reaction produces different types of molecules. One of the molecules is pentane. The first part of its name tells us that it has five carbon atoms (*pent-*). The last part of its name (*-ane*) shows that it is an alkane. Like all other alkanes, pentane is a **saturated hydrocarbon** – its molecule has as much hydrogen as possible in it.

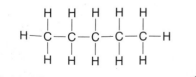

Figure 2 A molecule of pentane

The other molecules in this reaction have names that end slightly differently. They end in -ene. We call this type of molecule an **alkene**. The different ending tells us that these molecules are **unsaturated** because they contain a *double bond* between two of their carbon atoms. Look at Figure 3:

Alkenes with one double bond have this general formula, C_nH_{2n}.

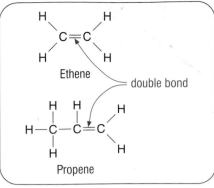

Figure 3 A molecule of propene and a molecule of ethene. These are both alkenes – each molecule has a carbon–carbon double bond in it.

PRACTICAL

Cracking

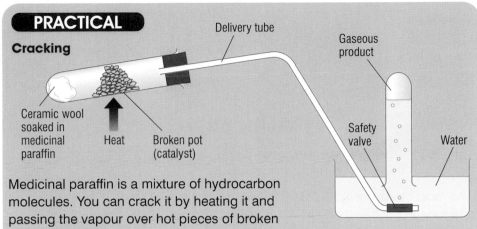

Medicinal paraffin is a mixture of hydrocarbon molecules. You can crack it by heating it and passing the vapour over hot pieces of broken pot. The products that you make in this reaction are insoluble gases, so you can collect them by bubbling them through water.

If you carry out this practical, collect at least two test tubes of gas. Test one by putting a lighted splint into it. Test the other by adding a few drops of bromine water to it.

● Why must you remove the end of the delivery tube from the water when you stop heating?

A simple experiment like the one described above shows that alkenes burn. They also react with bromine water (which is orange-yellow) – the products of this reaction are colourless. This means that we have a good test to see if a hydrocarbon is unsaturated:

unsaturated hydrocarbon + bromine water → products
(colourless)   (orange-yellow)   (colourless)

saturated hydrocarbon + bromine water → no reaction
(colourless)   (orange-yellow)   (orange-yellow)

DID YOU KNOW...

... that ethene is a really important chemical? Although we know over 500 ways of making ethene, the only way that it is currently made commercially is from oil.

SUMMARY QUESTIONS

1 Copy and complete using the words below:

alkenes   catalyst   cracking   double   heating   unsaturated

We can break down large hydrocarbon molecules by ...... them and passing them over a .... This is called .... Some of the molecules produced when we do this contain a ...... bond – they are called ...... hydrocarbons and we call them ....

2 Cracking a hydrocarbon makes two new hydrocarbons, A and B. When bromine water is added to A, nothing happens. Bromine water added to B loses its colour. Which hydrocarbon is unsaturated?

3 a) An alkene molecule with one double bond contains 7 carbon atoms. How many hydrogen atoms does it have? Write down its formula.

   b) Decane (with 10 carbon atoms) is cracked into octane (with 8 carbon atoms) and ethene. Write a balanced equation for this reaction.

KEY POINTS

1 We can split large hydrocarbon molecules up into smaller molecules by heating them and passing the gas over a catalyst,

2 Cracking produces unsaturated hydrocarbons, which we call alkenes.

3 Alkenes burn, and also react with bromine water, producing colourless products.

Making polymers from alkenes

LEARNING OBJECTIVES

1 What are monomers and polymers?
2 How do we make polymers from alkenes?

Refining crude oil produces a huge range of hydrocarbon molecules which are very important to our way of life. Oil products are all around us. We simply cannot imagine life without them.

Figure 1 All of these products were manufactured using chemicals made from oil

The most obvious way that we use hydrocarbons from crude oil is as fuels. We use fuels in our transport and at home. We also use them to make electricity in oil-fired power stations.

Then there are the chemicals we make from crude oil. We use them to make things ranging from margarines to medicines, from dyes to explosives. But one of the most important ways that we use chemicals from oil is to make plastics.

Plastics

Plastics are made up of huge molecules made from lots of small molecules that have joined together. We call the small molecules **monomers**. We call the huge molecules they make **polymers** – *mono* means 'one' and *poly* means 'many'. We can make different types of plastic which have very different properties by using different monomers.

NEXT TIME YOU...

... pick up a plastic bag, think how tiny monomers have joined together to make the huge polymer molecules that the bag is made from.

a) List three ways that we use fuels.
b) What are the small molecules that make up a polymer called?

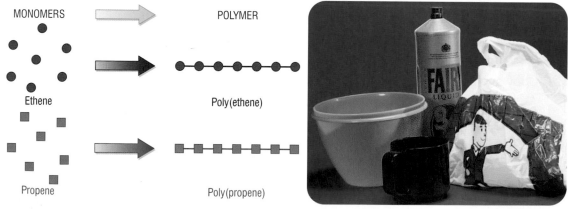

Figure 2 Polymers produced from oil are all around us and are part of our everyday lives

Ethene **(C₂H₄)** is the smallest unsaturated hydrocarbon molecule. We can turn it into a polymer known as poly(ethene) or polythene. Polythene is a really useful plastic. It is easy to shape, strong and transparent (unless we add colouring material to it). 'Plastic' bags, plastic drink bottles, dustbins and clingfilm are all examples of polythene that are very familiar to us in everyday life.

Propene (C_3H_6) is another alkene. We can also make polymers with propene as the monomer. The plastic formed is called poly(propene) or polypropylene. Poly(propene) is a very strong, tough plastic. We can use it to make many things, including milk crates and ropes.

c) Is ethene an alkane or an alkene?

d) Which plastic can we make from the monomer called propene?

How do monomers join together?

When alkene molecules join together, the double bond between the carbon atoms in each molecule 'opens up'. It is replaced by single bonds as thousands of molecules join together. This is an example of an *addition reaction*. Because a polymer is made, we call it *addition polymerisation*.

Ethene monomers → Poly(ethene)

We can also write this much more simply:

Many single ethene monomers → Long chain of poly(ethene)

where n is a large number

GET IT RIGHT!

The double $C=C$ bond in ethene (an alkene) makes it much more reactive than ethane (an alkane).

PRACTICAL

Modelling polymerisation

Use a molecular model kit to show the polymerisation of ethene to form poly(ethene).

Make sure you can see how the equation shown above represents the polymerisation reaction you have modelled.

● Describe what happens to the bonds in the reaction.

e) Think up a model to demonstrate the polymerisation of ethene, using people in your class as monomers. Evaluate the ideas of other groups.

SUMMARY QUESTIONS

1 Copy and complete using the words below:

> **addition   ethene   monomers   polymers**

Plastics are made out of large molecules called .... We make these by joining together lots of small molecules called .... One example of a plastic is poly(ethene), made from .... Poly(ethene) is formed as a result of an ...... reaction.

2 Why is ethene the smallest unsaturated hydrocarbon molecule?

3 a) Draw a propene molecule.

   b) Draw structures to show how propene molecules join together to form poly(propene).

   c) Explain the polymerisation reaction in b).

KEY POINTS

1 Plastics are made of polymers.

2 Polymers are large molecules made when monomers (small molecules) join together.

The properties of plastics

As you have just seen on pages 186 and 187, we can make plastics from chemicals made from crude oil. Small molecules called monomers join together to make much bigger molecules called polymers. As the monomers join together they produce a tangled web of very long chain molecules.

The atoms in these chains are held together by very strong chemical bonds. This is true for all plastics. But the size of the forces **between** polymer molecules in different plastics can be very different.

We call the forces between molecules **intermolecular forces**. The size of the intermolecular forces between the polymer molecules in a plastic depends partly on:

• the monomer used, and
• the conditions we choose to carry out polymerisation.

a) How are the atoms in polymer chains held together?
b) What do we call the forces between polymer chains?

In some plastics the forces between the polymer molecules are weak. When we heat the plastic, these weak intermolecular forces are broken and the plastic becomes soft. When we cool the plastic, the intermolecular forces bring the polymer molecules back together, and the plastic hardens again. We call plastics which behave like this **thermosoftening** plastics.

Poly(ethene), poly(propene) and poly(chloroethene) are all examples of thermosoftening plastics.

Figure 1 The forces between the molecules in poly(ethene) are relatively weak. This means that this plastic softens fairly easily when heated.

Figure 2 We usually call poly(chloroethene) by its more common name – polyvinylchloride, or PVC for short. We make lots of everyday plastic articles from PVC.

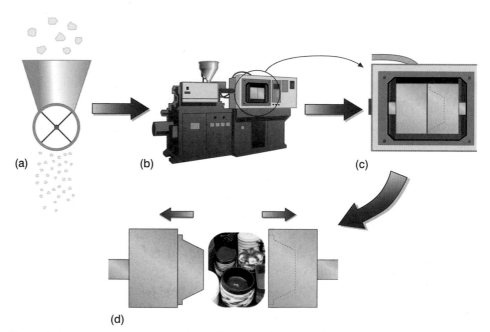

(a)   (b)   (c)

(d)

Figure 3 A very common way of making things out of polymers is to use a thermosoftening plastic that can be shaped in a mould: **(a)** chunks of monomer are ground into small pieces. **(b)** These are heated to melt them and then . . . **(c)** the molten plastic is forced into a mould. **(d)** The mould is separated to release the finished article.

There are two types of poly(ethene) – high density (HDPE) and low density poly(ethene) (LDPE). Both are made from ethene monomers but are formed under different conditions.

Using very high pressures and a trace of oxygen, ethene forms LDPE. The polymer chains are branched and they can't pack closely together.

Using a catalyst at 50°C and a slightly raised pressure gives us HDPE. This is made up of straight poly(ethene) molecules. They can pack closely together. Therefore forces between molecules (intermolecular forces) are stronger. HDPE has a higher softening temperature and is stronger than LDPE.

We cannot soften all plastics. Some monomers also make chemical bonds between the polymer chains when they are first heated in order to shape them. These bonds are strong, and they stop the plastic from softening when we heat it in the future. We call this type of polymer a **thermosetting plastic**.

c) What do we call a plastic that softens when we heat it?
d) What do we call a plastic that does not soften again once it has been made?

FOUL FACTS

Our body makes slimy chemicals (in our mouths and noses for example!) from natural polymers. These polymer molecules have charged parts, which repel other similar molecules, so they move past one another easily and the polymer feels slippery!

Figure 4 Plastic kettles are made out of thermosetting plastics

PRACTICAL

Modifying a polymer

... add a few drops of borax solution

Stir well for about 2 minutes

Slime

Warm solution of PVA glue

Take some PVA glue ...

- How could you investigate if the properties of slime depend on how much borax you add?

The glue becomes slimy because the borax makes the long polymer chains in the glue link together to form a jelly-like substance.

SUMMARY QUESTIONS

1 Copy and complete using the words below:

     bonds   thermosetting   tangled   weak

The polymer chains in a plastic form a …… web. In the chain, the atoms are held together by strong …… . If the intermolecular forces between the polymer chains are …… , the plastic softens at a relatively low temperature. Some polymers have strong bonds between the chains – we call these …… polymers.

2 Why do we use thermosetting plastics to make plastic kettles?

3 Polymer A starts to soften at 400°C while polymer B softens at 150°C.

Explain this statement using ideas about intermolecular forces.

KEY POINTS

1 Monomers affect the properties of the polymers that they produce.
2 Changing reaction conditions can also change the type of polymer that is produced.

New and useful polymers

Figure 1 Plastic drinks bottles are made from a plastic called poly(etheneterephthalate), or PET for short

You have probably heard the question – 'which came first, the chicken or the egg?' Polymers and the way that we use them are a bit like this. That's because sometimes we use a polymer to do a job because of its properties. But then at other times we might design a polymer with special properties so that it can do a particular job.

The bottles that we buy fizzy drinks in are a good example of a polymer that we use because of its properties. These bottles are made out of a plastic called PET.

The polymer it is made from is ideal for making drinks bottles. It produces a plastic that is very strong and tough, and which can be made transparent. The bottles made from this plastic are much lighter than glass bottles. This means that they cost less to transport.

a) Why is the plastic called PET used to make drinks bottles?
b) Why do drinks in PET bottles cost less to transport than drinks in glass bottles?

Now, rather than choosing a polymer because of its properties, materials scientists are designing new polymers with special properties. These are polymers that have the right properties to do a certain job.

c) What do we mean by a 'designer polymer'?

Medicine is one area where we are beginning to see big benefits from these 'polymers made to order'.

Figure 2 A sticking plaster is often needed when we cut ourselves. Getting hurt isn't much fun – and sometimes taking the plaster off can be painful too.

We all know how uncomfortable pulling a plaster off your skin can be. But for some of us taking off a plaster is really painful. Both very old and very young people have quite fragile skin. But now a group of chemists has made a plaster where the 'stickiness' can be switched off before the plaster is removed. The plaster uses a light sensitive polymer. Look at Figure 3.

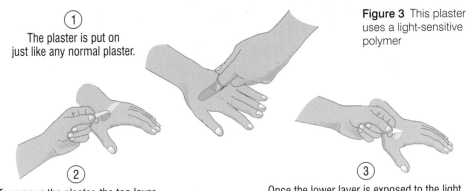

① The plaster is put on just like any normal plaster.

Figure 3 This plaster uses a light-sensitive polymer

② To remove the plaster, the top layer is peeled away from the lower layer which stays stuck to the skin.

③ Once the lower layer is exposed to the light, the adhesive becomes less sticky, making it easy to peel off your skin.

NEXT TIME YOU...

. . . take a plaster off your skin, think about the technologies used to create it. Did it hurt to pull it off? Did it leave a mark on your skin? Are there ways the plaster could be made even better?

New polymers can also come to our rescue when we are cut badly enough to need stitches. A new shape memory polymer is being developed by doctors which will make stitches that keep the sides of a cut together. Not only that, but the polymer will also dissolve once it has done its job. So there will be no need to go back to the doctor to have the stitches out.

Figure 4 When a shape memory polymer is used to stitch a wound loosely, the temperature of the body makes the thread tighten and close the wound, applying just the right amount of force. This is an example of a '**smart polymer**' i.e. one that changes in response to changes around it. In this case a change in temperature causes the polymer to change its shape. Later, after the wound is healed, the material is designed to dissolve and is harmlessly absorbed by the body.

PRACTICAL

Evaluating plastics

Carry out an investigation to compare the suitability of different plastics for a particular use.

For example, you might look at treated and untreated fabrics for waterproofing and 'breatheability' (gas permeability) or different types of packaging.

SUMMARY QUESTIONS

1 Copy and complete using the words below:

    cold   hot   PET   properties   shape   strong   transparent

We choose a polymer for a job because it has certain .... For example, we make drinks bottles out of a plastic called ...... because it is ...... and ....
Scientists can also design 'smart' polymers, for example ...... memory polymers. These change their shape when they are ...... or ....

2 a) The polymer in some sticking plasters is switched off by light because light makes bonds form between the polymer chains. Suggest why this may make the polymer less sticky.
  b) Design a leaflet for a doctor to give to a patient, explaining how stitches made from smart polymers work.

KEY POINTS

1 New polymers are being developed all the time. They are designed to have properties that make them specially suited for certain uses.
2 Smart polymers may have their properties changed by light, temperature or by other changes in their surroundings.

Plastics, polymers and packaging food

C1b 4.5

Shopping for food

Buying things in a shop like this probably took longer than shopping in the supermarket. But buying food already packed isn't just quicker – the packaging material used to wrap the food is designed to keep it fresh for longer, and to help to stop dirt and germs getting into it. And new packaging materials that we are developing will do this job even better – for example, they'll allow air in and out of the package without letting the food dry out.

Look at this shop – it's a bit different to the supermarket where we do our shopping each week! For a start, everything needs weighing out before you can buy it. Shopping like this must have taken ages!

Yeah – but my Gran says her Mum used to go shopping nearly every day, not just once a week! My Mum buys everything already wrapped up from the supermarket – except cheese. She always gets that from the deli at the supermarket where they weigh it out for you – I don't know why.

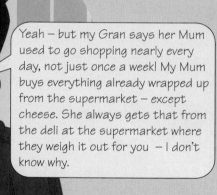

ACTIVITY

To wrap or not to wrap?

You are going to think about whether we should buy food that has already been wrapped – or whether it would be better if we had it weighed and wrapped when we bought it.

Start off by working on your own. Draw up a table like the one below, and use it to list reasons for buying food already wrapped and for buying food that needs to be wrapped.

When you have drawn up your lists, compare them with someone else in your class. Discuss any differences in your tables.

Buying food that has not been wrapped	Buying food that is already wrapped

'Would you like that in a bag, sir?'

When you buy something in a shop, you need to get it home safely. If you're a well-organised kind of person you may have taken a bag with you – but most of us are quite happy to say 'yes' when the shop assistant asks us 'Would you like a bag for that?'And when we get home – what happens to the bag?

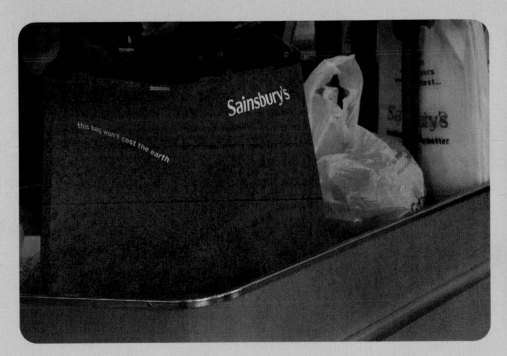

So just how many bags do we use in one year? How many bags does one supermarket give out in a year? Is this a problem?

Are the bags *biodegradable* (will they be broken down in nature by microorganisms when we throw them away)? Or will they take up valuable space in land-fill sites for years to come?

ACTIVITY

How many bags?

You are going to carry out a survey to find out how many carrier bags people use.
Design a questionnaire to answer questions like:

- how many carrier bags do you have at home?
- how many carrier bags do you bring home from the supermarket each week?
- do you use re-usable bags for your shopping?
- what happens to the carrier bags once you have used them?

When you have collected the answers to your questions, present the information using tables, graphs and charts.
Discuss the questions below:

- What do you think about the number of carrier bags used by people? Should we try to reduce this? Why (not)? How?
- Is your data representative of your local area? How could you be more certain?

What factors, if any, might affect the number of carrier bags used in different parts of the country?
How could you collect data to best reflect the picture in the whole country?

SUMMARY QUESTIONS

1 Write simple definitions for the following words:
   a) cracking
   b) distillation
   c) refining
   d) saturated hydrocarbon
   e) unsaturated hydrocarbon.

2 Propene is a hydrocarbon molecule containing three carbon atoms and six hydrogen atoms.
   a) What is meant by the term **hydrocarbon**?
   b) Draw the structural formula of propene.
   c) Is propene a **saturated** molecule or an **unsaturated** molecule? Explain your answer.
   d) Test tubes A and B In the diagram look identical. One test tube contains propane, while the other test tube contains propene.

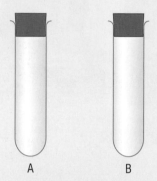

A   B

Explain clearly what test you could carry out to find out which tube contains propene, stating clearly the results obtained in each case.
   e) Propene molecules will react together to form long chains.
      i)   What do we call this type of reaction?
      ii) What properties does this new substance have?
      iii) Why does the widespread use of substances like this present an environmental problem?

3 a) What is a **polymer**?
   b) How do **monomers** join together to make a polymer?
   c) What factors can affect the way that a polymer behaves?

4 'Of all the problems facing us when disposing of our waste, the problem of plastics is the greatest.' Write a paragraph setting out arguments both agreeing and disagreeing with this statement.

EXAM-STYLE QUESTIONS

1 This question is about polymers.
   Match the words **A**, **B**, **C** and **D** with the spaces **1** to **4** in the sentences.
   A transparent
   B light sensitive
   C thermosetting
   D thermosoftening
   A plastic that can be remoulded is made from a ......**1**...... polymer.
   The plastic used to make a handle for a grill pan would be best made from a polymer that is ......**2**......
   The polymers used to make food wrappings for use in a supermarket are best if they are ......**3**......
   Some new types of sticking plaster can be removed easily because they are made from ......**4**......polymers.
   (4)

2 Match the words **A**, **B**, **C** and **D** with the descriptions **1** to **4** in the table.
   A bromine
   B butene
   C poly(ethene)
   D propane

	Description
1	An alkane
2	A polymer
3	Used to make polymers
4	Used to test for unsaturation

   (4)

3 Match the words **A**, **B**, **C** and **D** with the spaces **1** to **4** in the sentences about plastics.
   A chemical bonds
   B intermolecular forces
   C monomers
   D polymers
   Plastics are made of many ......**1**...... joined together in long chains to form ......**2**...... . In thermosoftening plastics the long chains are held together by ......**3**...... which are overcome by heating. In thermosetting plastics ......**4**...... form between the long chains when they are heated and the plastic sets hard.   (4)

4 An alkane, $C_{12}H_{26}$, was cracked. The reaction that took place is represented by the equation:

$$C_{12}H_{26} \rightarrow C_6H_{14} + C_4H_8 + \ldots\ldots$$

(a) The formula of the missing compound is …

A CH_4 **B** C_2H_4

C C_2H_6 **D** C_3H_8

(b) The compound C_4H_8 is …

A an alkane   **B** an alkene

C a poly(alkane)   **D** a poly(alkene)

(c) Addition polymers can be made from …

A $C_{12}H_{26}$ **B** C_6H_{14}

C C_4H_8 **D** C_2H_6

(d) The structure of C_4H_8 could be …

A $H-\overset{\displaystyle H}{\underset{\displaystyle H}{C}}=C=\overset{\displaystyle H}{C}-\overset{\displaystyle H}{\underset{\displaystyle H}{C}}-H$ **B** $H-\overset{\displaystyle H}{\underset{\displaystyle H}{C}}-\overset{\displaystyle H}{C}=\overset{\displaystyle H}{C}-\overset{\displaystyle H}{\underset{\displaystyle H}{C}}-H$

C $H-\overset{\displaystyle H}{\underset{\displaystyle H}{C}}-\overset{\displaystyle H}{\underset{\displaystyle H}{C}}-\overset{\displaystyle H}{C}=\overset{\displaystyle H}{C}-H$ **D** $\overset{\displaystyle H}{\underset{\displaystyle H}{C}}-\overset{\displaystyle H}{\underset{\displaystyle H}{C}}-\overset{\displaystyle H}{\underset{\displaystyle H}{C}}-\overset{\displaystyle H}{\underset{\displaystyle H}{C}}$ (4)

5 Read the passage about 'Slime' and use the information to help you answer the questions.

'Slime' has some of the properties of a liquid and some of the properties of a solid. It can be poured but it bounces if dropped on the floor. 'Slime' is made by mixing a solution containing a polymer called PVA with borax. When the substances are mixed the borax forms cross-links between the polymer chains. Some of the cross-links are chemical bonds and some are intermolecular forces involving water molecules. Lots of water molecules are held between the polymer chains and these give 'Slime' its flexibility and fluidity.

(a) Describe a molecule of a typical polymer. (2)

(b) Suggest why 'Slime' has the properties of both a solid and a liquid. (3)

(c) Suggest one method that you could use to modify the properties of 'Slime'. (1)

(d) A student tested different types of 'Slime' by measuring how far they stretched before they broke.
  (i) What was the independent variable in the investigation? (1)
  (ii) What type of variable was the dependent variable – categoric, ordered, discrete or continuous? (1)

HOW SCIENCE WORKS QUESTIONS

Biodegradable plastics could be used for growing crops

Non-biodegradable plastic has been used for many years for growing melons. The plants are put into holes in the plastic and their shoots grow up above the plastic. The melons are protected from the soil by the plastic and grow with very few marks on them. Biodegradable plastic has been tested – to reduce the amount of non-recycled waste plastic.

In this investigation two large plots were grown. One using biodegradable plastic, the other using normal plastic. The results were as follows:

Plastic used	Early yield (kg/hectare)	Total yield (kg/hectare)	Average melon weight (kg)
Normal	210	4 829	2.4
Biodegradable	380	3 560	2.2

a) This was a field investigation. Describe how the experimenter would have chosen the two plots. (3)

b) The hypothesis was that the biodegradable plastic would produce less fruit than the normal plastic. Is the hypothesis supported or refuted, or should a new hypothesis be considered? Explain your answer. (2)

c) How could the accuracy of this investigation be improved? (1)

d) How could the reliability of these results be tested? (1)

e) How would you view these results if you were told that they were funded by the manufacturer of the normal plastic? (1)

Extracting vegetable oils

Plants use the Sun's energy to produce glucose from carbon dioxide and water:

$$6CO_2 + 6H_2O \xrightarrow{\text{energy (from sunlight)}} C_6H_{12}O_6 + 6O_2$$

Plants then turn glucose into other chemicals using more chemical reactions. In some cases these other chemicals can be very useful to us – for example the **vegetable oils** that some plants make.

Farmers plant crops like oilseed rape, which is an important source of vegetable oil. We find the oils in the seeds of the rape plant. Once it has flowered and set seed, the farmer collects the seeds using a combine harvester.

The seeds are then taken to a factory where they are crushed, then pressed to extract the oil in them. The impurities are removed from the oil, and we can process it to make it into important foods, as we shall see later in this chapter.

Figure 1 Oilseed rape is a common sight in our countryside. As its name tells us, it is a good source of vegetable oil.

We extract other vegetable oils using steam. For example, we can extract lavender oil from lavender plants by distillation. The plants are put into boiling water and the oil from the plant evaporates. It is then collected by condensing it, when the water and other impurities can be removed to give pure lavender oil.

Figure 2 Norfolk lavender oil is extracted from lavender plants using steam

PRACTICAL

Extracting plant oil by distillation

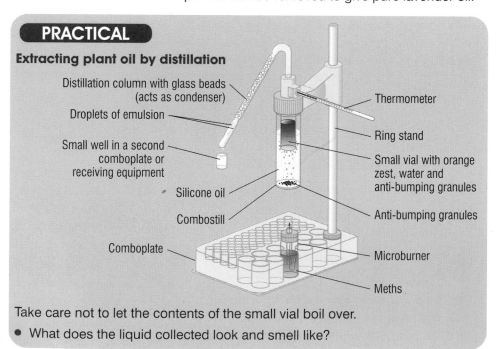

Distillation column with glass beads (acts as condenser)

Droplets of emulsion

Small well in a second comboplate or receiving equipment

Silicone oil

Combostill

Comboplate

Thermometer

Ring stand

Small vial with orange zest, water and anti-bumping granules

Anti-bumping granules

Microburner

Meths

Take care not to let the contents of the small vial boil over.

● What does the liquid collected look and smell like?

a) Write down **two** ways that can be used to extract vegetable oils from plants.

Vegetable oils as foods

Vegetable oils are very important foods because they contain a great deal of energy, as the table shows.

There are lots of different vegetable oils. Each vegetable oil has slightly different molecules. However, all vegetable oils have molecules which contain chains of carbon atoms with hydrogen atoms attached to them:

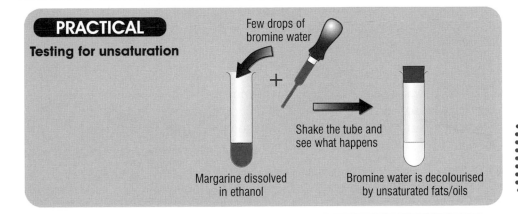

In some vegetable oils there are carbon atoms joined to each other by carbon–carbon double bonds (C=C). We call these **unsaturated oils**. We can detect the double bonds in unsaturated oils with bromine solution, as you did with the double bonds in alkene molecules. (See page 185.) They also react with iodine solution.

This provides us with an important way of detecting unsaturated oils:

unsaturated oil + bromine solution (yellow/orange) → colourless solution

unsaturated oil + iodine solution (violet/reddy brown depending on solvent) → colourless solution

b) What will you see if you test a polyunsaturated margarine with iodine water?

Energy in vegetable oil and other foods	
Food	**Energy in 100 g (kJ)**
vegetable oil	3900
sugar	1700
animal protein (meat)	1100

Figure 3 Vegetable oils have a high energy content

PRACTICAL

Testing for unsaturation

Few drops of bromine water

+

Shake the tube and see what happens

Margarine dissolved in ethanol

Bromine water is decolourised by unsaturated fats/oils

DID YOU KNOW...

... that no more than 20% of the energy in your diet should come from fats?

SUMMARY QUESTIONS

1 Copy and complete using the words below:

bromine   distillation   energy   iodine   pressing   unsaturated

We can extract vegetable oils from some plants by ...... or ....
Vegetable oils are particularly important as foods because they contain a lot of .... Some vegetable oils contain carbon–carbon double bonds – we call these ...... vegetable oils. They can be detected by reacting them with ...... or ...... solution.

2 Why might a diet containing too much vegetable oil make you fat?

3 A sample of vegetable oils is tested with iodine solution. The solution is decolourized. Which of the following statements is true?

a) The sample contains *only* unsaturated oils.
b) The sample contains *only* saturated oils.
c) The sample may contain a mixture of saturated and unsaturated oils.

Explain your answer.

KEY POINTS

1 Vegetable oils can be extracted from plants by pressing or by distillation.
2 Vegetable oils are important foods.
3 Unsaturated oils contain carbon–carbon double bonds. We can detect them using bromine or iodine solutions.

Cooking with vegetable oils

1 Why do we cook with vegetable oils?
2 What does it mean when we 'harden' vegetable oils?
3 How do we turn vegetable oils into spreads?

The temperature that a liquid boils at depends on the size of the forces between its molecules. The bigger these forces are, the higher the liquid's boiling point.

The molecules in vegetable oils are much bigger than water molecules. This makes the forces between the molecules in vegetable oils much larger. So the boiling points of vegetable oils are much higher than the boiling point of water.

When we cook food we heat it to a temperature where chemical reactions cause permanent changes to happen to the food. When we cook food in vegetable oil the result is very different to when we cook it in water. This is because vegetable oils can be used at a much higher temperature than boiling water.

So the chemical reactions that take place are very different. The food cooks more quickly, and very often the outside of the food turns a different colour, and becomes crisper.

a) How does the boiling point of vegetable oils compare to the boiling point of water?

Cooking food in oil also means that the food absorbs some of the oil. As you know, vegetable oils contain a lot of energy. This can make the energy content of fried food much higher than that of the same food cooked by boiling it in water. This is one reason why too much fried food can be bad for you!

Figure 1 An electric fryer like this one enables vegetable oil to be heated safely to a high temperature.

Figure 2 Boiled potatoes and fried potatoes are very different. One thing that probably makes chips so tasty is the contrast of crisp outside and soft inside, together with the different taste produced by cooking at a higher temperature. The different colour may be important too.

PRACTICAL

Investigating cooking

Compare the texture and appearance of potato pieces after equal cooking times in water and oil.

You might also compare the cooking times for boiling, frying and oven baking chips.

If possible carry out some taste tests in hygienic conditions.

b) How is food cooked in oil different to food cooked in water?

Unsaturated vegetable oils are usually liquids at room temperature. This is because the carbon–carbon double bonds in their molecules stop the molecules fitting together very well. This reduces the size of the forces between the molecules.

The boiling and melting points of these oils can be increased by adding hydrogen to the molecules. The reaction replaces some or all of the carbon–carbon double bonds with carbon–carbon single bonds. This allows the molecules to fit next to each other better. So the size of the forces between the molecules increases and the melting point is raised.

With this higher melting point, the liquid oil becomes a solid at room temperature. We call changing a vegetable oil like this **hardening** it.

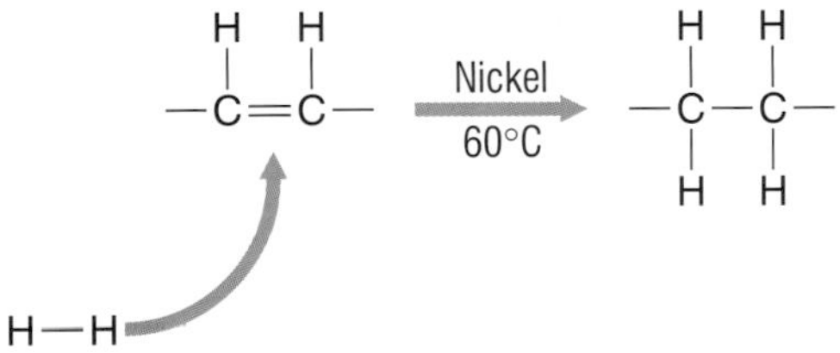

Figure 3 We harden a vegetable oil by reacting it with hydrogen. To make the reaction happen, we must use a nickel catalyst, and carry it out at about 60°C.

c) What do we call it when we add hydrogen to a vegetable oil?

Oils that we have treated like this are sometimes called *hydrogenated oils*. Because they are solids at room temperature, it means that they can be made into spreads to be put on bread. We can also use them to make cakes, biscuits and pastry.

Figure 4 We can use hydrogenated vegetable oils in cooking to make a huge number of different, and delicious, foods!

SUMMARY QUESTIONS

1 Copy and complete using the words below:

energy   hardening   higher   hydrogen   melting   tastes

The boiling points of vegetable oils are …… than the boiling point of water. This means that food cooked in oil …… different to boiled food. It also contains more …… .
The boiling and …… points of oils may also be raised by adding …… to their molecules. We call this …… the oil.

2 a) Why are hydrogenated vegetable oils more useful than oils that have not been hydrogenated?
   b) Explain how we harden vegetable oils and why the melting point is raised.

GET IT RIGHT!

No chemical bonds are broken when vegetable oils melt or boil – these are physical changes.
When oils are hardened with hydrogen, a chemical change takes place, producing margarine (which has a higher melting point than the original oil).

KEY POINTS

1 Vegetable oils are useful in cooking because of their high boiling points.
2 Vegetable oils are hardened by reacting them with hydrogen to increase their boiling and melting points.

Everyday emulsions

Figure 1 Smooth food has a good texture and looks as if it will taste nice – but it is not always easy to make, or to keep it smooth

The texture of food – what it feels like in your mouth – is a very important part of foods. Smooth foods like ice cream and mayonnaise are examples of food that might seem simple. In fact, they are quite complicated and difficult to make.

Smooth foods are made from a mixture of oil and water. Everyone knows that oil and water don't mix. Just try it by pouring a little cooking oil into a glass of water.

But we can persuade these two very different substances to mix together by making the oil into very small droplets. These spread out throughout the water and produce a mixture called an **emulsion**.

A good example of this is milk. Milk is basically made up of small droplets of animal fat dispersed in water.

Figure 2 Milk is an emulsion made up of animal fat and water, together with some other substances

PRACTICAL

A closer look at milk

You can view different types of milk – skimmed, semi-skimmed and full fat – under a microscope.

● What differences do you think you will see? Try it out.

Emulsions often behave very differently to the things that we make them from. For example, mayonnaise is made from ingredients that include oil and water. Both of these are runny – but mayonnaise is not!

Another very important ingredient in mayonnaise is egg yolks. Apart from adding a nice yellow colour, egg yolks have a very important job to do in mayonnaise. They stop the oil and water from separating out into layers. Food scientists have a word for this type of substance – they call them **emulsifiers**.

a) What do we mean by 'an emulsifier'?

Emulsifiers make sure that the oil and water in an emulsion cannot separate out. This means that the emulsion stays thick and smooth. Any creamy sauce needs an emulsifier. Without it we would soon find blobs of oil or fat floating around in the sauce. This might make us think that something had gone wrong in the cooking process. It may even make you feel quite sick!

b) How does an emulsifier help to make a good creamy sauce?

One very popular emulsion is *ice cream*. Everyday ice cream is usually made from vegetable oils, although luxury ice cream may also use animal fats.

Ice cream is one of the most complicated foods to make. To make the ice cream light, a great deal of air needs to be beaten into it to make a *foam*. The air is held in this foam by adding **stabilisers** and *thickeners*. These are substances with molecules that produce large 'cages' full of air when they are mixed with water.

Emulsifiers then keep the oil and water mixed together in the ice cream while we freeze it. Without them, the water in the ice cream freezes separately, producing crystals of ice. It makes the ice cream crunchy rather than smooth. This happens if you allow ice cream to melt and then put it back in the freezer.

Figure 3 Ice cream is a very complicated mixture of chemicals

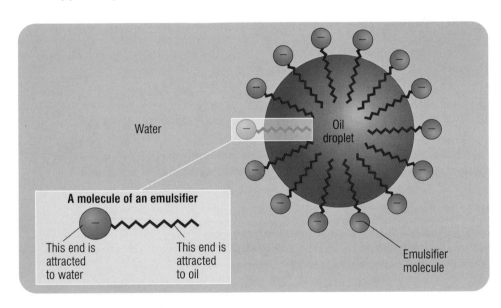

Figure 4 An emulsifier is a molecule with 'a tail' that 'likes' oil and 'a head' that 'likes' water. The charged oil droplets repel each other, keeping them spread throughout the water.

SUMMARY QUESTIONS

1 Copy and complete using the words below:

> **emulsifier   emulsion   ice cream   mayonnaise**
> **mix   separating   small**

Oil and water do not ...... together, but if the oil droplets can be made very ...... it is possible to produce a mixture of oil and water called an .... To keep the oil and water from ...... we can use a chemical called an .... Important examples of food made like this include ...... and ....

2 Salad cream is an emulsion made from vegetable oil and water. In what ways is salad cream different from both oil and water?

3 a) Why do we need to add an emulsifier to an emulsion like salad cream?
b) Explain how emulsifier molecules do their job.

KEY POINTS

1 Oils do not dissolve in water.
2 Oils can be used to produce emulsions which have special properties.
3 Emulsions made from vegetable oils are used in many foods, such as salad dressings and ice creams.

What is added to our food?

LEARNING OBJECTIVES

1 What are food additives and why are they put in our food?
2 How can we detect food additives?

DID YOU KNOW...

... that lemon juice can prevent a cut apple turning brown, thanks to the antioxidant action of vitamin C?

People have always needed to find ways of making food last longer – to *preserve* it. For hundreds of years we have added substances like salt or vinegar to food in order to keep it longer. As our knowledge of chemistry has increased we have used other substances too, to make food look or taste better.

We call a substance that is added to food to make it keep longer or to improve its taste or appearance a **food additive**. Additives that have been approved for use in Europe are given **E numbers** which identify them. For example E102 is a yellow food colouring called *tartrazine*, while E220 is the preservative *sulfur dioxide*.

Figure 1 Modern foods contain a variety of additives to improve their taste or appearance, and to make them keep longer

a) What is a food additive?

There are six basic types of food additives. Each group of additives is given an E number. The first digit of the number tells us what kind of additive it is.

E number	Additive	What the additive does	Example
E1 . . .	colours	Improve the appearance of the food. There are three main classes of colour in foods: *natural colours*, *browning colours*, which are produced during cooking and processing, and *additives*.	E150 – caramel, a brown colouring
E2 . . .	preservatives	Help food to keep longer. Many foods go bad very quickly without preservatives. Wastage of food between harvesting and eating is still a problem in many countries.	E211 – sodium benzoate
E3 . . .	antioxidants	Help to stop food reacting with oxygen. Oxygen in the air affects many foods badly, making it impossible to eat them. A good example of this is what happens when you cut an apple open – the brown colour formed is due to oxygen reacting with the apple.	E300 – vitamin C
E4 . . .	emulsifiers, stabilisers and thickeners	Help to improve the texture of the food – what it feels like in your mouth. Many foods need to be treated like this, for example, jam and the soya proteins used in veggieburgers.	E440 – pectin
E5 . . .	acidity regulators	Help to control pH. The acidity of foods is an essential part of their taste. All fruits contain sugar, but without acids they would be sickly and dull.	E501 – potassium carbonate
E6 . . .	flavourings	There are really only five flavours – *sweet*, *sour*, *bitter*, *salt* and *savoury*. What we call flavour is a subtle blend of these five, together with the smells that foods give off.	E621 – monosodium glutamate

Detecting additives

There is a wide range of chemical instruments that scientists can use to identify unknown chemical compounds, including food additives. Many of these are simply more sophisticated and automated versions of techniques that we use in the school lab.

One good example of a technique used to identify food additives is **chromatography**. This technique separates different compounds based on how well they dissolve in a particular solvent. Their solubility then determines how far they travel across a surface, like a piece of chromatography paper.

PRACTICAL

Detecting dyes in food colourings

Make a chromatogram to analyse various food colourings.

● What can you deduce from your chromatogram?

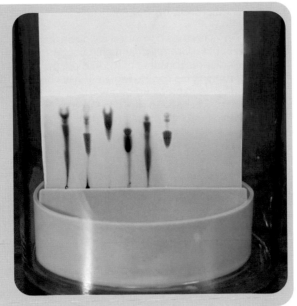

Figure 2 The technique of paper chromatography that we use at school. Although they are more complex, techniques used to identify food additives are often based on the same principles as the simple tests we do in the school science lab.

b) What happens when you make a paper chromatogram of food colourings?

Once the compounds in a food have been separated out using chromatography, they can be identified by comparing them with known substances. Alternatively they may be fed into another instrument – the **mass spectrometer**. This can be used for identifying both elements and compounds – it measures the relative formula mass of substances placed in it for analysis, which we can then use to identify the sample.

Figure 3 In the UK in 2005 a batch of red food colouring was found to be contaminated with a chemical suspected of causing cancer. This dye had found its way into hundreds of processed foods. All of these had to be removed from the shelves of our supermarkets and destroyed.

SUMMARY QUESTIONS

1 Copy and complete the table:

Additive	Reason	Additive	Reason
colouring	. . .	. . .	improving texture
. . .	help food keep longer	acidity regulators	. . .
antioxidants	. . .	. . .	changing flavour

2 a) Carry out a survey of some processed foods. Identify some examples of food additives and explain why they have been used.

   b) Describe how we can separate the dyes in a food colouring and identify them.

KEY POINTS

1 Additives may be added to food in order to improve its appearance, taste and how long it will keep (its shelf-life).
2 Food scientists can analyse foods to identify additives.

Vegetable oils as fuels

LEARNING OBJECTIVES

1 How can we use vegetable oils as a fuel?
2 What are the advantages of using vegetable oils as fuels?

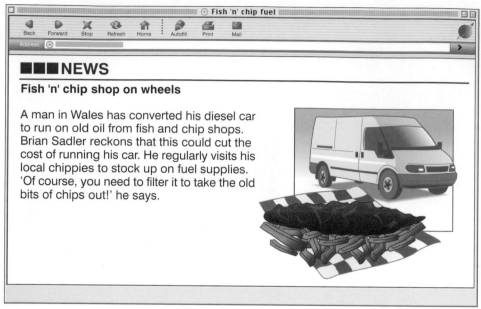

Figure 1 A true story of recycling

Running our cars on the old oil from fish and chip shops isn't realistic for most of us. But it is possible to make fuel for cars using vegetable oils, even used cooking oil. We just need to use a little bit of clever chemistry.

Biodiesel is the name we give to any fuel made from vegetable oils. We can use these fuels in any car or van that has a diesel engine.

Most modern biodiesel is made by treating vegetable oils to remove some unwanted chemicals. The biodiesel made like this can be used on its own, or mixed with diesel fuel made from refining crude oil.

When we make biodiesel we also produce other useful products. For example, we get a solid waste material that we can feed to cattle as a high-energy food. We also get glycerine which we can use to make soap.

a) What is biodiesel?

b) What is biodiesel made from?

There are some very big advantages in using biodiesel as a fuel. First, biodiesel is a very clean fuel. Biodiesel is much less harmful to animals and plants than diesel made from crude oil. If it is spilled, it breaks down about five times faster than 'normal' diesel. Also, when we burn biodiesel in an engine it burns much more cleanly. It makes very little sulfur dioxide and other pollutants.

DID YOU KNOW...

. . . that the largest factory for turning used cooking oil into biodiesel was opened in Scotland in 2005? The plant can produce up to 5% of Scotland's diesel fuels requirements.

But using biodiesel has one really big advantage over petrol and diesel. It is the fact that the crops used to make biodiesel absorb carbon dioxide gas as they grow. So biodiesel is effectively 'CO₂ neutral'. That means the amount of carbon dioxide given off is nearly balanced by the amount absorbed. Therefore, biodiesel makes little contribution to the greenhouse gases in our atmosphere.

Figure 2 This coach runs on biodiesel

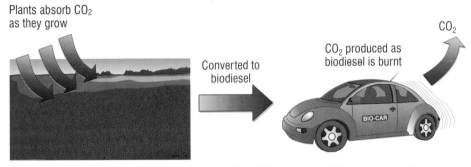

Figure 3 Cars run on biodiesel produce very little CO_2 overall, as CO_2 is absorbed by plants as the fuel is made

Using ethanol as a bio-fuel

Another bio-fuel is ethanol. We can make it by fermenting the sugar from sugar beet or sugar cane. In Brazil they can grow lots of sugar cane. They add the ethanol to petrol, saving our dwindling supplies of crude oil. The ethanol gives off carbon dioxide (a greenhouse gas) when it burns, but the sugar cane absorbs the gas during photosynthesis.

Ethanol can also be made from the ethene we get by cracking the heavier fractions from crude oil (see page 184):

$$C_2H_4 + H_2O \xrightarrow[\text{high pressure}]{\text{catalyst}} C_2H_5OH$$

ethene   steam   ethanol

c) Why is ethanol from sugar cane known as a 'bio-fuel' whereas ethanol from ethene isn't?

Figure 4 Ethanol can be made from sugar cane

SUMMARY QUESTIONS

1 Copy and complete using the words below:

   carbon dioxide   cattle   diesel   plants   soap

   Biodiesel is a fuel made from .... Making the fuel also produces some other useful products, including ...... and food for .... Biodiesel produces less pollution than ...... , and absorbs nearly as much .... when it is made as it does when it burns.

2 Where does the energy in biodiesel come from?

3 a) How is ethene converted into ethanol? Include a balanced equation in your answer.
   b) When can we describe ethanol as a bio-fuel?
   c) Write an article for a local newspaper describing the arguments for using biodiesel instead of other fuels made from crude oil.

KEY POINTS

1 Vegetable oils can be burned as fuels.
2 Vegetable oils are a renewable source of energy that could be used to replace some fossil fuels.

Vegetable oils

GO FASTER!

Plant-powered performance

BIO BLEND DIESEL Cheaper Eco Friendly Fuel SOLD HERE

Whatever will those ace scientists come up with next? Just when you think you've seen everything, the idea of filling up your car with plant power comes along! The latest idea by those who want us all to save the planet is to take oil made from oilseed rape plants (you know, that yellow stuff you see growing all over the countryside in springtime), mix it with diesel and then shove it in the tank of your motor! And they call it ... *biodiesel!*

So why would anyone want to do this? Well, mainly, because biodiesel is better for the environment. Not only does it burn just like ordinary diesel fuel, but it doesn't produce as much pollution. And what's more, it's a 'green fuel' which doesn't contribute as much to global warming – though we couldn't quite see why ... there's a year's free subscription on offer to anyone who can explain it to us!

ACTIVITY

Imagine that you are a presenter on a popular television programme about cars and driving. Write and present an article for this programme about biodiesel.

Make your article fun and informative. Include as much factual information as you can so that people can decide whether they would like to use this fuel in their car.

Look after your family's hearts

Everyone knows the benefits of a healthy diet. But do you know the benefits of ensuring that you eat vegetable oils as part of your diet?

Scientists have found that eating vegetable oils instead of animal fats can do wonders for the health of your heart. The saturated fats you find in things like butter and cheese can make the blood vessels of your heart become clogged up with fat.

However, the unsaturated fats in vegetable oils (like olive oil and corn oil) are especially good for you. They help to keep your arteries clear and reduce the chance of you having heart disease.

The levels of a special fat called **cholesterol** in your blood give doctors an idea about your risk of heart disease. People who eat lots of vegetable oils tend to have a much healthier level of cholesterol in their blood.

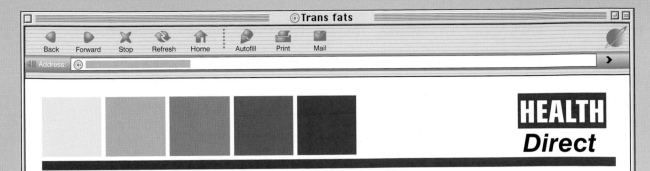

HEALTH
Direct

Online Health Encyclopaedia

Trans fats

Trans unsaturated fatty acids, or ***trans fats***, are solid fats. They are produced artificially by heating liquid vegetable oils with metal catalysts and hydrogen. Trans fats are made in huge quantities when we harden vegetable oils to make margarine.

The fats used to cook French fries and other fast foods usually contain trans fats. Concerns have been raised for several decades that eating trans fats might have contributed to the 20th century epidemic of coronary heart disease.

Studies have shown that trans fats have adverse effects on cholesterol in the blood – increasing 'bad' cholesterol while decreasing 'good' cholesterol. Trans fats have also been associated with an increased risk of coronary heart disease in other studies.

Changes in food labelling are very important. But many products, including fast food, often contain extremely high levels of trans fats. Yet these are exempt from labelling regulations and may have labels such as 'cholesterol-free' and 'cooked in vegetable oil'.

For example, a person eating one doughnut for breakfast and a large order of French fries for lunch would eat 10 g of trans fats, or 5 per cent of the total energy of an 1800-calorie diet. Thus, simple labelling changes alone will not be sufficient.

ACTIVITIES

a) Write an article for a family lifestyle magazine about 'feeding your family'. Include in this article reasons for including vegetable oils in a balanced diet and their effect on people's health.

b) Design a poster with the title 'Vegetable oils – good or bad?'

SUMMARY QUESTIONS

1 Write simple definitions for the following words:

   a) vegetable oils

   b) unsaturated oils

   c) saturated oils.

2 A vegetable oil removes the colour from bromine solution. When the oil has been hardened it does not react with bromine solution.
   Explain these observations.

3 a) Some ice cream is left standing out on a table during a meal on a hot day. It is then put back in the freezer again. When it is taken out of the freezer a few days later, people complain that the ice cream tastes 'crunchy'. Why is this?

   b) A recipe for ice cream says: 'Stir the ice cream from time to time while it is freezing.' Why must you stir ice cream when freezing it?

4 a) Why is it important to be able to identify food additives in food?

   b) What is the first number of the E number of the following additives?
      i) lecithin (an emulsifier)
      ii) quinoline yellow (a food colouring)
      iii) calcium oxide (an acidity regulator)
      iv) inosinic acid (a flavour enhancer)

5 Biodiesel is almost 'carbon neutral'. Explain, using your knowledge of the carbon cycle, whether it would ever be possible to produce biodiesel that produces less carbon dioxide when burnt than was absorbed by the plants which made it.

6 'Alternative fuels like biodiesel and gasohol are all very well for countries like Brazil and India – but no good in countries like the UK!' Do you agree with this statement? Give your reasons.

7 Some food are marketed as 'free from artificial additives'.

   a) Does this mean that the food contains no added substances?

   b) Give an example of a substance that is added to food, and outline the reasons for using it.

   c) Is 'additive-free food' better than food which contains additives?

EXAM-STYLE QUESTIONS

1 The energy values of chips depend on their fat content. Match the energy values **A**, **B**, **C** and **D** with the numbers **1** to **4** in the table.

   A 687 kJ/100 g

   B 796 kJ/100 g

   C 1001 kJ/100 g

   D 1174 kJ/100 g

	Description of type of chips	Fat content (g/100 g)
1	Fish and chip shop, fried in blended oil	12.4
2	French fries from burger outlet	15.5
3	Homemade fried in blended oil	6.7
4	Oven chips, frozen, baked	4.2

(4)

2 Match the words **A**, **B**, **C** and **D** with spaces **1** to **4** in the sentences.

   A cooking oils

   B emulsifiers

   C emulsions

   D hydrogenated oils

   Mayonnaise and salad dressings are ......**1**...... that are made by mixing oil and vinegar with other ingredients such as egg yolk.

   In mayonnaise the egg yolk contains ......**2**......that stop the oil and water separating.

   Vegetable oils can be converted into ......**3**...... by reacting with hydrogen and a catalyst.

   Biodiesel is a fuel that can be made from waste ......**4**...... . (4)

3 The table on the next page gives some information about four different vegetable oils.
   Smoke point is the temperature at which the oil begins to smoke when heated.
   Match descriptions **A**, **B**, **C** and **D** with numbers **1** to **4** in the table.

   A The oil that contains the most monounsaturated fat.

   B The oil that reacts with the largest volume of bromine water.

   C The oil with the highest melting point.

   D The oil with the widest range of smoke point.

Type of oil	1	2	3	4
	Corn oil	Olive oil	Sunflower oil	Rapeseed oil
Saturated fat (%)	14.4	14.3	12.0	6.6
Mono-unsaturated fat (%)	29.9	73.0	20.5	59.3
Poly-unsaturated fat (%)	51.3	8.2	63.3	29.3
Melting point (°C)	−15	−12	−18	5
Smoke point (°C)	229–268	204–210	229–252	230–240

(4)

4 Use the table of types of chips in question **1** to help you answer these questions.

(a) Why do chips contain fat? (1)

(b) Why do French fries contain most fat? (1)

(c) Why do oven chips contain least fat? (2)

(d) Why do all chips have a golden brown colour, but boiled potatoes remain white? (2)

(e) (i) How would you display the data in the table? (1)
    (ii) Explain your answer to part (i). (1)

5 Virgin olive oil is extracted by mechanical methods that do not modify its properties. If the temperature during extraction does not exceed 27°C the oil can be labelled as 'cold pressed'. Any olive oil that remains in the pressed pulp can be extracted by dissolving it in a solvent. The solvent is removed from the oil by evaporation. This type of oil is called pomace oil.

(a) Why is it important that the temperature does not exceed 27°C during extraction? (2)

(b) Suggest why some people prefer virgin olive oil to pomace oil. (2)

6 Some students made a solution of the colours in a soft drink. Describe how you could use paper chromatography to show how many colours were in the solution. (4)

HOW SCIENCE WORKS QUESTIONS

The teacher decided that her class should do a survey of different foods to find out the degree of unsaturated oils present in them. She chose five different oils and divided them amongst her students. This allowed one oil to be done twice, by two different groups. They were given strict instructions as to how to do the testing.

Bromine water was added to each oil from a burette. The volume added before the bromine was no longer colourless was noted.

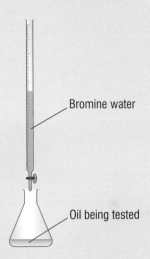
Bromine water
Oil being tested

The results are in this table.

Type of oil	Amount of bromine water added (cm³)	
	Group 1	Group 2
Ollio	24.2	23.9
Soleo	17.8	18.0
Spreo	7.9	8.1
Torneo	13.0	12.9
Margeo	17.9	17.4

a) Why was it important that the teacher gave strict instructions to all of the groups on how to carry out the tests? (1)

b) List some controls that should have been included in the instructions. (3)

c) Are there any anomalies? If so, state which results are anomalies. (1)

d) What evidence is there in the results that indicate that they are reliable? (1)

e) How might the accuracy be checked? (1)

f) How would you present these results? (1)

Structure of the Earth

The deepest mines go down to about 3500 m, while geologists have drilled down to more than 12 000 m in Russia. Although these figures seem large, they are tiny compared with the diameter of the Earth. The Earth's diameter is about 12 800 km. That's more than one thousand times the deepest hole ever drilled!

The Earth is made up of layers that formed many millions of years ago, early in the history of our planet. Heavy matter sank to the centre of the Earth while lighter material floated on top. This produced a structure consisting of a dense **core**, surrounded by the **mantle**. Outside the mantle there is a thin layer called the **crust**.

The uppermost part of the mantle and the crust make up the Earth's *lithosphere*.

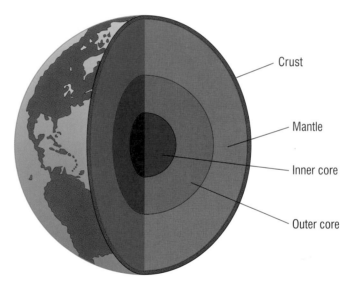

Figure 1 The structure of the Earth

We call the outer layer of the Earth the crust. This layer is very thin compared to the diameter of the Earth (thinner than the outer layer of a football!). Its thickness can vary from as thin as 5 km under the oceans to as much as 70 km under the continents.

Underneath the crust is the mantle. This layer is much, much thicker than the crust – nearly 3000 km. The mantle behaves like a solid, but it can flow in parts very slowly.

Finally, inside the mantle lies the Earth's core. This is about half the radius of the Earth, and is made of a mixture of nickel and iron. The core is actually two layers. The outer core is a liquid, while the inner core is solid.

a) What is the outer layer of the Earth called?
b) What is the next layer of the Earth called?
c) How many parts make up the Earth's core?

How do we know the structure inside the Earth if we have never seen it? Scientists use evidence from earthquakes. Following an earthquake, seismic waves travel through the Earth. The way in which seismic waves travel through the Earth is affected by the structure of the Earth. By observing how seismic waves travel, scientists have built up the detailed picture of the inside of the Earth described here.

Also, by making careful measurements, physicists have been able to measure the mass of the Earth, and to calculate its density. The density of the Earth as a whole is much greater than the density of the rocks found in the crust. This suggests that the centre of the Earth must be made from a different material to the crust. This material must have a much greater density than the material that makes up the crust.

Crust	Mantle	Core
Averages: about 6 km under the oceans about 35 km under continental areas	Starts underneath crust and continues to about 3000 km below Earth's surface Behaves like a solid, but is able to flow very slowly	Radius of about 3500 km Made of nickel and iron Outer core is liquid, inner core is solid

At one time scientists thought that features like mountain ranges on the surface of the Earth were caused by the crust shrinking as the early molten Earth cooled down. They thought of it rather like the skin on the surface of a bowl of custard shrinks then wrinkles as the custard cools down.

However, scientists now have a better explanation for the features on the Earth's surface, as we shall see later in this chapter.

Figure 2 All of the minerals that we depend on in our lives – iron, aluminium and copper, for example, as well as oil and gas – come from the thin crust of the Earth

SUMMARY QUESTIONS

1 Copy and complete using the words below:

    core   crust   mantle   slowly   solid   thin

The structure of the Earth consists of three layers – the ……, the …… and the …… . The outer layer of the Earth is very …… compared to its diameter. The layer below this is ……. but can flow in parts very ……..

2 Why do some people think that the mantle is best described as a 'very thick syrupy liquid'?

3 Why do scientists think that the core of the Earth is made of much denser material than the crust?

KEY POINTS

1 The Earth consists of a series of layers.
2 Scientists originally thought that the features on the Earth's surface were caused as the crust cooled and shrank.

The restless Earth

LEARNING OBJECTIVES

1 What are tectonic plates?
2 Why do they move?
3 Why is it difficult for scientists to predict when earthquakes and volcanic eruptions will occur?

Have you ever looked at the western coastline of Africa and the eastern coastline of South America on a map? If you have, you might have noticed that these edges of the two continents have a remarkably similar shape.

The fossils and rock structures that we find when we look in Africa and South America are also similar. Fossils show that the same reptiles and plants once lived in both continents. And the layers of rock in the two continents are arranged in the same sequence, with layers of sandstone lying above seams of coal.

Scientists now believe that they can explain the similarity in the shapes of the continents and of the rocks and fossils found there. They think that the two continents were once joined together as one land mass.

a) What evidence is there that Africa and South America were once joined?

Figure 2 shows the vast 'supercontinent' of Pangaea. This land mass is believed to have existed up until about 250 million years ago. Slowly Pangaea split in two about 160 million years ago. The land masses continued to move apart until about 50 million years ago. Then they began to closely resemble the map of the world we know today.

b) What was the name of the original 'supercontinent'?

Of course, the continents moved and split up very, very slowly – only a few centimetres each year. They moved because the Earth's crust and uppermost part of the mantle (its lithosphere) is cracked into a number of large pieces. We call these **tectonic plates**.

Deep within the Earth, radioactive decay produces vast amounts of energy. This heats up molten minerals in the mantle which expand. They become less dense and rise towards the surface and are replaced by cooler material. It is these **convection currents** which pushed the tectonic plates over the surface of the Earth.

Figure 1 *Glossopteris* was a tree-like plant growing about 230 million years ago. It had tongue-shaped leaves, and grew to a height of about 4 metres. Its fossils have been found in Africa and in South America.

Figure 2 The break up of Pangaea into Laurasia and Gondwanaland led eventually to the formation of the land masses we recognise today. Notice how, 100 million years ago, India is still moving northwards to take up the position it occupies today. The collision between India and the continent of Asia produced the mountain range we call the Himalayas.

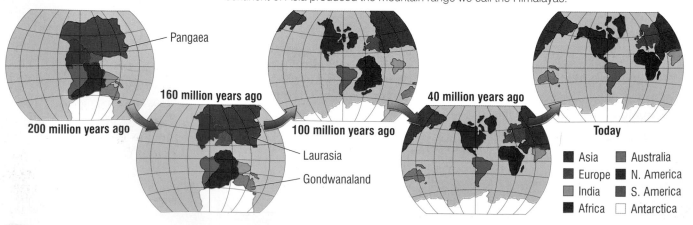

Where the boundaries of the plates meet, huge forces are exerted. These forces make the plates buckle and deform, and mountains may be formed. The plates may also move suddenly and very quickly past each other. These sudden movements cause earthquakes.

If earthquakes happen under the sea, they may cause huge tidal waves called **tsunamis**. However, it is difficult for scientists to know exactly where and when the plates will move like this. So predicting earthquakes is still a very difficult job.

The Earth's tectonic plates are made up of the crust and the upper part of the mantle (not just the crust).

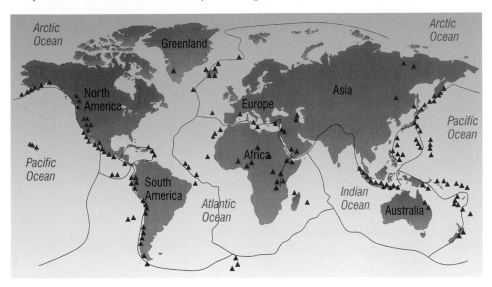

Figure 3 The distribution of volcanoes around the world largely follows the boundaries of the tectonic plates

Wegener's revolutionary theory

The idea that huge land masses once existed before the continents we know today was put forward in the late 19th century by the geologist Edward Suess. He thought that a huge southern continent had sunk. He suggested that this left behind a land bridge (since vanished) between Africa and South America.

The idea of continental drift was put first forward by Alfred Wegener in 1915. However, his fellow scientists found Wegener's ideas hard to accept. This was mainly because he could not explain **how** the continents had moved. So they stuck with their existing ideas.

His theory was finally shown to be right almost 50 years later. Scientists found that the sea floor is spreading apart in some places, where molten rock is spewing out between two continents. This led to a new theory, called plate tectonics.

SUMMARY QUESTIONS

1 Copy and complete using the words below:

   convection   earthquakes   mantle   tectonic   tsunamis   volcanoes

   The surface of the Earth is split up into a series of …… plates. These move across the Earth's surface due to …… currents in the …… . Where the plates meet or rub against each other …… and …… may form, and …… may happen.

2 a) Explain how tectonic plates move.
   b) Why are earthquakes and volcanic eruptions difficult to predict?
   c) Imagine that you are a scientist who has just heard Wegener talking about his ideas for the first time. Write a letter to another scientist explaining what Wegener has said and why you have chosen to reject his ideas.

KEY POINTS

1 The Earth's lithosphere is cracked into a number of pieces (tectonic plates) which are constantly moving.

2 The motion of the tectonic plates is caused by convection currents in the mantle, due to radioactive decay.

3 Earthquakes and volcanoes happen where tectonic plates meet. It is difficult to know when the plates may slip past each other. This makes it difficult to predict accurately when and where earthquakes will happen.

The Earth's atmosphere in the past

LEARNING OBJECTIVES

1   What was the Earth's atmosphere like in the past?
2   How were the gases in the Earth's atmosphere produced?
3   How was oxygen produced?

Scientists think that the Earth was formed about 4.5 billion years ago. To begin with it was a molten ball of rock and minerals. For its first billion years it was a very violent place. The Earth's surface was covered with volcanoes belching fire and smoke into the atmosphere.

Figure 1 Volcanoes moved chemicals from inside the Earth to the surface and the newly forming atmosphere

The volcanoes released carbon dioxide, water vapour and nitrogen gas, which formed the early atmosphere. Water vapour in the atmosphere condensed as the Earth gradually cooled down, and fell as rain. This collected to form the first oceans.

Comets also brought water to the Earth. As icy comets rained down on the surface of the Earth they melted, adding to the water supplies. Even today many thousands of tonnes of water fall onto the surface of the Earth from space every year.

So as the Earth began to stabilise, the early atmosphere was probably mainly carbon dioxide, with some water vapour, nitrogen and traces of methane and ammonia. There was very little or no oxygen. This is very like the atmospheres which we know exist today on the planets Mars and Venus.

Figure 2 The surface of one of Jupiter's moons called Io, with its small atmosphere and active volcanoes. This photograph probably gives us a reasonable glimpse of what our own Earth was like billions of years ago.

a) What was the main gas in the Earth's early atmosphere?
b) How much oxygen was there in the Earth's early atmosphere?

After the initial violent years of the history of the Earth, the atmosphere remained relatively stable. That's until life first appeared on Earth.

Scientists think that life on Earth began about 3.4 billion years ago, when simple organisms like bacteria appeared. These could make food for themselves, using the breakdown of other chemicals as a source of energy.

Later, bacteria and other simple organisms such as algae evolved. They could use the energy of the Sun to make their own food by photosynthesis, and oxygen was produced as a waste product.

By two billion years ago the levels of oxygen were rising steadily as algae and bacteria filled the seas. More and more plants evolved, all of them also photosynthesising, removing carbon dioxide and making oxygen.

$$\text{carbon dioxide} + \text{water} \xrightarrow{\text{(energy from sunlight)}} \text{sugar} + \text{oxygen}$$

As plants evolved and successfully colonised most of the surface of the Earth, the atmosphere became richer and richer in oxygen. Now it was possible for animals to evolve. These animals could not make their own food and needed oxygen to respire.

On the other hand, many of the earliest living microorganisms could not tolerate oxygen (because they had evolved without it). They largely died out, as there were fewer and fewer places where they could live.

Figure 3 Some of the first photosynthesising bacteria probably lived in colonies like these stromatolites. They grew in water and released oxygen into the early atmosphere.

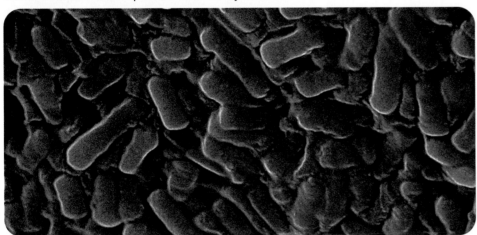

Figure 4 Bacteria such as these not only do not need oxygen – they die if they are exposed to it. But they can survive and breed in rotting tissue and other places where there is no oxygen.

SUMMARY QUESTIONS

1 Copy and complete using the words below:

   carbon dioxide   methane   oxygen   volcanoes   water

The Earth's early atmosphere probably consisted mainly of the gas .... .There could also have been ...... vapour and nitrogen, plus small amounts of ...... and ammonia. These gases were released by ...... as they erupted. Plants removed carbon dioxide from the atmosphere and produced ....

2 How was the Earth's early atmosphere formed?

3 Why was there no life on Earth for several billion years?

4 Draw a chart that explains the early development of the Earth's atmosphere.

KEY POINTS

1 The Earth's early atmosphere was formed by volcanic activity.
2 It probably consisted mainly of carbon dioxide. There may also have been water vapour together with traces of methane and ammonia.
3 As plants colonised the Earth, the levels of oxygen in the atmosphere rose.

Gases in the atmosphere

LEARNING OBJECTIVES

1 What are the main gases in the atmosphere?
2 What are their relative proportions?
3 What are the noble gases and why are they useful?

Figure 1 There is clear evidence in carbonate rocks of the organisms which lived millions of years ago, now preserved with their ancient carbon in the structure of our rocks

We think that the early atmosphere of the Earth contained a great deal of carbon dioxide. Yet the modern atmosphere of the Earth has only around 0.04% of this gas. Where has it all gone? The answer is mostly into living organisms and into materials formed from living organisms.

Carbon dioxide is taken up by plants during photosynthesis and the carbon can end up in new plant material. Then animals eat the plants and the carbon is transferred to the animal tissues, including bones, teeth and shells.

Over millions of years the dead bodies of huge numbers of these living organisms built up at the bottom of vast oceans. Eventually they formed sedimentary carbonate rocks like limestone.

Some of these living things were crushed by movements of the Earth and heated within the crust. They formed fossil fuels such as coal and oil. In this way much of the carbon from carbon dioxide in the ancient atmosphere became locked up within the Earth's crust.

a) Where has most of the carbon dioxide in the Earth's early atmosphere gone?

Carbon dioxide also dissolved in the oceans. It reacted and made insoluble carbonate compounds. These fell to the sea-bed and helped to form carbonate rocks.

At the same time, the ammonia and methane, from the Earth's early atmosphere, reacted with the oxygen formed by the plants. This got rid of these poisonous gases and increased the nitrogen and carbon dioxide levels:

$$CH_4 + 2O_2 \rightarrow CO_2 + 2H_2O$$
$$4NH_3 + 3O_2 \rightarrow 2N_2 + 6H_2O$$

By 200 million years ago the proportions of the different gases in the Earth's atmosphere were much the same as they are today.

Look at the pie chart in Figure 2.

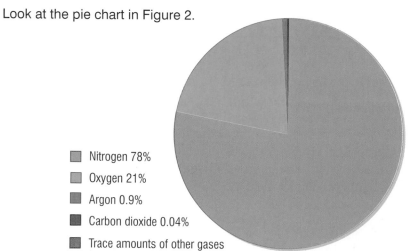

■ Nitrogen 78%
■ Oxygen 21%
■ Argon 0.9%
■ Carbon dioxide 0.04%
■ Trace amounts of other gases

Figure 2 The relative proportions of nitrogen, oxygen and other gases in the Earth's atmosphere

b) What gas did plants produce that changed the Earth's atmosphere?

The noble gases

The Earth's atmosphere contains tiny amounts of a group of gases that we call the **noble gases**. They are all found in Group 0 of the periodic table.

Helium, neon and argon, along with krypton, xenon and radon are the least reactive elements known. It is very difficult to make them react with any other elements. They don't even react with themselves to form molecules. They exist as single atoms. We say that they are monatomic.

Because the noble gases are very unreactive, we cannot use them to make useful materials. Instead, we use them in situations where they are useful because of their extreme lack of reactivity.

Uses of the noble gases

Helium is used in airships and in party balloons. Its low density means that balloons filled with the gas float in air. It is also safer than the only alternative gas, hydrogen, because its low reactivity means that it does not catch fire. We also use it with oxygen as a breathing mixture for deep-sea divers. The mixture reduces their chances of suffering from the 'bends'.

We use **neon** in electrical discharge tubes – better known as **neon lights**. When we pass an electrical current through the neon gas it gives out a bright light. Neon lights are familiar as street lighting and in advertising.

We use **argon** in a different type of lighting – the everyday light bulb (or filament lamp). The argon provides an inert atmosphere inside the bulb. When the electric current passes through the metal filament, the metal becomes white hot. If any oxygen was inside the bulb, it would react with the hot metal. However, no chemical reaction takes place between the metal filament and argon gas. This stops the filament from burning away and makes light bulbs last longer.

Figure 3 The noble gases are all found in Group 0 of the periodic table

Figure 4 These brightly coloured balloons are filled with helium, which makes them float upwards through the air

---

The carbon cycle

Over the past 200 million years the levels of carbon dioxide in the atmosphere have not changed much. This is due to the natural carbon cycle in which carbon moves between the oceans, rocks and the atmosphere.

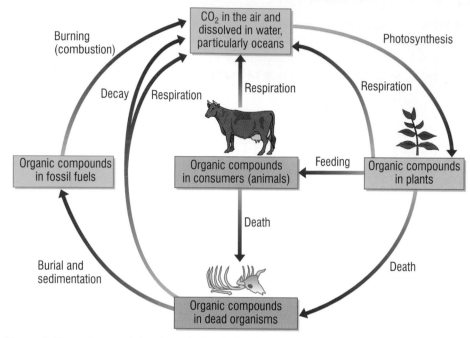

Figure 1 The carbon cycle has kept the level of carbon dioxide in the atmosphere steady for the last 200 million years

Left to itself, the carbon cycle is self-regulating. The oceans act as massive reservoirs of carbon dioxide. They absorb excess CO_2 when it is produced and release it when it is in short supply. Plants also soak up carbon dioxide from the atmosphere. We often call plants and oceans carbon dioxide 'sinks'.

Carbon dioxide moves back into the atmosphere when living things respire or when they die and decompose.

Then we also have the CO_2 that comes from volcanoes. Carbonate rocks are sometimes moved deep into the Earth by the movements of the Earth's crust. If that rock then becomes part of a volcano, heat causes the carbonates in the rock to break down. Then the carbon dioxide gas is released as the volcano erupts.

a) What has kept carbon levels stable over the last 200 million years?
b) What happens when carbonate rocks become part of a volcano?

PRACTICAL

The properties of carbon dioxide

Carry out a series of tests to find out the properties of carbon dioxide gas.

• Record your findings in a bullet-pointed list.

The changing balance

Over the last fifty years or so we have increased the amount of carbon dioxide released into the atmosphere tremendously. We burn fossil fuels to make electricity, heat our homes and drive our cars. This has enormously increased the amount of carbon dioxide we produce.

There is no doubt that the levels of carbon dioxide in the atmosphere are increasing.

We can record annual changes in the levels of carbon dioxide which are due to seasonal differences in the plants. (See the graph on page 124.) This shows how important plants are for removing CO_2 from the atmosphere. But the overall trend for the last 30 years has been ever upwards.

The balance between the carbon dioxide produced and the carbon dioxide absorbed by 'CO_2 sinks' is very important.

Think about what happens when we burn fossil fuels. Carbon, which has been locked up for hundreds of millions of years in the bodies of once living animals, is released as carbon dioxide into the atmosphere. For example:

$$\text{propane} + \text{oxygen} \rightarrow \text{carbon dioxide} + \text{water}$$
$$C_3H_8 + 5O_2 \rightarrow 3CO_2 + 4H_2O$$

As carbon dioxide levels in the atmosphere go up, so the reaction between carbon dioxide and sea water increases. This reaction makes insoluble carbonates (mainly calcium carbonate). These are deposited as sediment on the bottom of the ocean. It also produces soluble hydrogencarbonates – mainly calcium and magnesium – which simply remain dissolved in the sea water.

In this way the seas and oceans act as a buffer, absorbing excess carbon dioxide but releasing it if necessary. However this buffering system probably cannot cope with all the additional carbon dioxide that we are currently pouring out into the atmosphere.

Figure 2 Most of the electricity that we use in the UK is made by burning fossil fuels. This releases carbon dioxide into the atmosphere. One solution would be to pump the CO_2 deep underground to be absorbed into porous rocks. This would increase the cost of producing electricity by about 10%.

DID YOU KNOW...

. . . that scientists predict that global warming may mean that the Earth's temperature could rise by as much as 5.8°C by the year 2100?

SUMMARY QUESTIONS

1 Match up the parts of sentences:

a) Carbon dioxide levels in the Earth's atmosphere	A ...... carbon locked up long ago is released as carbon dioxide.
b) Plants and oceans are known as	B ...... were kept steady by the carbon cycle.
c) When we burn fossil fuels	C ...... the reaction between carbon dioxide and sea water increases.
d) As carbon dioxide levels rise	D ...... carbon dioxide sinks

2 Draw a labelled diagram to illustrate how boiling an electric kettle may increase the amount of carbon dioxide in the Earth's atmosphere.

3 Why has the amount of carbon dioxide in the Earth's atmosphere risen in the last 50 years?

KEY POINTS

1 Carbon moves into and out of the atmosphere due to plants, the oceans and rocks.
2 The amount of carbon dioxide in the Earth's atmosphere has risen due to the amount of fossil fuels we burn.

Earth issues

EARTHQUAKES

The tsunami that tore across the Indian Ocean on 26 December 2004 left nearly 300 000 people dead. It was caused when a huge earthquake – the second biggest ever recorded – lifted billions of tonnes of sea water over 20 metres upwards. This produced a huge wave that travelled thousands of miles at speeds of several hundred miles an hour.

The earthquake was detected thousands of miles away by the Pacific Tsunami Warning Centre, but no-one could tell that it would produce such devastation. In the aftermath of the disaster, people discussed the lessons that had been learnt.

DEATH TOLL FROM TSUNAMI MAY REACH **300,000**

six page report

MASSIVE EARTHQUAKE CAUSES TIDAL WAVE IN INDIAN OCEAN

Full report

What we need to do is to build an early warning system that would tell us when a tidal wave is coming. A system like this already exists in the Pacific Ocean – but now we need one in the Indian Ocean.

Rather than detect tidal waves it would be better to predict when an earthquake was going to happen – then we could help to protect people against earthquakes and tidal waves.

Any early warning system would cost millions of dollars. Tidal waves don't happen that often. The money would be better spent on feeding people who are starving and buying medicines to treat people who are sick.

ACTIVITY

Set up a discussion between three people on a TV news show or a radio phone-in. Each person should take one of the viewpoints described above, and should argue for their particular point of view.
Try to find as much factual information as you can to back up your view:

- whether you think a tsunami early warning system is needed, or
- that money is better spent on food and health, or
- that the effort should be put into detecting earthquakes.

THE CARBON PROBLEM

Our lifestyles all affect the Earth – and in particular, the amount of carbon dioxide we produce from the fuels we burn and the electricity we use. Here are four different people – you might even know one or two of them . . .

It's sooo difficult just keeping up these days. The right clothes, makeup and hair – not to mention being seen at all the right places. Fortunately Mummy has lots of time to take me to where I have to be in her new car ... cool isn't it? Must dash – I hear they've got some new jeans in at that lovely little shop, only 30 miles drive away! Mummy ... !

It's alright round here I suppose – but not a lot to do except watch telly and play computer games. I'd like to go up to London to see the football at the weekend but I can't afford it and Dad says it's too far for him to drive. S'pose I'll just have to listen to it on the radio ...

It's amazing what cheap air tickets you can get now ... I'm just off to see a mate in California! Spent time in Spain, Morocco and Eastern Europe so far this year – and if I can get the money together, I might even manage to get to Australia. Brilliant gap year!

Haven't got much time to stop and talk – just on my way to the 'save the whale' rally. It's 20 miles away, but it won't take me long to get there on my trusty bike. Like my new jeans – cool label, but from the charity shop! Wanna share a tofu sandwich ... !

ACTIVITY

Look at each of these characters. Choose **one** of them and write a 'diary' for what they do and where they go in a typical week. When you have done this, think about the amount of fossil fuels this character uses, not just in travel but in everything they do and buy. Make a list of ways that they could reduce their fossil fuel consumption – and so reduce the amount of carbon dioxide they produce.

SUMMARY QUESTIONS

1 Write simple definitions for the following words describing the structure of the Earth:

a) mantle

b) core

c) lithosphere

d) tectonic plate.

2 Write down *three* pieces of evidence that suggest that South America and Africa were once joined together.

3 The pie charts show the atmosphere of a planet shortly after it was formed (A) and then millions of years later (B).

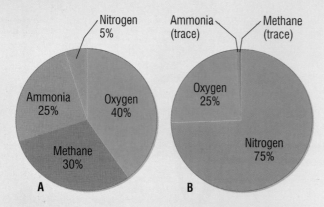

a) How did the atmosphere of the planet change?

b) What might have caused this change?

c) Copy and complete the word equations showing the chemical reactions that may have taken place in the atmosphere.

   i) methane + ...   carbon dioxide + ...

   ii) ammonia + ...   nitrogen + ...

4 Wegener suggested that all the Earth's continents were once joined in a single land mass.

a) Describe the evidence for this idea, and explain how the single land mass separated into the continents we see today.

b) Why were other scientists slow to accept Wegener's ideas?

5 The Earth and its atmosphere are constantly changing. Design a poster to show this, that would be suitable for displaying in a classroom with children aged 10–11 years.
Use diagrams and words to describe and explain ideas and to communicate them clearly to the children.

EXAM-STYLE QUESTIONS

1 Match words **A**, **B**, **C** and **D** with the numbers **1** to **4** in the table.

A atmosphere

B core

C crust

D mantle

	Description
1	Almost entirely solid, but can flow very slowly
2	Contains mainly the elements nitrogen and oxygen
3	Has an average thickness of about 6 km under oceans and 35 km under continents
4	Part liquid and part solid, with a radius of about 3 500 km

(4)

2 Match words **A**, **B**, **C** and **D** with the spaces **1** to **4** in the sentences.

A believed

B dismissed

C produced

D published

In 1912 Alfred Wegener ......**1**...... a theory that a single land mass had split apart into continents that moved to their current positions.

At the time geologists ......**2**...... that the continents moved up and down – not sideways.

Wegener's theory was ......**3**...... by geologists because he could not explain how the continents moved.

In 1944 an English geologist explained that heat from radioactivity ......**4**...... convection currents strong enough to move continents. (4)

3 Match words **A**, **B**, **C** and **D** with spaces **1** to **4** in the sentences.

A ammonia

B carbon dioxide

C noble gases

D oxygen

The Earth's early atmosphere consisted mainly of ......**1**...... with some nitrogen, water vapour, methane and ......**2**...... .

The Earth's atmosphere now contains 78% nitrogen, 21% ......**3**......, about 1% ......**4**....... and 0.04% carbon dioxide. (4)

4 This question is about three of the noble gases, helium, neon and argon.

(a) Why is helium used in balloons and airships rather than hydrogen? (2)

(b) Explain how argon allows you to use an electric light bulb for many hours. (2)

(c) Explain how neon is used for advertising. (2)

5 The graph shows the percentage of carbon dioxide in the atmosphere in recent years.

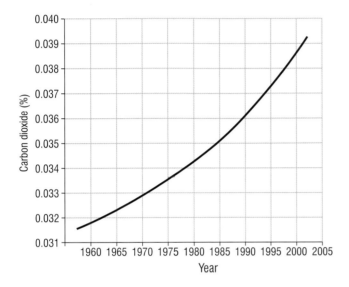

(a) (i) Could the 'percentage of carbon dioxide in the atmosphere' be described as a categoric, discrete or continuous variable? (1)

   (ii) There is considerable variation in the percentage of carbon dioxide within each 5 year period. What do we call the line that 'smooths out' these variations on the graph? (1)

(b) By how much has the percentage of carbon dioxide increased from 1960 to 2005? (1)

(c) What is this increase as a percentage of the 1960 figure? (1)

(d) Suggest **two** reasons for this increase. (2)

(e) What natural processes remove carbon dioxide from the atmosphere? (3)

(f) Why should we be concerned about the increase in carbon dioxide? (2)

Core samples have been taken of the ice from Antarctica. The deeper the sample the longer it has been there. It is possible to date the ice and to take air samples from it. The air was trapped when the ice was formed. It is possible therefore to test samples of air that have been trapped in the ice for many thousands of years.

This table shows some of these results. The more recent results are from actual air samples taken from a Pacific island.

Year	CO$_2$ concentration (ppm)	Source
2005	379	Pacific island
1995	360	Pacific island
1985	345	Pacific island
1975	331	Pacific island
1965	320	Antarctica
1955	313	Antarctica
1945	310	Antarctica
1935	309	Antarctica
1925	305	Antarctica
1915	301	Antarctica
1905	297	Antarctica
1895	294	Antarctica
1890	294	Antarctica

a) If you have access to a spreadsheet, enter this data and produce a line graph. (3)

b) Draw a line of best fit. (1)

c) What pattern can you detect? (2)

d) What conclusion can you make? (1)

e) Should the fact that the data came from two different sources affect your conclusion? Explain why. (2)

Science A

See pages 186–9

1   Match the polymers **A**, **B**, **C** and **D** with the numbers **1** to **4** in the table.

   A   high density poly(ethene)

   B   low density poly(ethene)

   C   poly(chloroethene)

   D   poly(propene)

	Density (g/cm³)	Monomer
1	0.90	C_3H_6
2	0.92	C_2H_4
3	0.96	C_2H_4
4	1.30	C_2H_3Cl

(4 marks)

See pages 212–13

2   Alfred Wegener first suggested his theory of continental drift in 1915. He suggested that the continents had once been a single large land mass that had split apart. He showed that fossils and rocks were similar on the parts of America and Africa that fitted together. He produced a lot of evidence to support his theory but other scientists did not accept his ideas for over 50 years.

   (a)   One of the main reasons why other scientists did not accept his ideas was . . .

      A   America and Africa are too far apart.

      B   Fossils are similar all over the Earth.

      C   He could not explain how the continents moved.

      D   He had no evidence that the continents were moving.

   (b)   Geologists did not like his ideas because . . .

      A   they could not understand his fossil evidence.

      B   they did not know enough about the rocks of Africa and America.

      C   they did not like German scientists.

      D   they would have to change all their own long established ideas.

   (c)   What new evidence was found in the 1960s?

      A   Land bridges had existed between the continents.

      B   The sea floor was spreading on either side of deep ocean ridges.

      C   The polar ice-caps were shown to be melting.

      D   The Earth had been shrinking since it had been formed.

   (d)   What new theory was developed using Wegener's ideas?

      A   convection currents

      B   plate tectonics

      C   radioactive decay

      D   seismic activity

(4 marks)

Science B

1   Read the information in the box and use it to help you to answer the questions.

See pages 190–1

> Smart polymers can be used to switch enzymes on and off. The polymers are described as 'smart' because they alter their properties when conditions such as light, temperature or acidity change. Tiny smart polymer chains are attached to enzymes next to the active sites. The chains extend or contract depending on the conditions and block or unblock the site. This switches the enzyme off or on. One application already in use is for drugs that need to remain inactive until they reach a particular place in the body.

(a)   What is a polymer?   *(2 marks)*

(b)   Why are the polymers in the article 'smart'?   *(2 marks)*

(c)   Suggest another application for smart polymers that is not mentioned in the article.   *(1 mark)*

2   Read the information in the box and use it to help you to answer the questions.

> Many plant oils are drying oils, which restricts their use as fuels. Drying happens when double bonds in the molecules are oxidised and form cross-links so the oil polymerises into a plastic-like solid. The process is accelerated at high temperatures and engines quickly become gummed up. Oils with high iodine values have more double bonds. An iodine value of less than 25 is required if the oil is to be used in unmodified diesel engines.

(a)   What is a drying oil?   *(2 marks)*

(b)   In what way are the polymers formed like thermosetting plastics?   *(1 mark)*

See page 189

(c)   Suggest why drying oils are used in oil paints.   *(2 marks)*

(d)   Describe how you could find the volume of iodine solution that reacted with a plant oil. Include how you would make your results as precise as possible.   *(4 marks)*

See pages 197, 209

(e)   Iodine values can be lowered by hydrogenation.

See page 199

(i)   How could this be done?   *(2 marks)*

(ii)   Give one disadvantage of hydrogenation.   *(1 mark)*

See page 207

GET IT RIGHT!

Do not worry if the questions are about something you have not met before or something you have not studied during the course. The information that you need to answer the questions will be given on the paper and the questions will be testing your understanding. Read the information carefully and be sure to use it in your answers.

P1a | Energy and energy resources

Figure 1 North sea oil – our reserves of fossil fuels are running out

Sunlight

Apples

Runner

Figure 2 Renewable energy?

What you already know

Here is a quick reminder of previous work that you will find useful in this unit:

- Fuels store energy from the Sun.
  Coal, oil and natural gas are examples of **fossil fuels**. These fuels were formed over millions of years from dead plants and marine animals.

- We generate most of our electricity in power stations that burn fossil fuels. The Earth's reserves of fossil fuels will run out sooner or later.

- We can use renewable energy sources, such as wind, waves and running water, to generate electricity. Renewable energy sources never run out because they do not burn fuel.

Heat energy is transferred from high to low temperatures. Heat transfer takes place in three different ways:

1. **Conduction** happens in solids, liquids and gases.
   - Metals are good conductors of heat.
   - Gases and most other non-metals are poor conductors.
   - Poor conductors, such as glass, are called thermal insulators.

2. **Convection** only happens in liquids and gases.
   - The flow of the liquid or gas due to convection is called a **convection current**.
   - Convection is more important than conduction in liquids and gases.

3. *Radiation* is energy carried by waves.
   - The Sun radiates energy into space.
   - The Earth absorbs only a tiny fraction of the energy radiated by the Sun.
   - Plants need sunlight for photosynthesis.

RECAP QUESTIONS

1. a) Which fuels listed below are fossil fuels?

   coal   natural gas   hay   oil   wood

   b) List two renewable sources of energy that do not need water.

2. a) Which one of the following renewable energy resources does not depend on energy from the Sun?

   **solar heating   tidal power
   wind power   wave power**

   b) Which of the above renewable energy resources is most reliable? Explain your answer.

3. Complete the sentences below.

   a) Heat transfer in a solid is due to ....
   b) Heat transfer in a liquid is due to ...... and ....
   c) Heat transfer through a vacuum is due to ....

4. a) Why is natural gas not a renewable source of energy?
   b) Why is natural gas not suitable as a source of energy for road vehicles?
   c) Give two reasons why natural gas is a very suitable fuel for cooking?

Making connections

Energy for everyone

Life is great if you can get energy at the flick of a switch. Sadly, there are many people in poor countries who can't get energy as easily as we can. Better fuel supplies and electricity would improve their lives. Yet people like us in rich countries are using up the world's reserves of fuel. Will there be enough energy for everyone in the future? This module will give you some of the answers but will also raise more questions. Can everyone live like we do?

I like to have a good time with my friends. We like the bright lights and the buzz when we go out to town.

We're trying to develop cheaper and better solar cells. We could use these to supply electricity to people in remote areas.

I can't do without my car. I hate these petrol queues every time there's a petrol shortage. I blame the government.

I'm tired and hungry. I've had to walk miles to gather this wood. I need to get home because my family need it for cooking our meals. I wish we had electricity.

ACTIVITY

Working as a small group, put together a two-minute radio slot to make people think about the issues raised above. Choose someone to be the host of the programme and someone for each person in the pictures above. Give everyone the same amount of time to speak.

We need lots of renewable energy resources like wind to reduce global warming.

Chapters in this unit

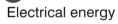

Heat transfer   Using energy   Electrical energy   Generating electricity

P1a 1.1

Thermal radiation

LEARNING OBJECTIVES

1 What is thermal radiation?
2 Do all objects give off thermal radiation?
3 How does it depend on the temperature of an object?

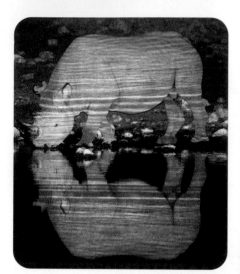

Figure 1 Keeping watch in darkness

Seeing in the dark

We can use special TV cameras to 'see' animals and people in the dark. These cameras detect thermal radiation. Every object gives out (emits) thermal radiation. The hotter an object is, the more thermal radiation it emits.

Look at the photo in Figure 1. The rhino is hotter than the ground.

a) Why is the water darker than the rhino?
b) Why is there an image of the rhino?

PRACTICAL

Detecting thermal radiation

You can use a thermometer with a blackened bulb to detect thermal radiation. Figure 2 shows how to do this.

Figure 2 Detecting infra-red radiation

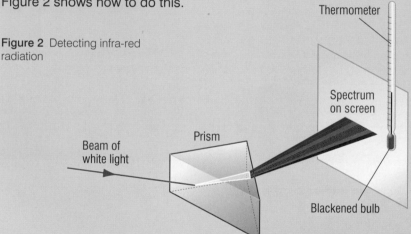

- The glass prism splits a narrow beam of light into the colours of the spectrum.
- The thermometer reading rises when it is placed just beyond the red part of the spectrum. Some of the thermal radiation in the beam goes there. Our eyes cannot detect it but the thermometer can.
- Thermal radiation is also called **infra-red radiation**. This is because it is beyond the red part of the visible spectrum.

● What would happen to the thermometer reading if the thermometer were moved away?

The electromagnetic spectrum

Radio waves, microwaves, infra-red radiation and visible light are part of the electromagnetic spectrum. So too are ultraviolet rays and X-rays. Electromagnetic waves are electric and magnetic waves that travel through space. You will learn more about electromagnetic waves on page 282.

Energy from the Sun

The Sun emits radiation in all parts of the electromagnetic spectrum. Fortunately for us, the Earth's atmosphere blocks most of the radiation, such as ultraviolet rays, that would harm us. But it doesn't block thermal radiation from the Sun.

Figure 3 shows a solar furnace. This is a giant reflector that focuses sunlight.

The temperature at the focus can reach thousands of degrees. That's almost as hot as the surface of the Sun, which is at 5500°C.

The greenhouse effect

The Earth's atmosphere acts like a greenhouse made of glass. In a greenhouse, shorter wavelength radiation from the Sun can pass through the glass. However, longer wavelength thermal radiation is trapped inside by the glass. So the greenhouse stays warm.

Gases in the atmosphere, such as water vapour, methane and carbon dioxide, act like the glass. They trap the thermal radiation given off from the Earth. These gases make the Earth warmer than it would be if it had no atmosphere.

But the Earth is becoming too warm. If the polar ice caps melt, it will cause sea levels to rise. Cutting back our use of fossil fuels will help to reduce 'greenhouse gases'.

Figure 3 A solar furnace in the Eastern Pyrenees, France

FOUL FACTS

If we don't combat global warming, sea levels will rise. London's sewage system wouldn't be able to cope with this – there would be rats, disease and nasty smells everywhere!

DID YOU KNOW?

Solar heating panels heat water by absorbing solar energy directly. Solar cells turn energy from the Sun directly into electricity. Lots of solar panels on roof tops would reduce our use of fossil fuels. (See pages 270 and 271 for more about solar panels.)

● Why do solar panels in Britain always face south?

SUMMARY QUESTIONS

1 Complete the table to show if the object emits infra-red radiation or light or both.

Object	Infra-red	Light
A hot iron		
A light bulb		
A TV screen		
The Sun		

2 How can you tell if an electric iron is hot without touching it?

3 a) Explain why penguins huddle together to keep warm.
  b) Design an investigation to model the effect of penguins huddling together. You could use beakers of hot water to represent the penguins.

KEY POINTS

1 Thermal radiation is energy transfer by electromagnetic waves.
2 All objects emit thermal radiation.
3 The hotter an object is, the more thermal radiation it emits.

P1a 1.2

Surfaces and radiation

LEARNING OBJECTIVES

1 Which surfaces are the best emitters of radiation?
2 Which surfaces are the best absorbers?

Figure 1 A thermal blanket in use

Which surfaces are the best emitters of radiation?

Rescue teams use thermal blankets to keep accident survivors warm. A thermal blanket has a light, shiny outer surface. This emits much less radiation than a dark, matt surface. The colour and smoothness of a surface affects how much radiation it emits.

Dark, matt surfaces emit more radiation than light, shiny surfaces.

PRACTICAL

Testing different surfaces

To compare the radiation from two different surfaces, you can measure how fast two beakers (or cans) of hot water cool. One beaker needs to be wrapped with shiny metal foil and the other with matt black paper. Figure 2 shows the idea. At the start, the volume and temperature of the water in each beaker need to be the same.

- Why should the volume and temperature of the water be the same at the start?
- Which one will cool faster?

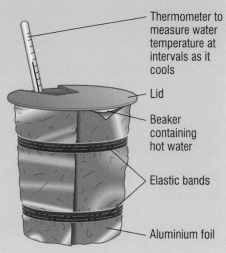

Thermometer to measure water temperature at intervals as it cools

Lid

Beaker containing hot water

Elastic bands

Aluminium foil

Figure 2 Testing different surfaces

Which surfaces are the best absorbers of radiation?

When you use a photocopier, why are the copies warm? This is because thermal radiation from a lamp dries the ink on the paper. Otherwise, the copies will be smudged. The black ink absorbs thermal radiation more easily than the white paper.

- A dark surface absorbs radiation better than a light surface.
- A matt surface absorbs radiation better than a shiny surface because it has lots of cavities. Look at Figure 3. It shows why these cavities trap and absorb the radiation.

Dark, matt surfaces absorb radiation better than light, shiny surfaces.

a) Why does ice on a road melt faster in sunshine if sand is sprinkled on it?
b) Why are solar panels painted matt black?

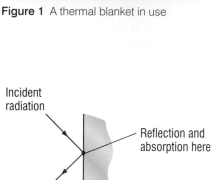

Incident radiation

Reflection and absorption here

Smooth surface

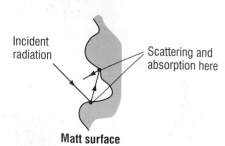

Incident radiation

Scattering and absorption here

Matt surface

Figure 3 Absorbing infra-red radiation

PRACTICAL

Absorption tests

To compare absorption by different surfaces, you can use identical beakers with cold water in. The beakers need to be coated with paint of different colours and different textures.

Figure 4 Testing absorbers

- The volume and the starting (initial) temperature of the water in each beaker must be the same.
- Place the beakers in a sunlit room in the sunlight.
- Use a thermometer to see which beaker warms up fastest.

● Why is it important to use the same volume of water in each beaker?
● Which beaker in Figure 4 do you think would warm up fastest? Give a reason for your answer.

SUMMARY QUESTIONS

1 Explain the following:

a) Houses in hot countries are usually painted white.
b) Solar heating panels are painted black.

2 A metal cube filled with hot water was used to compare the heat radiated from its four vertical faces, A, B, C and D.

An infra-red sensor was placed opposite each face at the same distance, as shown in Figure 5. The sensors were connected to a computer. The results of the test are shown in the graph below.

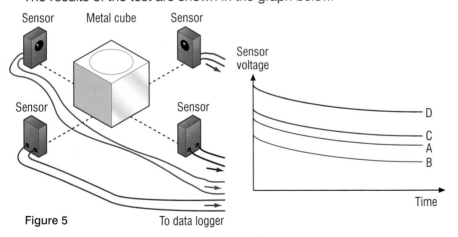

Figure 5

a) Why was it important for the distance from each sensor to the face to be the same?
b) One face was silvery and shiny, one was silvery and matt, one was black and shiny and one was matt black.
   Which face A, B, C or D was i) silvery and shiny, ii) matt and black?
c) Which face radiated i) most heat, ii) least heat?
d) What are the advantages of using data logging equipment to collect the data in this investigation?

3 Explain any advantages and disadvantages of wearing black clothing in a hot country.

KEY POINTS

1 Dark matt surfaces are better emitters of thermal radiation than light shiny surfaces.
2 Dark matt surfaces are better absorbers of thermal radiation than light shiny surfaces.

P1a 1.3

Conduction

LEARNING OBJECTIVES

1 What materials make the best conductors and insulators?
2 Why are metals good conductors?
3 Why are non-metals poor conductors?

Figure 1 At a barbecue – the steel cooking utensils have wooden or plastic handles

Conductors and insulators

When you have a barbecue, you need to know which materials are good thermal conductors and which are good insulators. If you can't remember, you are likely to burn your fingers!

Testing rods of different materials as conductors

The rods need to be the same width and length for a fair test. Each rod is coated with a thin layer of wax near one end. The uncoated ends are then heated together.

Look at Figure 2.

The wax melts fastest on the rod that conducts heat best.

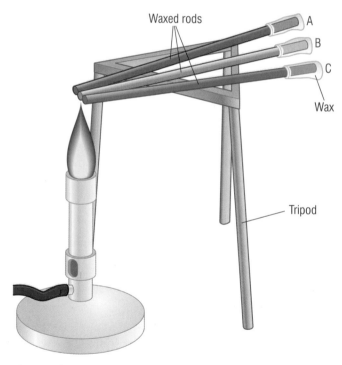

Figure 2 Comparing conductors

- Metals conduct heat better than non-metals.
- Copper is a better conductor of heat than steel.
- Wood conducts heat better than glass.

a) Why do steel pans have handles made of plastic or wood?
b) Name the independent and the dependent variables investigated in Figure 2.
(See page 7.)

PRACTICAL

Testing sheets of materials as insulators

Use different materials to insulate identical cans (or beakers) of hot water. The volume of water and its temperature at the start should be the same. Use a thermometer to measure the water temperature after the same time. The results should tell you which insulator was best.

The table gives the results of comparing two different materials using the method above.

c) Which material, felt or paper, was the best thermal insulator?

d) Which variable shown in the table was controlled to make this a fair test?

Material	Starting temperature (°C)	Temperature after 300 s (°C)
paper	40	32
felt	40	36

Conduction in metals

Metals contain lots of **free electrons**. These electrons move about at random inside the metal and hold the positive ions together. They collide with each other and with the positive ions. (Ions are charged particles.)

- Ion
- Electron
- Atom

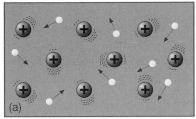 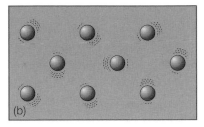

Figure 4 Energy transfer a) in a metal, b) in a non-metal

When a metal rod is heated at one end, the free electrons at the hot end gain kinetic energy and move faster.

- These electrons **diffuse** (i.e. spread out) and collide with other free electrons and ions in the cooler parts of the metal.
- As a result, they transfer kinetic energy to these electrons and ions.

So energy is transferred from the hot end of the rod to the colder end.

In a non-metallic solid, all the electrons are held in the atoms. Energy transfer only takes place because the atoms vibrate and shake each other. This is much less effective than energy transfer by free electrons. This is why metals are much better conductors than non-metals.

SUMMARY QUESTIONS

1 Choose the best insulator or conductor from the list for a), b) and c).

        fibreglass   plastic   steel   wood

  a) …… is used to insulate a house loft.

  b) The handle of a frying pan is made of …… or …… .

  c) A radiator in a central heating system is made from …… .

2 a) Choose a material you would use to line a pair of winter boots? Explain your choice of material.

  b) How could you carry out a test on 3 different lining materials?

3 Explain why metals are good conductors of heat.

DID YOU KNOW?

Materials like wool and fibreglass are good thermal insulators. This is because they contain air trapped between the fibres. Trapped air is a good insulator. We use insulators like fibreglass for loft insulation and for lagging water pipes.

Figure 3 Insulating a loft. The air trapped between fibres make fibreglass a good thermal insulator.

KEY POINTS

1 Conduction in a metal is due mainly to free electrons transferring energy inside the metal.

2 Non-metals are poor conductors because they do not contain free electrons.

3 Materials such as fibreglass are good insulators because they contain pockets of trapped air.

P1a 1.4

Convection

Figure 1 A natural glider – birds use convection currents to soar high above the ground

Gliders and birds know how to use convection to stay in the air. Convection currents of warm air can keep them high above the ground for hours.

Convection happens whenever we heat **fluids**. A fluid is a gas or a liquid.

Look at the diagram in Figure 2. It shows a simple demonstration of convection.

The hot gases from the burning candle go straight up the chimney above the candle. Cold air is drawn down the other chimney to replace the air leaving the box.

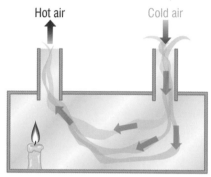

Figure 2 Convection

Using convection

(1) Hot water at home

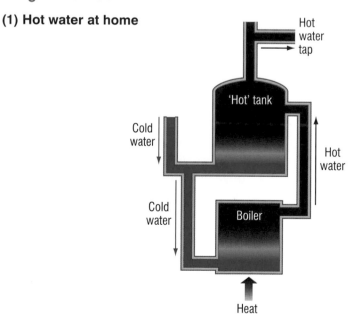

Figure 3 Hot water at home

Many homes have a hot water tank. Hot water from the boiler rises and flows into the tank where it rises to the top. Figure 3 shows the system. When you use a hot water tap at home, you draw off hot water from the top of the tank.

a) What would happen if we connected the hot taps to the bottom of the tank?

(2) Sea breezes

Sea breezes keep you cool at the seaside. On a sunny day, the ground heats up faster than the sea. So the air above the ground warms up and rises. Cooler air from the sea flows in as a 'sea breeze' to take the place of the warm air.

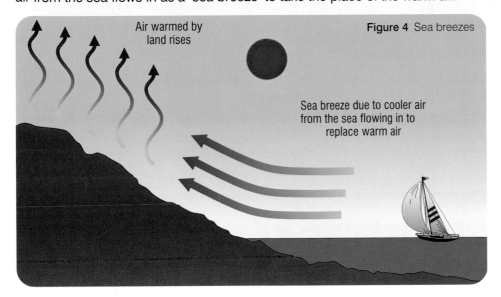

Air warmed by land rises

Figure 4 Sea breezes

Sea breeze due to cooler air from the sea flowing in to replace warm air

How convection works

Convection takes place:

- only in liquids and gases (i.e. fluids)
- due to circulation (convection) currents within the fluid.

The circulation currents are caused because fluids rise where they are heated (as heating makes them less dense). Then they fall where they cool down (as cooling makes them more dense). Convection currents transfer thermal energy from the hotter parts to the cooler parts.

So why do fluids rise when heated?

Most fluids expand when heated. This is because the particles move about more, taking up more space. Therefore the density decreases because the same mass of fluid occupies a bigger volume. So heating part of a fluid makes it less dense and it therefore rises.

GET IT RIGHT!

When you explain convection, remember it is the hot fluid that rises, NOT 'heat'.

SUMMARY QUESTIONS

1 Use each word from the list *once* only to complete the following sentences.

cools   falls   mixes   rises

When a fluid is heated, it ...... and ...... with the rest of the fluid. The fluid circulates and ...... then it ....

2 Figure 5 shows a convector heater. It has an electric heating element inside and a metal grille on top.

a) What does the heater do to the air inside the heater?

b) Why is there a metal grille on top of the heater?

c) Where does air flow into the heater?

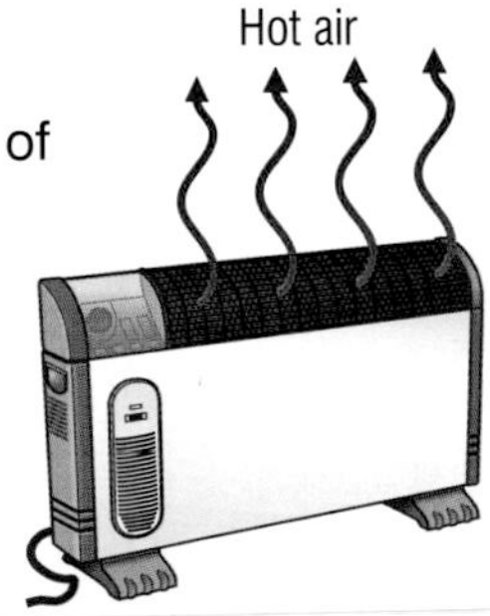

Hot air

Figure 5 A convector heater

3 Describe how you could demonstrate convection currents in water using a strongly coloured crystal. Explain in detail what you would see.

KEY POINTS

1 Convection takes place only in liquids and gases (fluids).

2 Heating a liquid or a gas makes it less dense.

3 Convection is due to a hot liquid or gas rising.

P1a 1.5

Heat transfer by design

LEARNING OBJECTIVES

1 What factors affect the rate of heat transfer from a hot object?
2 What can we do to keep things hot?
3 What can we do to cut down heat losses from a house?

Figure 1 A car radiator helps to transfer heat from the engine

Cooling by design

Lots of things can go wrong if we don't control heat transfer. For example, a car engine that overheats can go up in flames.

The cooling system of a car engine transfers thermal energy from the engine to a radiator. The radiator is flat so it has a large surface area. This increases heat loss through convection in the air and through radiation.

Most cars also have a cooling fan that switches on when the engine is too hot. This increases the flow of air over the surface of the radiator.

a) Why are car radiators painted black?
b) What happens to the rate of heat transfer when the cooling fan switches on?

The vacuum flask

If you are outdoors in cold weather, a hot drink from a vacuum flask keeps you warm. In the summer the same vacuum flask is great for keeping your drinks cold.

The liquid is in the double-walled glass container.

● The vacuum between the two walls of the container cuts out heat transfer by conduction and convection between the walls.
● Glass is a poor conductor so there is little heat conduction through the glass.
● The glass surfaces are silvery to reduce radiation from the outer wall.

c) List the other parts of the flask that are good insulators. What would happen if they weren't good insulators?

PRACTICAL

Investigating heat transfer

Carry out investigations to find out:

– how the starting temperature of a hot object affects its rate of cooling
– how the size of a hot object affects how quickly it cools.

● Name the independent variable in each investigation. (See page 7.)
● What are your conclusions in each case?

Plastic cap

Double-walled plastic container

Plastic protective cover

Hot or cold liquid

Sponge pad (for protection)

Inside surfaces silvered to stop radiation

Vacuum prevents conduction and convection

Plastic spring for support

Figure 2 A vacuum flask

Reducing heat losses at home

Home heating bills can be expensive. Figure 4 shows how we can reduce heat losses at home and cut our home heating bills.

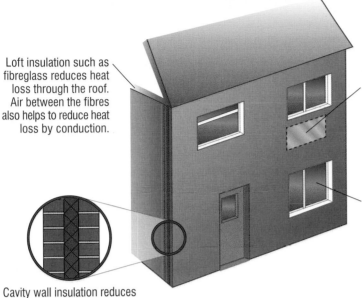

Loft insulation such as fibreglass reduces heat loss through the roof. Air between the fibres also helps to reduce heat loss by conduction.

Aluminium foil between a radiator panel and the wall reflects heat radiation away from the wall.

A double glazed window has two glass panes with dry air or a vacuum between the panes. Dry air is a good insulator so it cuts down heat conduction. A vacuum cuts out heat transfer by convection as well.

Cavity wall insulation reduces heat loss through the walls. We place insulation between the two layers of brick that make up the walls of a house.

Figure 4 Saving money

NEXT TIME YOU...

... see a little dog with a coat on, don't laugh! Small animals lose heat at a faster rate than large animals. This is because they have a larger surface area in relation to their volume than large animals.

Figure 3 A hot dog!

d) Why is cavity wall insulation better than air in the cavity between the walls of a house?

SUMMARY QUESTIONS

1 Hot water is pumped through a radiator like the one in Figure 5.

Complete the sentences below about the radiator.

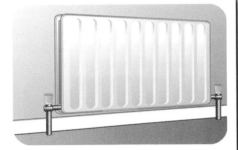

Figure 5 A central heating radiator

a) Heat transfer through the walls of the radiator is due to ....
b) Hot air in contact with the radiator causes heat transfer to the room by ....
c) Heat transfer to the room takes place directly due to ....

2 Some double-glazed windows have a plastic frame and a vacuum between the panes.

a) Why is a plastic frame better than a metal frame?
b) Why is a vacuum between the panes better than air?
c) Design a test to show that double glazing is more effective at preventing heat loss than single glazing.

3 Describe, in detail, how the design of a vacuum flask reduces the rate of thermal energy transfer.

KEY POINTS

1 A radiator has a large surface area so it can lose heat easily.
2 Small objects lose heat more easily than large objects.
3 Heat loss from a building can be reduced using:
   - aluminium foil behind radiators.
   - cavity wall insulation.
   - double glazing.
   - loft insulation.

P1a 1.6   Hot issues

Pay back time!

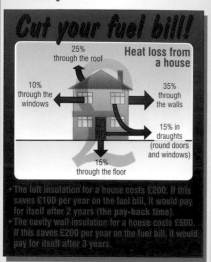

Cut your fuel bill!

Heat loss from a house

25% through the roof

10% through the windows

35% through the walls

15% in draughts (round doors and windows)

15% through the floor

• The loft insulation for a house costs £200. If this saves £100 per year on the fuel bill, it would pay for itself after 2 years (the pay-back time).
• The cavity wall insulation for a house costs £600. If this saves £200 per year on the fuel bill, it would pay for itself after 3 years.

Figure 1 Heat loss from a house

ACTIVITIES

a) i) Figure 1 shows the heat losses from a house. Choose the two best ways to cut heat losses from this house.

 ii) A double-glazed window costs £200. It saves £10 per year on the fuel bill. How long is the pay-back time?

b) Produce a poster to encourage people to save energy in the home.

c) What proportion of the cost of energy-saving measures should be paid by householders and by the government. Should grants be made available? Who should receive them? Discuss these issues in a small group and come up with a few suggestions for MPs to consider.

Thermal imaging

We use thermal scanners in hospitals to spot tumours just below the skin. Tumours are warmer than the normal tissue. So they warm the nearby skin and cause it to radiate more energy.

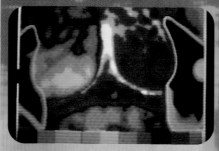

Figure 2 A thermal scan

QUESTIONS

1 Why are thermal scanners unable to detect tumours deep inside the body?

2 Do you think 'high-tech' devices are a good thing in hospitals? Why?

3 Why might some hospitals have more of the latest technology than others? What costs are involved in introducing new medical equipment to hospitals?

4 Use these questions to write a two-minute script for a slot for a weekly TV programme on medical issues.

Colourful chemicals

Imagine if objects changed colour when their temperature changed. A room could be a cool pale blue in summer and a warm shade of red in winter.

Engineers have developed **heat-smart paint** to show if a component gets too hot. The paint changes colour if its temperature goes above a certain level and stays that colour when it cools down. The change is not reversible. If it were, you couldn't tell later if the painted object got too hot.

We use **temperature strips** in hospitals, in industry and in food packaging. Reversible strips change colour and show the corresponding temperature. Non-reversible strips can show the highest or the lowest temperature.

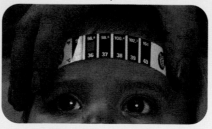

What about reversible heat-smart paints? Scientists in China have invented a reversible heat-smart paint that changes colour and helps to keep buildings warmer in winter and cooler in summer.

Figure 3 Temperature strips

QUESTIONS

5 What would be the best colour for the outside of a building
  a) in summer,   b) in winter?

6 What type of temperature strip, reversible or non-reversible, would you use:
  a) in a freezer at home,
  b) on an oven door to tell you if the oven is too hot?

Who am I?
A life-saving invention

I was already a famous scientist when I invented my lamp in 1813. My invention saved the lives of countless coal miners. Before my lamp, coal miners used candles to see underground. The candle flames often caused fatal accidents. They ignited pockets of methane gas, causing explosions in the mine.

As professor of chemistry at the Royal Institution in London, I made important discoveries about gases. I investigated flames and discovered that wire gauze could stop a flame spreading. The wire conducted heat away from the flame and lowered its temperature so the flame couldn't go through it.

QUESTIONS

7 a) What would happen if the gauze had a hole in it?

b) Why was it important to inspect a lamp before it was taken down a mine?

c) What replaced this lamp?

8 Do some research to find out more about the life of the above scientist. Present the main events as a series of bullet-pointed statements. Colour code those connected to his personal life and those connected to his work in science.

Sir Humphry Davy

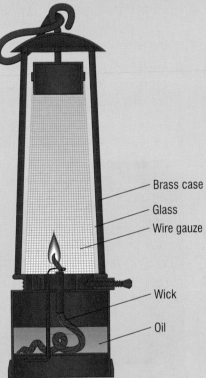

Figure 4 My lamp

- Brass case
- Glass
- Wire gauze
- Wick
- Oil

Science and safety

TRAGEDY ON CAMPUS!

Two students were found dead in their flat yesterday. Police said they had been overcome by fumes from a poorly ventilated gas fire. A spokesman for the gas supplier warned anyone with a gas fire at home to make sure that the room is well-ventilated and that the gas fire is serviced regularly.

Gas heaters and boilers need good ventilation. Figure 5 shows how a gas heater works. Hot air and gases produced by the flames rise and escape to the outside through a ventilation duct. Colder air is drawn in at the bottom.

If the ventilation is poor, carbon monoxide gas is produced. As this is lethal to inhale, people can die as a result of poor ventilation. Gas heaters and boilers need servicing regularly.

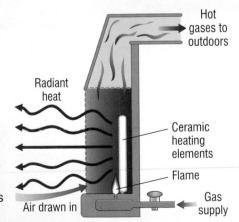

Hot gases to outdoors

Radiant heat

Ceramic heating elements

Flame

Gas supply

Air drawn in

Figure 5 A gas heater

ACTIVITY

Produce a leaflet to make people aware of the dangers of poorly-ventilated gas heaters and boilers. Make sure you include scientific information, explained so that most people will understand.

SUMMARY QUESTIONS

1 a) Why does a matt surface in sunshine get hotter than a shiny surface?

b) What type of surface is better for a flat roof – a matt dark surface or a smooth shiny surface? Explain your answer.

c) A solar heating panel is used to heat water. Why is its top surface painted matt black?

d) Why is a car radiator painted matt black?

2 Choose the correct word from the list for each of the four spaces in the sentences below.

collide   electrons   ions   vibrate

a) Heat transfer in a metal is due to particles called …… moving about freely inside the metal. They transfer energy when they …… with each other.

b) Heat transfer in a non-metallic solid is due to particles called …… inside the non-metal. They transfer energy because they …… .

3 A heat sink is a metal plate or clip fixed to an electronic component to stop it overheating.

a) When the component becomes hot, how does heat transfer from where it is in contact with the plate to the rest of the plate?

b) Why does the plate have a large surface area?

4 Complete each of the sentences using words from the list below.

conduction   convection   radiation

a) …… cannot happen in a solid or through a vacuum.

b) Heat transfer from the Sun is due to …… .

c) When a metal rod is heated at one end, heat transfer due to …… takes place in the rod.

d) …… is energy transfer by electromagnetic waves.

EXAM-STYLE QUESTIONS

1 The transfer of thermal energy through space from the Sun to the Earth is by

A Conduction

B Convection

C Condensation

D Radiation   (1)

2 Infra-red radiation is absorbed, reflected and emitted to different extents by different surfaces.
Which of the following statements best applies to **dark, matt** surfaces?

A They are good emitters of infra-red radiation.

B They are good reflectors of infra-red radiation.

C They are poor absorbers of infra-red radiation.

D They are poor emitters of infra-red radiation.   (1)

3 The diagram shows water being heated in a saucepan on a hotplate.
Match words from the list with the numbers **1** to **4** in the sentences.

A conduction

B convection

C radiation

D insulator

Heat energy is transferred through the base of the saucepan by ……**1**……

Heat energy is transferred through the water in the saucepan by ……**2**……

Some energy is transferred from the hotplate to the air by ……**3**……

The handle of the saucepan is made from wood because wood is a good ……**4**……. (4)

4 The diagram shows some ways of reducing energy loss from a house.

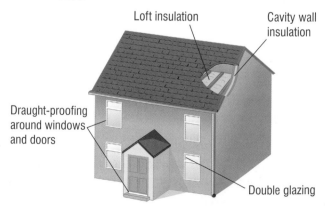

The table shows the cost of fitting, and the annual savings on energy bills, for different methods of reducing energy loss from a house.

Method of reducing energy loss	Cost of fitting	Annual saving
Cavity wall insulation	£800	£100
Double glazing	£4000	£80
Draught proofing	£60	£60
Loft insulation	£300	£60

(a) Which method pays for itself in the shortest time?

    A cavity wall insulation   **B** double glazing

    C draught proofing   **D** loft insulation   (1)

(b) Which method reduces energy loss by the greatest amount?

    A cavity wall insulation   **B** double glazing

    C draught proofing   **D** loft insulation   (1)

(c) The time it takes for the saving on energy bills to equal the cost of installing the insulation is called the pay-back time. What is the pay-back time for loft insulation?

    A 1 year   **B** 2 years

    C 3 years   **D** 5 years   (1)

(d) Fitting double glazing reduces energy loss by . . .

    A conduction   **B** convection

    C evaporation   **D** radiation   (1)

5   A cook at a barbecue has baked some potatoes. He takes them off the barbecue.

(a) The hot potatoes radiate thermal energy. The amount of thermal energy radiated by a potato depends upon the nature of its surface. Name two other variables that affect the amount of thermal energy radiated by a potato.   (2)

(b) The cook wraps the hot potatoes in clean, shiny foil. Explain why.   (2)

(c) Explain what the difference would be if the cook wrapped the potatoes in black, dull foil.   (2)

(d) Suggest why:

    (i) the outside of the potatoes cooks before the inside.   (2)

    (ii) some cooks put metal skewers through the potatoes before they put them on the barbecue.   (2)

HOW SCIENCE WORKS QUESTIONS

Tamsin was asked to investigate how quickly different metals conducted heat. She was given four metal rods. They were made of brass, steel, iron and copper. She was told to stick pins to the metal rods, using a jelly that easily melted. She thought carefully about how to do the investigation. She decided to stick the pins to the metal rods and then heat the rods with a Bunsen burner. She would time how long it took for the pins to fall off.

Tamsin is going to need some help getting reliable and valid results.

a) How should she arrange these pins on the metal rods? Explain your ideas.   (2)

b) How should the pins be attached to the metal rods?   (1)

c) How should she heat the metal rods? She only has one Bunsen burner.   (1)

d) How could she make the results more accurate? Explain your ideas.   (2)

e) Tamsin will need a table to record her results. Produce a table for Tamsin.   (3)

f) What type of graph should Tamsin use to present her results?   (1)

g) Once Tamsin has had a go at the investigation, she suggests that you try it to see if you get the same results. How might this show the reliability of Tamsin's results?   (1)

h) In another lesson Tamsin investigated which of two materials was the better thermal insulator. She wrapped a copper beaker in each material, A and B, then filled the beaker with hot water.
Here are her results:

Material	Temp. of water at start (°C)	Temp. of water after 10 minutes (°C)
A	91.8	72.3
B	79.3	70.8

She concluded that Material B is a better thermal insulator than Material A.
Comment on the validity of her conclusion drawn from this data.   (2)

P1a 2.1

Forms of energy

1 What forms of energy are there?
2 How can we describe energy changes?

On the move

Cars, buses, planes and ships all use energy from fuel. They carry their own fuel. Electric trains use energy from fuel in power stations. Electricity transfers energy from the power station to the train.

Figure 1 The French TGV (Train à grande vitesse) electric train can reach speeds of more than 500 km/hour

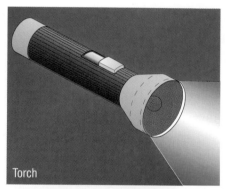
Torch

Skier

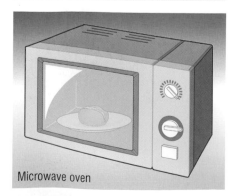
Microwave oven

Figure 2 Energy changes

We describe energy stored or transferred in different ways as *forms of energy*.

Here are some examples of forms of energy:

● *Chemical energy* is energy stored in fuel (including food). This energy is released when chemical reactions take place.
● *Kinetic energy* is the energy of a moving object.
● *Gravitational potential energy* is the energy of an object due to its position.
● *Elastic (or strain) energy* is the energy stored in a springy object when we stretch or squash it.
● *Electrical energy* is energy transferred by an electric current.
● *Thermal (heat) energy* of an object is energy due to its temperature. This is partly because of the random kinetic energy of the particles of the object.

a) What form of energy is supplied to the train in Figure 1?
b) What does TGV mean?

We say that energy is *transformed* when it changes from one form into another.

242

In the torch in Figure 2, the torch's battery pushes a current through the bulb. This makes the torch bulb emit light and it also gets hot. We can show the energy changes using a flow diagram.

Look at the example below:

chemical energy in the battery → electrical energy → light energy + thermal energy

c) What happens to the thermal energy of the torch bulb?

PRACTICAL

Energy changes

When an object starts to fall freely, it gains kinetic energy because it speeds up as it falls. So its gravitational potential energy changes to kinetic energy as it falls.

Look at Figure 3. It shows a box that hits the floor with a thud. All of its kinetic energy changes to heat and sound energy at the point of impact. The proportion of kinetic energy transformed to sound is much smaller than that changed to heat.

- Draw an energy flow diagram to show the changes in Figure 3.

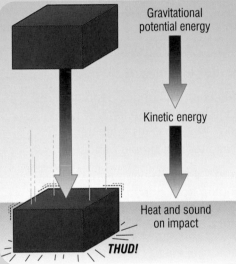

Gravitational potential energy

Kinetic energy

Heat and sound on impact

THUD!

Figure 3 An energetic drop

SUMMARY QUESTIONS

1 Copy and complete a) and b) using the words below:

  electric   kinetic   gravitational potential   thermal

  a) When a ball falls in air, it loses …… energy and gains …… energy.
  b) When an electric heater is switched on, it changes …… energy into …… energy.

2 a) List two different objects you could use to light a room if you have a power cut. For each object, describe the energy changes that happen when it produces light.
  b) Which of the two objects in a) is:
     i) easier to obtain energy from,
     ii) easier to use?

3 Read the 'Science @ Work' box at the top of this page.
  Explain the energy changes involved in using a pile driver.

d) Where does the energy supplied to the hammer come from?

KEY POINTS

1 Energy exists in different forms.
2 Energy can change (transform) from one form into another form.

P1a 2.2

Conservation of energy

LEARNING OBJECTIVES

1 What energy changes happen on a roller coaster ride?
2 What do we mean by 'conservation of energy'?
3 Why is conservation of energy a very important idea?

Figure 1 On a roller coaster – having fun with energy transformations!

At the fun-fair

Fun-fairs are very exciting places because lots of energy changes happen quickly. A roller coaster gains gravitational potential energy when it climbs. Then it loses gravitational potential energy when it races downwards.

As it descends:

its gravitational potential energy $\rightarrow$ kinetic energy + sound + thermal energy due to air resistance and friction

a) When a roller coaster gets to the bottom of a descent, what energy transformations happen if:
   i) we apply the brakes to stop it?
   ii) it goes up and over a second 'hill'?

PRACTICAL

Investigating energy changes

When energy changes happen, does the total amount of energy stay the same? We can investigate this question with a simple pendulum.

Figure 2 shows a pendulum bob swinging from side to side.

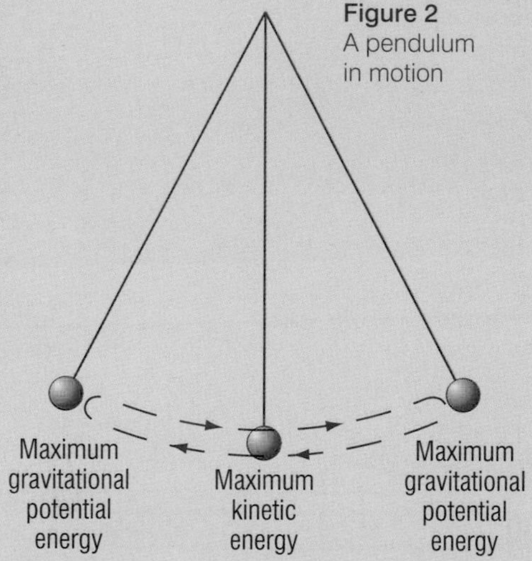

Figure 2
A pendulum in motion

- As it moves towards the middle, its gravitational potential energy changes to kinetic energy.
- As it moves away from the middle, its kinetic energy changes back to gravitational potential energy. You should find that the bob reaches the same height on each side.
- What does this tell you about the energy of the bob at its maximum height on each side?
- Why is it difficult to mark the exact height the pendulum bob rises to? How could you make your judgement more accurate?

Maximum gravitational potential energy   Maximum kinetic energy   Maximum gravitational potential energy

Conservation of energy

Scientists have done lots of tests to find out if the total energy after a change is the same as the energy before the change. All the tests so far show it is the same.

This important result is known as the **conservation of energy**.

It tells us that **energy cannot be created or destroyed.**

Bungee jumping

What energy changes happen to a bungee jumper after jumping off the platform?

- Some of the gravitational potential energy of the bungee jumper changes to kinetic energy as the jumper falls with the rope slack.
- Once the slack in the rope has been used up, the rope slows the bungee jumper's fall. Most of the gravitational potential energy and kinetic energy of the jumper is changed into elastic (strain) energy.
- After reaching the bottom, the rope pulls the jumper back up. As the jumper rises, most of the elastic (strain) energy of the rope changes back to gravitational potential energy and kinetic energy of the jumper.

The bungee jumper doesn't return to the same height as at the start. This is because some of the initial gravitational potential energy has been changed to heat energy as the rope stretched then shortened again.

Figure 3 Bungee jumping

b) What happens to the gravitational potential energy lost by the bungee jumper?

c) Draw a flow diagram to show the energy changes.

PRACTICAL

Bungee jumping

You can try out the ideas about bungee jumping using the experiment shown in Figure 4.

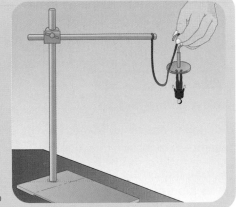

Figure 4 Testing a bungee jump

SUMMARY QUESTIONS

1 a) Complete the sentences below using the words below (one option is used twice):

   electrical   gravitational potential   thermal

   A person going up in a lift gains …… energy. The lift is driven by electric motors. Some of the …… energy supplied to the motors is changed to …… energy instead of …… energy.

2 a) A ball dropped onto a trampoline returns to the same height after it bounces. Describe the energy change of the ball from the point of release to the top of its bounce.

   b) What can you say about the energy of the ball at the point of release compared with at the top of its bounce?

   c) You could use the test above to see which of three trampolines was the bounciest.

      i) Name the independent variable in this test. (See page 7.)

      ii) Is this variable categoric, discrete or continuous? (See page 4.)

3 One exciting fairground ride acts like a giant catapult. The capsule, in which you are strapped, is fired high into the sky by rubber straps. Explain the energy changes taking place in the ride.

KEY POINTS

1 Energy can be transformed from one form to another or transferred from one place to another.

2 Energy cannot be created or destroyed.

P1a 2.3

Useful energy

LEARNING OBJECTIVES

1 What do we mean by 'useful' energy?
2 What causes some energy to be 'wasted'?
3 What eventually happens to wasted energy?

Energy for a purpose

Where would we be without machines? We use washing machines at home. We use machines in factories to make the goods we buy. We use them in the gym to keep fit and we use them to get us from place to place.

a) What happens to all the energy you use in a gym?

A machine transfers energy for a purpose. Friction between the moving parts of a machine causes the parts to warm up. So not all of the energy supplied to a machine is usefully transferred. Some energy is wasted.

- **Useful energy** is energy transferred to where it is wanted in the form it is wanted.
- **Wasted energy** is energy that is not usefully transferred or transformed.

b) What happens to the kinetic energy of a machine when it stops?

PRACTICAL

Investigating friction

Friction in machines always causes energy to be wasted. Figure 2 shows two examples of friction in action. Try one of them out.

- In **A**, friction acts between the drill bit and the wood. The bit becomes hot as it bores into the wood. Some of the electrical energy supplied to the bit changes into thermal energy of the drill bit (and the wood).
- In **B**, when the brakes are applied, friction acts between the brake blocks and the wheel. This slows the bicycle and the cyclist down. Some of the kinetic energy of the bicycle and the cyclist changes into thermal energy of the brake blocks (and the bicycle wheel).

Figure 1 Using energy

DID YOU KNOW?

Crashing in a car is all about energy transfer. The faster you travel the more kinetic energy the car has and the more it has to lose before stopping. If the car crashes, all of that kinetic energy is quickly turned into strain energy on the car distorting its shape and then thermal energy heating up the metal. There is usually quite a lot of sound energy too! Scientists crash a lot of cars in tests to try to improve their safety.

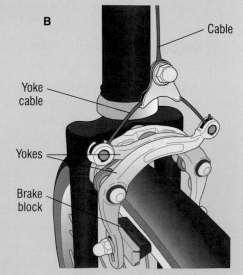

Cable

Yoke cable

Yokes

Brake block

Figure 2 Friction in action. A) Using a drill, B) braking.

Figure 3 Disc brakes

NEXT TIME YOU...

. . . are in a car slowing down at traffic lights, think about what is making the car stop. Figure 3 shows how the disc brakes of a car work. When the brakes are applied, the pads are pushed onto the disc in each wheel. Friction between the pads and each disc slows the wheel down. Some of the kinetic energy of the car changes into thermal energy of the disc pads and the discs.

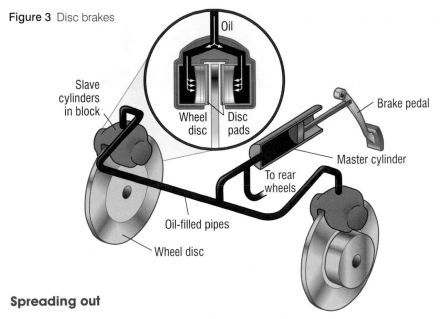

Spreading out

- Wasted energy spreads out to the surroundings.
  For example, the gears of a car get hot when the car is running. So thermal energy transfers from the gear box to the surrounding air.
- Useful energy eventually transfers to the surroundings too.
  For example, the useful energy supplied to the road wheels of a car changes into thermal energy of the tyres, the road and the surrounding air.
- Energy becomes less useful, the more it spreads out.
  For example, the hot water from the cooling system of a CHP (combined heat and power) power station gets used to heat nearby buildings. The thermal energy supplied to the buildings will eventually be lost to the surroundings.

c) The hot water from many power stations flows into rivers or lakes. Why is this wasteful?

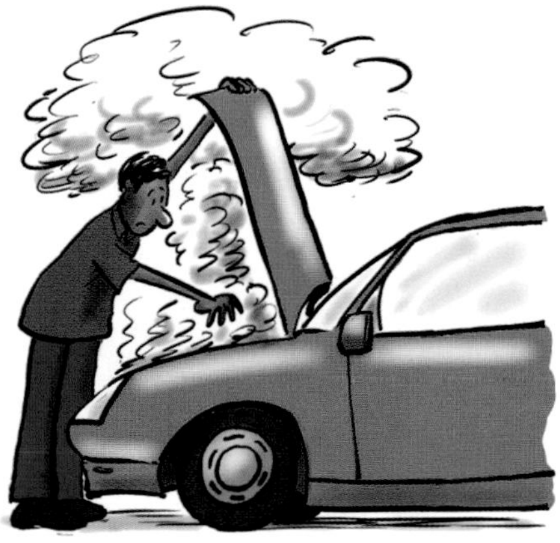

Figure 4 Energy spreading out

SUMMARY QUESTIONS

1 Copy and complete the table below. It should show what happens to the energy transferred in each case.

Energy transfer by	Useful energy	Wasted energy
a) an electric heater		
b) a television		
c) an electric kettle		
d) headphones		

2 What would happen to:

a) a gear box that was insulated so it could not lose thermal energy to the surroundings?
b) a jogger wearing running shoes, which are well-insulated?
c) a blunt electric drill if you use it to drill into hard wood?

3 Explain why a swinging pendulum eventually stops.

KEY POINTS

1 Useful energy is energy in the place we want it and in the form we need it.
2 Wasted energy is energy that is not useful energy.
3 Useful energy and wasted energy both end up being transferred to the surroundings, which become warmer.
4 As energy spreads out, it gets more and more difficult to use for further energy transfers.

P1a 2.4

Energy and efficiency

LEARNING OBJECTIVES

1 What do we mean by efficiency?
2 How can we make machines more efficient?

When you lift an object, the useful energy from your muscles goes to the object as gravitational potential energy. This depends on its weight and how high it is raised.

- Weight is measured in **newtons (N)**. The weight of a 1 kilogram object on the Earth's surface is about 10 N.
- Energy is measured in **joules (J)**. The energy needed to lift a weight of 1 N by a height of 1 metre is equal to 1 joule.

Your muscles get warm when you use them so they must waste some energy.

a) You lower a weight. What happens to its gravitational potential energy?

Figure 1 represents the energy transfer through a device. It shows how we can represent any energy transfer where energy is wasted. This type of diagram is called a **Sankey diagram**.

Because energy cannot be created or destroyed,

energy supplied = useful energy delivered + energy wasted

For any device that transfers energy,

$$\text{efficiency} = \frac{\text{useful energy transferred by the device}}{\text{total energy supplied to the device}}$$

For example, a light bulb with an efficiency of 0.15 would radiate 15 J of energy as light for every 100 J of electrical energy we supply to it.

The percentage efficiency of the light bulb is 15% (= 0.15 × 100%).

b) How much energy is wasted for every 100 J of electrical energy supplied?
c) What happens to the wasted energy?

Energy transfer per second INTO machine

MACHINE OR APPLIANCE

Energy wasted per second

Useful energy transfer per second OUT of machine

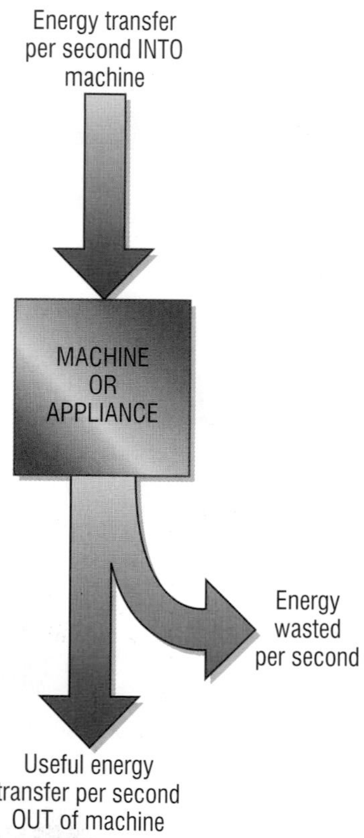

Figure 1 Energy transfer shown on a Sankey diagram

Worked example

An electric motor is used to raise an object. The object gains 60 J of gravitational potential energy when the motor is supplied with 200 J of electrical energy. Calculate the efficiency of the motor.

Solution

Total energy supplied to the device = 200 J

Useful energy transferred by the device = 60 J

$$\text{Efficiency of the motor} = \frac{\text{useful energy transferred by the motor}}{\text{total energy supplied to the motor}}$$

$$= \frac{60\,\text{J}}{200\,\text{J}} = 0.30$$

PRACTICAL

Investigating efficiency

Figure 3 shows how you can use an electric winch to raise a weight. You can use the joulemeter to measure the electrical energy supplied.

– If you double the weight for the same increase in height, do you need to supply twice as much electrical energy to do this task?

– The gravitational potential energy gained by the weight = weight in newtons × height increase in metres

● Use this equation and the joulemeter measurements to work out the efficiency of the winch.

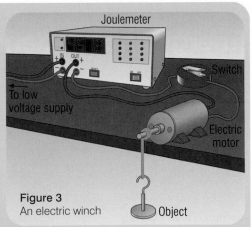

Figure 3
An electric winch   Object

Improving efficiency

	Why machines waste energy	How to reduce the problem
1	Friction between the moving parts causes heating.	Lubricate the moving parts to reduce friction.
2	The resistance of a wire causes the wire to get hot when a current passes through it.	Use wires in circuits with as little electrical resistance as possible.
3	Air resistance causes energy transfer to the surroundings.	Streamline the shapes of moving objects to reduce air resistance.
4	Sound created by machinery causes energy transfer to the surroundings.	Cut out noise (e.g. tighten loose parts to reduce vibration).

d) Which of the above solutions would hardly reduce the energy supplied?

SUMMARY QUESTIONS

1 Complete the sentences below using words from the list.

supplied to   wasted by

a) The useful energy from a machine is always less than the total energy …… it.

b) Friction between the moving parts of a machine causes energy to be …… the machine.

c) Because energy is conserved, the energy …… a machine is the sum of the useful energy from the machine and the energy …… by the machine.

2 An electric motor is used to raise a weight. When you supply 60 J of electrical energy to the motor, the weight gains 24 J of gravitational potential energy.
Work out:
a) the energy wasted by the motor,
b) the efficiency of the motor.

3 A machine is 25% efficient. If the total energy supplied to the machine is 3200 J, how much useful energy can be transferred?

GET IT RIGHT!

Efficiency and percentage efficiency are numbers without units. The maximum efficiency is 1 or 100%, so if a calculation produces a number greater than this it must be wrong.

KEY POINTS

1 Energy is measured in joules.

2 The efficiency of a device =

$$\frac{\text{useful energy transferred by the device}}{\text{total energy supplied to the device}}$$

3 Wasted energy causes inefficiency.

P1a 2.5   Energy and efficiency issues

People in poor countries want more energy to raise their standard of living. People like us in rich countries use much more energy than people in poor countries. What can we do to help them?

Walk to and from school or 'car share'?

Switch things off when they are not in use?

Ideas about energy

Recycle materials more?

ACTIVITY

Discuss this as a group and think up as many ideas as you can. Then choose one idea to develop for a short presentation.

The bouncy ball test

Release a ball above a hard floor and see if it rebounds to the same height. Repeat your measurements as often as necessary.

If the ball does bounce back to the same height, it hasn't lost any energy.

What if it doesn't? Some of its energy must have changed to sound and heat energy when it hit the floor.

Figure 1 A bouncy test

A BURNING ISSUE!

By Jack Daniels

– row over Council plans causes chaos

Last night's Council meeting had to be abandoned when protestors objected to plans for a local incinerator.

The Council want to use it to burn waste paper and to heat their buildings. Nearby residents claim it will produce smoke and noise 24 hours a day and will ruin their lives.

A Council officer said 'Money will be saved on fuel bills. We can't let protestors tell us what to do.'

MORE ON PAGE 2

ACTIVITY

Imagine you were at the meeting as a radio reporter. Prepare a 3-minute slot for local radio. Include two short interviews with people representing each 'side' of the issue. Also interview a scientist from the local university about her plans to test the environmental impact of the incinerator.

QUESTIONS

1 **a)** Compare the rebound height with the original height.
   b) Use the comparison to work out what fraction of the initial gravitational potential energy is lost in the impact.
2 How could you check the accuracy of your measurements?
3 Comment on the reliability of your measurements. (See page 3.)
4 List the variables that might affect your answer to 1b) in other tests. Label each variable as categoric, discrete or continuous. (See page 4.)
5 Comment on the conclusion drawn in 1b). Is it a powerful generalisation? Or does it have limitations? How would you extend the investigation given time?

James Joule VIP!

The unit of energy is named after James Joule. In 1840, he found he could heat water by making a falling weight turn a paddle in the water.

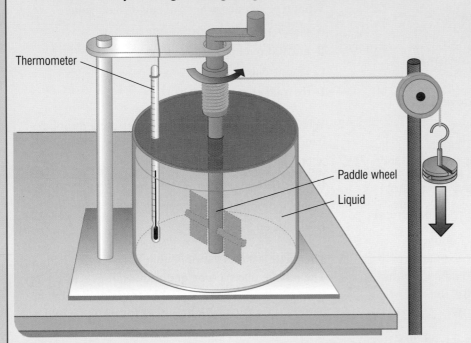

Thermometer

Paddle wheel

Liquid

Figure 2 Heating water

Figure 3
James Joule

The water gains thermal energy and the weight loses gravitational potential energy. Joule showed that the water gained the same amount of energy as the weight lost. He had discovered a very important principle:

Energy cannot be created or destroyed.

He repeated his tests with different weights and liquids and reached the same conclusion in each experiment. He realised that the total amount of energy is unchanged every time.

Hybrid cars

My car doesn't use petrol so it doesn't cause fumes. It's also really quiet!

But you can't go far without recharging and you're too slow!

I have a petrol engine and a battery-powered motor. I use less fuel than you and in slow traffic the petrol engine switches off and the motor takes over. When I accelerate, the engine switches on again. Mine is the best car!

ELECTRIC PETROL HYBRID

ACTIVITY

a) How did James Joule improve the strength of the conclusions he came to? Why would this be important when presenting his new ideas to other scientists?
b) Imagine you are James Joule. Someone wants to patent your discovery as a water heater. They've sent you a drawing. Write a letter saying why their patent idea is not likely to catch on.

QUESTIONS

1 Why is a hybrid car better than
   a) an ordinary car?
   b) an electric car?

2 In a road test, a hybrid car used 2 litres of fuel to travel a distance of 50 km. A petrol car in the same test used 4 litres of the same fuel to travel the same distance.
   a) Which car is more efficient?
   b) The fuel cost 85 p per litre. A typical motorist drives 300 km each week. How much would such a motorist save each week using a hybrid car instead of a petrol-only car?

ACTIVITY

Design an advert for the latest hybrid car. It could be a storyboard for a TV advert or a full-page advert in a newspaper. Use your answers to questions 1 and 2 to help you persuade people to buy the hybrid car.

SUMMARY QUESTIONS

1 The devices listed below transfer energy in different ways.

   1. Car engine   2. Electric bell
   3. Electric light bulb   4. Gas heater

The next list gives the useful form of energy the devices are designed to produce.

Match words A, B, C and D with the devices numbered 1 to 4.

  A Heat (thermal energy)   B Light
  C Movement (kinetic energy)   D Sound

2 Use words from the list to complete the sentences:

useful   wasted   thermal   light   electrical

When a light bulb is switched on, ... energy is changed into ...... energy and into ...... energy of the surroundings. The energy that radiates from the light bulb is ...... energy. The rest of the energy supplied to the light bulb is ...... energy.

3 You can use an electric motor to raise a load. In a test, you supply the winch with 10 000 J of electrical energy and the load gains 1500 J of gravitational potential energy.

a) Calculate its efficiency.

b) How much energy is wasted?

c) Copy and complete the energy transfer diagram below for the winch.

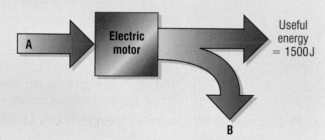

4 A ball gains 4.0 J of gravitational potential energy when it is raised to a height of 2.0 m above the ground. When it is released, it rebounds to a height of 1.5 m.

a) How much kinetic energy did it have just before it hit the ground? Assume air resistance is negligible.

b) How much gravitational potential energy did it have at the top of the rebound?

c) How much energy did it lose in the rebound?

d) What happened to the energy it lost on impact?

EXAM-STYLE QUESTIONS

1 On a building site a machine is used to lift a bag of sand from the ground to the top of a building.
What type of energy has the bag of sand gained?

  A elastic potential energy

  B gravitational potential energy

  C kinetic energy

  D thermal energy   (1)

2 What type of energy is stored in a stretched rubber band?

  A chemical energy

  B elastic strain energy

  C gravitational potential energy

  D kinetic energy   (1)

3 The picture shows a mobile that hangs over a baby's cot. The mobile plays a tune and rotates. It gets its energy from a battery. The electrical energy supplied by the battery is transformed into other forms of energy.

Match words from the list with the numbers **1** to **4** in the sentences.

  A kinetic energy

  B light

  C sound

  D thermal energy

A motor makes the mobile go round. The motor transforms electrical energy mainly into ......**1**....... When the mobile is switched on it becomes warm after a short while. This is because some of the electrical energy is transformed into ......**2**...... Speakers in the mobile transform electrical energy into ......**3**...... There is a 'power on' indicator on the mobile that transforms electrical energy into ......**4**...... .   (4)

4 An electric fan is used to move air around a room.

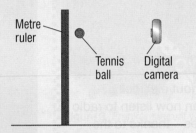

(a) The fan **usefully** transforms electrical energy into

    A elastic energy

    B heat energy

    C kinetic energy

    D sound   (1)

(b) Energy that is not usefully transformed by the fan is **wasted** as

    A heat energy and sound

    B heat energy only

    C kinetic energy and sound

    D sound energy only   (1)

(c) Which of the following statements about the energy wasted by the fan is **not** true?

    A It makes the surroundings warmer.

    B It can no longer be transformed in useful ways.

    C It becomes very thinly spread out.

    D It makes the surroundings cooler.   (1)

(d) A second design of fan transforms useful energy at the same rate but wastes less of the energy supplied to it. This means that the second fan

    A is 100% efficient.   **B** is less efficient.

    C is more efficient.   **D** has the same efficiency.

    (1)

5 A chair lift carries skiers to the top of a mountain.

(a) When the skiers get to the top of the mountain they have gained gravitational potential energy. As they ski back down the mountain what type of energy is this transformed into?   (1)

(b) The chair lift is powered by an electric motor. What useful energy transformation takes place in the motor?   (2)

(c) Some of the electrical energy supplied to the motor is wasted as heat. Why does this happen?   (1)

(d) The energy required to lift two skiers to the top of the mountain is 240 000 J. The energy supplied to the motor is 800 000 J. Calculate the efficiency of the motor.   (2)

HOW SCIENCE WORKS QUESTIONS

Whilst watching a tennis match I wondered why, when they asked for a new set of balls, they were fetched from a fridge. Could it be that they behave differently when they are hot? I decided to test this idea and set up a controlled investigation.

The tennis balls were heated to different temperatures and then dropped from the same height. I used a digital camera to photograph the bounce so that I could get an accurate reading of how high each ball bounced.

My prediction was that as the temperature increased the ball would bounce higher.

My results, and the manufacturer's results, are in this table.

Temperature (°C)	Height bounced (cm)	
	My results	**Manufacturer's**
3	11.0	14.0
10	37.3	40.5
19	53.9	57.1
29	64.1	67.0
40	70.2	73.4
51	74.5	77.5
60	74.3	79.3

a) Plot a graph of my results, including a line of best fit.   (3)

b) What is the sensitivity of the instrument used to measure the height of the bounce?   (1)

c) What is the pattern in the results?   (2)

d) What do these results suggest about my prediction?   (1)

e) Is there any evidence for a random error in my investigation? Explain your answer.   (2)

f) Is there any evidence for a systematic error in my investigation? Explain your answer.   (2)

g) What is the importance of these results to the professional tennis player?   (1)

P1a 3.2   **Electrical power**

Powerful machines

When you use a lift to go up, a powerful electric motor pulls you and the lift upwards. The lift motor transforms energy from electrical energy to gravitational potential energy when the lift is raised. We also get electrical energy transformed to wasted thermal energy and sound energy.

Figure 1 A lift motor

- The energy we supply per second to the motor is the **power** supplied to it.
- The more powerful the lift motor is, the faster it is able to move a particular load.

In general we can say that:

the more powerful a device, the faster the rate at which it transforms energy.

We measure the power of a device in watts (W) or kilowatts (kW).

For any device,

- its input power is the energy per second supplied to it.
- its output power is the useful energy per second transferred by it.

1 watt is a rate of transfer of energy of 1 joule per second (J/s).

1 kilowatt is equal to 1000 watts.

Power (in watts, W) = rate of transfer of energy

$$= \frac{\text{energy transferred (in joules, J)}}{\text{time taken (in seconds, s)}}$$

Worked example

A motor transfers 10 000 J of energy in 25 s. What is its power?

Solution

Power (in watts, W) = $\frac{\text{energy transferred (in joules, J)}}{\text{time taken (in seconds, s)}}$

$$\text{Power} = \frac{10\,000\,\text{J}}{25\,\text{s}} = 400\,\text{W}$$

---

a) What is the power of a lift motor that transfers 50 000 J of energy from the electricity supply in 10 s?

Power ratings

Here are some typical values of power ratings for different energy transfer 'devices':

Device	Power rating
A torch	1 W
An electric light bulb	100 W
An electric cooker	10 000 W = 10 kW (where 1 kW = 1000 watts)
A railway engine	1 000 000 W = 1 megawatt (MW) = 1 million watts
A Saturn V rocket	100 MW
A very large power station	10 000 MW
World demand for power	10 000 000 MW
The Sun	100 000 000 000 000 000 000 MW

b) How many 100 W electric light bulbs would use the same amount of power as a 10 kW electric cooker?

Figure 2 Rocket power

Muscle power

How powerful is a weight lifter?

A 30 kg dumbell has a weight of 300 N. Raising it by 1 m would give it 300 J of gravitational potential energy. A weight lifter could lift it in about 0.5 seconds. The rate of energy transformation would be 600 J/s (= 300 J/0.5 s). So the weight lifter's power output would be about 600 W in total!

c) An inventor has designed an exercise machine that can also generate 100 W of electrical power. Do you think people would buy this machine in case of a power cut?

Figure 3 Muscle power

SUMMARY QUESTIONS

1 a) Which is more powerful?
   ii) A torch bulb or a mains filament lamp.
   iii) A 3 kW electric kettle or a 10 000 W electric cooker.
   b) There are about 20 million homes in Britain. If a 3 kW electric kettle was switched on in 1 in 10 homes at the same time, how much power would need to be supplied?

2 The input power of a lift motor is 5000 W. In a test, it transforms 12 000 J of electrical energy to gravitational potential energy in 20 seconds.

   a) How much electrical energy is supplied to the motor?
   b) What is its efficiency in the test?

3 Choose one of the energy transfer devices listed at the top of this page. Carry out some research and describe how it works.

4 A machine has a power rating of 100 kW. If the machine runs for 2 minutes, how much energy does it transfer?

KEY POINTS

1 The unit of power is the watt (W), equal to 1 J/s.
2 1 kilowatt (kW) = 1000 watts
3 Power (in watts) =

$$\frac{\text{energy transferred (in joules)}}{\text{time taken (in seconds)}}$$

P1a 3.3

Using electrical energy

LEARNING OBJECTIVES

1 How can we work out the energy used by a mains device?
2 How is the cost of mains electricity worked out?

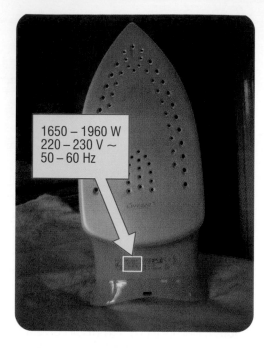

Figure 1 Mains power

1650 – 1960 W
220 – 230 V ~
50 – 60 Hz

GET IT RIGHT!

Remember that a kilowatt-hour (kWh) is a unit of energy.

How much electrical energy is transferred from the mains when you use an electric heater? You can work this out if you know its power and how long you use it for.

A 1 kW heater uses the same amount of electrical energy in 1 hour as a 2 kW heater would use in half-an-hour. For ease, we say that:

the energy supplied to a 1 kW device in 1 hour is **1 kilowatt-hour (kWh)**.

We use the kilowatt-hour as the unit of energy supplied by mains electricity. You can work out the energy, in kilowatt-hours, used by a mains device in a certain time using this equation:

$$\begin{array}{c}\textbf{Energy transferred} \\ \textbf{(kilowatt-hours, kWh)}\end{array} = \begin{array}{c}\textbf{power of device} \\ \textbf{(kilowatts, kW)}\end{array} \times \begin{array}{c}\textbf{time in use} \\ \textbf{(hours, h)}\end{array}$$

For example,

• a 1 kW heater switched on for 1 hour uses 1 kWh of electrical energy (= 1 kW × 1 hour).
• a 1 kW heater switched on for 10 hours uses 10 kWh of electrical energy (= 10 kW × 1 hour).
• a 0.5 kW heater switched on for 6 hours uses 3 kWh of electrical energy (= 0.5 kW × 6 hours).

a) How many kWh are used by a 100 W lamp in 24 hours?

How many joules are there in 1 kilowatt-hour?

One kilowatt-hour is the amount of electrical energy supplied to a 1 kilowatt device in 1 hour. So 1 kilowatt-hour = 1000 joules/second × 60 × 60 s = 3 600 000 J.

NEXT TIME YOU…

… leave the TV on overnight, think about how much it is costing you. It's still using energy in standby mode.

Paying for electrical energy

The **electricity meter** in your home measures the amount of electrical energy your family uses. It records the total energy supplied, no matter how many devices you all use. It gives us a reading of the number of kilowatt-hours (kW h) of energy supplied by the mains.

Figure 2 An electricity meter

NELEB

L. Jones
26 Homewood Road
Otwood M51 9YZ

Meter readings		units	pence per unit	amount	VAT %
present	previous				
31534	30092	1442	5.79	83.49	Zero
					07.30
Standing charge					
TOTAL NOW DUE					90.79
PERIOD ENDED					31.3.06

Figure 3 Checking your bill

In most houses, somebody reads the meter every three months. Look at the electricity bill in Figure 3.

The difference between the two readings is the number of kilowatt-hours (or units) supplied since the last bill.

b) Check for yourself that 1442 kW h of electrical energy is supplied in the bill shown.

We use the kilowatt-hour to work out the cost of electricity. For example, a cost of 7p per kWh (or 7p per unit) means that each kilowatt-hour of electrical energy costs 7p. Therefore,

total cost = number of kWh used × cost per kWh

c) Work out the cost of 1442 kW h at 7p per kWh.

SUMMARY QUESTIONS

1 Use words from the list to complete the sentences below.

hour   kilowatt   kilowatt-hours

a) The ...... is a unit of power.
b) Electricity meters record the mains electrical energy transformed in units of ....
c) One ...... is the energy transformed by a 1 ......device in 1 ....

2 a) Work out the number of kW h transformed in each case below.
   i) A 3 kilowatt electric kettle is used 6 times for 5 minutes each time.
   ii) A 1000 watt microwave oven is used for 30 minutes.
   iii) A 100 watt electric light is used for 8 hours.
b) Calculate the total cost of the electricity used in a) if the cost of electricity is 7.0p per kW h.

3 An electric heater is left on for 3 hours.
During this time it uses 12 kWh of electrical energy. What is the power of the heater?

KEY POINTS

1 Energy transferred (kilowatt-hours, kWh) = power of device (kilowatts, kW) × time in use (hours, h)

2 Total cost = number of kWh used × cost per kWh

P1a 3.4   The National Grid

LEARNING OBJECTIVES

1 Why is there a National Grid for electricity?
2 How does electricity from power stations reach our homes?

SCIENCE @ WORK

The cables of the National Grid system are well-insulated from each other and from the ground. The insulators used on electricity pylons need to be very effective as insulators – or else the electricity would short-circuit to the ground. In winter, ice on the cables can cause them to snap. Teams of electrical engineers are always on standby to deal with sudden emergencies.

Figure 1 Electricity pylons carry the high voltage cables of the National Grid

Your electricity supply at home reaches you through the **National Grid**. This is a network of cables connecting power stations to homes and other buildings. The network also contains transformers. Step-up transformers are used at power stations and step-down transformers are used at sub-stations near homes.

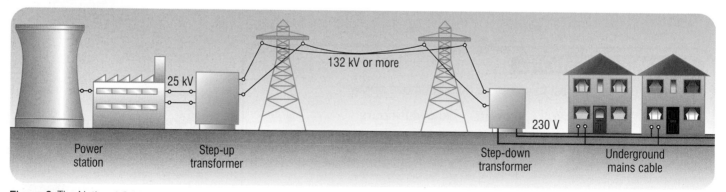

132 kV or more

25 kV

230 V

Power station   Step-up transformer   Step-down transformer   Underground mains cable

Figure 2 The National Grid

GET IT RIGHT!

Remember that step-up transformers are used at power stations and step-down transformers are used at sub-stations near homes.

The National Grid's voltage is 132 000 volts or more. This is because transmitting electricity at a high voltage reduces power loss, making the system more efficient.

Power stations produce electricity at a voltage of 25 000 volts.

● We use *step-up* **transformers** to step this voltage up to the grid voltage.
● We use *step-down* **transformers** at local sub-stations to step the grid voltage down to 230 volts for use in homes and offices.

DEMONSTRATION

Watch a demonstration of the effect of a transformer using this apparatus.

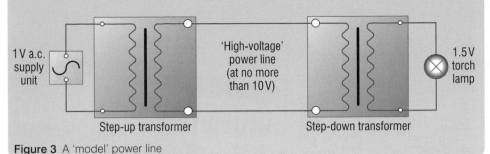

Figure 3 A 'model' power line

Power and the grid voltage

By making the grid voltage as high as possible, the energy losses are reduced to almost zero. This is because less current passes through the cables (for the same power delivered) so its heating effect is less.

a) What difference would it make if we didn't step up the grid voltage?

Underground or overground?

Lots of people object to electricity pylons. They say they spoil the landscape or they affect their health. Electric currents produce electric and magnetic fields that might affect people.

Why don't we bury all cables underground?

Underground cables would be:

- much more expensive,
- much more difficult to repair,
- difficult to bury where they cross canals, rivers and roads.

What's more, overhead cables are high above the ground. Underground cables could affect people more because the cables wouldn't be very deep.

b) Give two reasons why underground cables are more difficult to repair?

DID YOU KNOW?

Figure 4 Engineers at work on the Grid. They certainly need a head for heights!

The National Grid was set up in 1926. Before then, every town had its own power station. The voltages in nearby towns were often different. If there was a sudden demand for electricity in one town, nearby towns couldn't help because they had different voltages. The UK government decided electricity would be supplied to homes at 240 volts. This was lowered to 230 volts in 1994.

SUMMARY QUESTIONS

1 Complete the sentences below using words from the list.

bigger   down   smaller   up

a) Power stations are connected to the National Grid using step- ......
transformers. This type of transformer makes the voltage ....

b) Homes are connected to the National Grid using step- ......
transformers. This type of transformer makes the voltage ....

2 Would you buy a house next to an electricity sub-station? Find out why some people would be worried. What advice would you offer them?

3 a) Why is electrical energy transferred through the National Grid at a much higher voltage than it is generated in a power station?

b) Why are transformers needed to connect local sub-stations to the National Grid?

KEY POINTS

1 The National Grid is a network of cables and transformers.

2 We use step–up transformers to step up power stations' voltages to the grid voltage,

3 We use step-down transformers to step the grid voltage down for use in our homes.

4 A high grid voltage reduces energy loss and makes the system more efficient.

P1a 3.5   Essential electricity issues

On holiday

Before you go on holiday abroad, be sure to find out what the voltage is in the country you are visiting.

• In North America, the mains voltage is 115 V not 230 V. A 230 V hairdryer won't work very well at 115 V. Some devices have a dual-voltage switch to enable it to work at either voltage. But make sure you reset the switch to 230 V when you get home!

• In most EU countries, the mains voltage is 230 V, the same as in the UK. But take an adaptor because the mains sockets are different to those in the UK.

QUESTION

1 Why would it be very dangerous to use a 115 V device on 230 V?

19th November 1878 1d   Cyprus under British Administration - page 2

Floodlit Football - A Brilliant Success!

Successful Trial Of Bright Idea

Supporters of Blackburn Rovers and Accrington Stanley will remember yesterday after being treated to one of the first floodlit football matches in Britain.   Six thousand fans paid to enter the ground and a further twenty thousand watched from Coronation Park.   The players, attired in picturesque costumes, amused the crowds before the match by playing leapfrog and racing.   The leather ball, which was painted white so the players could see it, was kicked off at about a quarter to eight.   Although the players' faces could not be distinguished at a distance, the ball could be seen very well and the crowd had no difficulty following the match.   Blackburn had a resounding victory – three goals to nil!

Turn to Page 12

Lighting The Way - London Gets Electric Street Lighting!

QUESTION

2 A football ground has 8 towers of floodlights, each using 50 kW of electrical power.

a) How much electrical energy, in kWh, is used when the lights are on for 4 hours?

b) How much does the electricity cost at 7 p per kWh?

Bygone times – before electricity

• Many towns used lamp lighters before they had electric street lamps installed. The lamp lighters would go round the streets at dusk and light all the gas lamps. Then they would go round and turn them off at dawn.

• A steam train is a sight to see at a distance but not if you get too close. The ash and smoke from a steam engine are very unpleasant if it blows in your face. Steam trains on the London Underground were replaced by electric trains long ago. Travelling on the Underground would be very unpleasant if the carriages were pulled by steam engines.

QUESTIONS

3 What did people use to wash clothes before the invention of the washing machine?

4 How would your life be different without electricity? Write an account of 'A day in my life without electricity'.

Shop around?

You can get your electricity from any one of a number of different electricity companies. You can even get your gas and electricity from the same company. Companies offer different deals to attract new customers.

	Homepower	Power Co	PowerGreen
First 100 kW h of electricity	9p per kW h	10p per kW h	
Electricity above 100 kW h	7p per kW h	5p per kW h	8p per kW h

Go online now to find a better deal!

QUESTION

5 Look at the table opposite and work out which would be the best deal if

a) you use 100 kW h of electricity per month on average,

b) you use 300 kW h of electricity per month on average.

A not-so brilliant idea!

Nikolai Tesla was a brilliant electrical engineer. He made important inventions and discoveries about electric motors and generators. He even discovered how to supply electricity using radio waves. No cables were needed to supply the electricity. But no supply company would take his idea up. Why? – because anyone could tap into a radio power grid for free – just by putting an aerial up. The company wouldn't know who was using their electricity.

ACTIVITY

Think up an invention that could make you rich using Nikolai's idea.

SUMMARY QUESTIONS

1 a) Name a device that transforms electrical energy into:
   i) light and sound energy,
   ii) kinetic energy.

   b) Complete the sentences below.
   i) In an electric bell, electrical energy is transformed into useful energy in the form of ...... energy and ...... energy.
   ii) In a washing machine, electrical energy is transformed into useful energy in the form of ...... energy and sometimes as ...... energy.

2 a) Which two units in the list below can be used to measure energy?

   joule   kilowatt   kilowatt-hour   watt

   b) Rank the electrical devices below in terms of energy used from highest to lowest,
   A a 0.5 kW heater used for 4 hours,
   B a 100 W lamp left on for 24 hours,
   C a 3 kW electric kettle used 6 times for 10 minutes each time,
   D a 750 W microwave oven used for 10 minutes.

3 a) The readings of an electricity meter at the start and the end of a month are shown below.

0	9	3	7	2		0	9	6	1	5

   i)   Which is the reading at the end of the month?
   ii) How many units of electricity were used during the month?
   iii) How much would this electricity cost at 7p per kWh?

   b) A pay meter in a holiday home supplies electricity at a cost of 10p per kWh.
   i) How many kW h would be supplied for £1.00?
   ii) How long could a 2 kW heater be used for after £1 is put in the meter slot?

4 An escalator in a shopping centre is powered by a 50 kW electric motor. The escalator is in use for a total time of 10 hours every day.

   a) How much electrical energy in kW h is supplied to the motor each day?

   b) The electricity supplied to the motor costs 7p per kW h. What is the daily cost of the electricity supplied to the motor?

   c) How much would be saved each day if the motor was replaced by a more efficient 40 kW motor?

EXAM-STYLE QUESTIONS

1 The devices shown transform electrical energy into other forms of energy.

The list gives the useful form of energy the devices are designed to produce. Match the words in the list with the devices numbered **1** to **4**.

   A kinetic energy   **B** light
   C sound   **D** thermal energy   (4)

2 A 3 kW electric motor is switched on for 15 minutes. How much energy, in kilowatt hours, does it transfer during this time?
   A 0.0075 kWh   **B** 0.075 kWh
   C 0.75 kWh   **D** 7.50 kWh   (1)

3 Which of the following does **not** represent a unit of energy?
   A J   **B** kJ
   C kW   **D** kWh   (1)

4 The diagram shows the readings on a household electricity meter, in kWh, at the beginning and end of one week. Each kWh of electricity costs 8p.

1	8	2	4	2		1	8	5	1	1

   At the beginning of the week   At the end of the week

   (a) How many kWh of electricity were used during the week?
   A 242   **B** 269
   C 511   **D** 753   (1)

   (b) On one day 30 kWh of electricity were used. How much would this electricity cost?
   A 24p   **B** 30p
   C £2.40   **D** £3.00   (1)

   (c) During the week a 2 kW iron was used for 2.5 hours. How much energy was transformed by the iron?
   A 0.50 kWh   **B** 0.75 kWh
   C 5.00 kWh   **D** 7.50 kWh   (1)

(d) How much does it cost to use a 9 kW shower for half an hour?

   A 3.6p   **B** 4.5p

   C 36p   **D** 45p   (1)

5 A student uses an electric iron.

(a) What useful energy transformation takes place in the iron? (1)

(b) The iron has a power of 1.2 kW. What is meant by 'power'? (1)

(c) Electricity cost 8 p per kWh. How much does it cost the student to use the iron for 30 minutes? (3)

6 Each town in Britain used to have its own power station. Now electricity is supplied by a system called the National Grid.

(a) Why is the National Grid system better than each town having its own supply? (2)

(b) Electricity in power stations is generated at 25 000 volts. Explain why:

  (i) it is transmitted across the National Grid system at 132 000 volts.

  (ii) it is supplied to homes at 230 volts. (2)

(c) What is the name of the device used to change the potential difference of the mains supply from 25 000 volts to 132 000 volts before transmission across the National Grid? (2)

(d) Suggest why the cables of the National Grid are carried high above the ground rather than being buried underground. (2)

HOW SCIENCE WORKS QUESTIONS

Josh set up an investigation to measure the efficiency of a small electric motor. He set the motor up to lift different masses. By measuring the voltage and the current used, he could calculate the energy used by the motor to lift different masses to the same height.

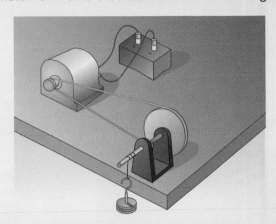

Josh hypothesised that heavier masses would reduce the efficiency of a motor. His prediction was that, if he increased the mass lifted by the motor, it would use much more energy than the energy gained by the masses. He thought this would happen because the motor would require more energy and more heat would be lost. His results are in the table below.

Mass lifted (g)	Efficiency (%)
50	10
100	16
150	24
200	22
250	21
300	19
350	15

a) Plot a graph of these results. (3)

b) Describe the pattern shown by these results. (3)

c) Do these results wholly support, partly support or refute Josh's hypothesis? (1)

d) Why might these results not be reliable? (1)

e) How could Josh improve the reliability of these results? (1)

P1a 4.1

Fuel for electricity

LEARNING OBJECTIVES

1 How is electricity generated in a power station?
2 What fossil fuels do we burn in power stations?
3 How do we use nuclear fuels in power stations?

Figure 2 Inside a gas-fired power station

SCIENCE @ WORK

When a popular TV programme ends, lots of people decide to put their kettles on. The national demand for electricity leaps as a result. Engineers meet these surges in demand by switching gas turbine engines on in gas-fired power stations.

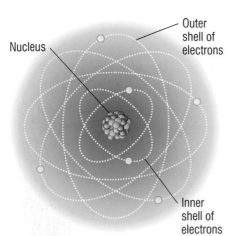

Figure 3 The structure of the atom

In a power station

Almost all the electricity you use is generated in power stations.

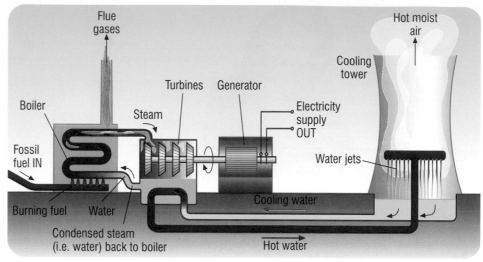

Figure 1 Inside a fossil fuel power station

- In a *coal* or *oil-fired power station*, the burning fuel heats water in a boiler to produce steam. The steam drives a turbine that turns an electricity generator.

a) What happens to the steam after it has been used?
b) What happens to the thermal energy of the steam after it has been used?

- In a *gas-fired power station*, we burn natural gas directly in a gas turbine engine. This produces a powerful jet of hot gases and air that drives the turbine. A gas-fired turbine can be switched on very quickly.

PRACTICAL

Turbines

See how we can use water to drive round the blades of a turbine.

- Why is steam better than water?

Nuclear power

Do you know what's inside an atom? Figure 3 shows you.

- Every atom contains a positively charged nucleus surrounded by electrons.
- The nucleus is composed of two types of particles, neutrons and protons.
- Atoms of the same element can have different numbers of neutrons in the nucleus.

How is electricity obtained from a nuclear power station?

The fuel in a nuclear power station is uranium. The uranium fuel is contained in sealed cans in the core of the reactor. The nucleus of a uranium atom is unstable and can split in two. Energy is released when this happens. We call this process **nuclear fission**. Because there are lots of uranium atoms in the core, it becomes very hot.

The thermal energy of the core is taken away by a fluid (called the 'coolant') that is pumped through the core. The coolant is very hot when it leaves the core. It flows through a pipe to a 'heat exchanger', then back to the reactor core. The thermal energy of the coolant is used to turn water into steam in the heat exchanger. The steam drives turbines which turn electricity generators.

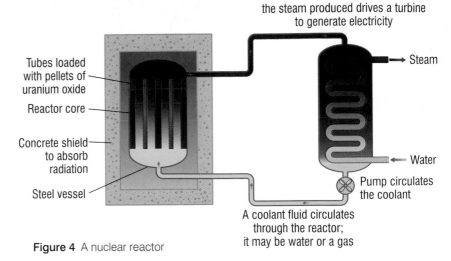

The heated fluid is used to boil water; the steam produced drives a turbine to generate electricity

Tubes loaded with pellets of uranium oxide

Reactor core

Concrete shield to absorb radiation

Steel vessel

Steam

Water

Pump circulates the coolant

A coolant fluid circulates through the reactor; it may be water or a gas

Figure 4 A nuclear reactor

Comparing nuclear power and fossil fuel power

	Nuclear power station	Fossil fuel power station
Fuel	uranium	coal, oil or gas
Energy released per kg of fuel	1 000 000 kW h (= about 10 000 × energy released per kg of fossil fuel)	100 kW h
Waste	radioactive waste that needs to be stored for many years	non-radioactive waste
Greenhouse gases	no – because uranium releases energy without burning	yes – because fossil fuels produce gases like carbon dioxide when they burn

SUMMARY QUESTIONS

1 Complete the sentences below using words from the list.

coal   gas   oil   uranium

a) …… is not a fossil fuel.
b) Power stations that use …… as the fuel can be switched on very quickly.
c) Steam is used to make the turbines rotate in a power station that uses coal, …… or…… as fuel.

2 a) State one advantage and one disadvantage of:
   i) an oil-fired power station compared with a nuclear power station,
   ii) a gas-fired power station compared with a coal-fired power station.
   b) Look at the table above:
      How many kilograms of fossil fuel would give the same amount of energy as 1 kg of uranium fuel?
   c) Some people think that more nuclear power stations are the only way to reduce greenhouse gases significantly. Explain your views on this.

KEY POINTS

1 Electricity generators in power stations are driven by turbines.
2 Much more energy is released per kilogram from uranium than from fossil fuel.

P1a 4.2   Energy from wind and water

Wind power

A wind turbine is an electricity generator at the top of a narrow tower. The force of the wind drives the turbine's blades around. This turns a generator. The power generated increases as the wind speed increases.

a) What happens if the wind stops blowing?

Wave power

Figure 1 A wind farm – why do some people oppose these developments?

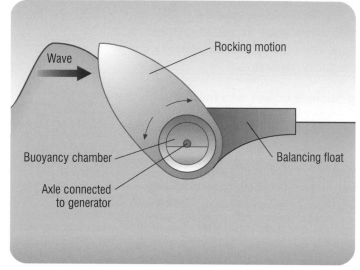

Figure 2 Energy from waves

A wave generator uses the waves to make a floating section move up and down. This motion drives a turbine which turns a generator. A cable between the generator and the shore delivers electricity to the grid system.

Wave generators need to withstand storms and they don't produce a constant supply of electricity. Also, lots of cables (and buildings) would be needed along the coast to connect the wave generators to the electricity grid. This would spoil areas of coastline. In addition, tidal flow patterns might be changed, affecting the habitats of marine life and birds.

b) What could happen if the waves get too high?

PRACTICAL

Pumped storage

When electricity demand is low, we can use wind and wave power to pump water uphill. We can use the energy stored in this way later.

See how we can pump water 'uphill' and then use it to generate electricity.

● What would happen if the upper 'reservoir' is too high?

Hydroelectric power

We can generate hydroelectricity when rainwater collected in a reservoir (or water in a pumped storage scheme) flows downhill. The flowing water drives turbines that turn electricity generators at the foot of the hill.

c) Where does the energy for hydroelectricity come from?

Tidal power

Tidal power stations trap the water from each high tide behind a barrage. We can then release the high tide into the sea through turbines. The turbines drive generators in the barrage. One of the most promising sites in Britain is the Severn estuary. This is because the estuary becomes narrower as you move 'up-river' away from the open sea. So it 'funnels' the incoming tide and makes it higher than elsewhere.

Figure 4 A tidal power station

Figure 3 A hydroelectric scheme

d) Why is tidal power more reliable than wind power?

SUMMARY QUESTIONS

1 Complete the following sentences below using words from the list.

> **hydroelectric   tidal   wave   wind**

a) …… power does not need water.
b) …… power does not need energy from the Sun.
c) …… power is obtained from water running downhill.
d) …… power is obtained from water moving up and down.

2 a) Use the table below for this question.
   i) How many wind turbines would give the same total output as a tidal power station?
   ii) How many kilometres of wave generators would give the same total output as a hydroelectric power station?

b) Use the words below to fill in the location column in the table.

> **coastline   estuaries   hilly or coastal areas   mountain areas**

	Output	Location
Hydroelectric power station	500 MW per station	
Tidal power station	2000 MW per station	
Wave power generators	20 MW per kilometre of coastline	
Wind turbines	2 MW per wind turbine	

c) Imagine you are a government adviser on alternative energy sources. Put the four methods in the table into an order to prioritise government spending. Explain your choices.

KEY POINTS

1 A wind turbine is an electricity generator on top of a tall tower.
2 A wave generator is a floating generator turned by the waves.
3 Hydroelectricity generators are turned by water running downhill.
4 A tidal power station traps each high tide and uses it to turn generators.

P1a 4.3

Power from the Sun and the Earth

LEARNING OBJECTIVES

1 What are solar cells and what do we use them for?
2 What are the advantages and disadvantages of solar cells?
3 What is geothermal energy and how can we use it to generate electricity?

Figure 1 Energy from the Sun

Figure 3 A solar-powered vehicle. Think of some advantages and disadvantages of this car.

Solar power

Solar radiation transfers energy to you from the Sun – sometimes more than you want if you get sunburnt. But we can use the Sun's energy to generate electricity using **solar cells**. We can also use it to heat water directly in **solar heating panels**.

a) Which generates electricity – a solar cell or a solar heating panel?

PRACTICAL

Solar cells

Use a solar cell to drive a small motor.

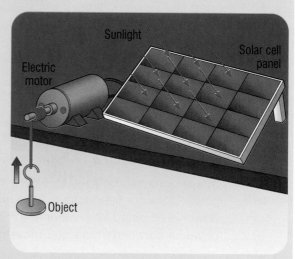

Figure 2 Solar cells at work

● What happens if you cover the solar cells with your hand?

(1) **Solar cells** in use now convert less than 10% of the solar energy they absorb into electrical energy. We connect them together to make solar cell panels.

● They are useful where we only need small amounts of electricity (e.g. in watches and calculators) or in remote places (e.g. on small islands in the middle of an ocean).
● They are very expensive to buy even though they cost nothing to run.
● We need lots of them to generate enough power to be useful – and plenty of sunshine!

(2) A **solar heating panel** heats water that flows through it. Even on a cloudy day in Britain, a solar heating panel on a house roof can supply plenty of hot water.

b) If the water stopped flowing through a solar heating panel, what would happen?

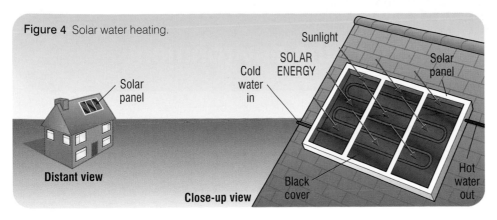

Figure 4 Solar water heating.

Distant view

Close-up view

Sunlight
SOLAR ENERGY
Cold water in
Solar panel
Solar panel
Black cover
Hot water out

Geothermal energy

Geothermal energy comes from energy released by radioactive substances, deep within the Earth.

● The energy released by these radioactive substances heats the surrounding rock.

● As a result, heat is transferred towards the Earth's surface.

We can build geothermal power stations in volcanic areas or where there are hot rocks deep below the surface. Water gets pumped down to these rocks to produce steam. Then the steam produced drives electricity turbines at ground level.

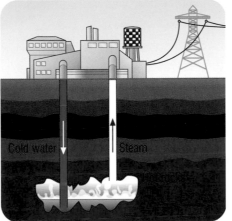

Cold water   Steam

Hot rocks

Figure 5 A geothermal power station and how it works.

c) Why don't geothermal power stations need energy from the Sun?

SUMMARY QUESTIONS

1 Match words from the list below with the spaces in the sentences.

   geothermal energy   solar energy   radiation   radioactivity

   a) The best energy resource for a calculator is …… .
   b) …… inside the Earth releases …… energy.
   c) …… from the Sun generates electricity in a solar cell.

2 A satellite in space uses a solar cell panel for electricity. The panel generates 300 W of electrical power and has an area of 10 m².

   a) Each cell generates 0.2 W. How many cells are in the panel?
   b) The satellite carries batteries that are charged by electricity from the solar cell panels. Why are batteries carried as well as solar cell panels?

3 Discuss the advantages and disadvantages of using solar and geothermal energy to generate electricity.

P1a 4.4   Energy and the environment

Can we get energy without creating any problems? Look at the chart in Figure 1.

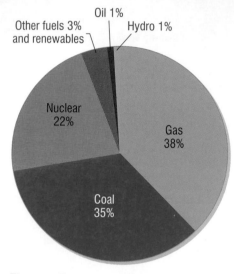

Figure 1 Energy sources for electricity

It shows the energy sources we use at present to generate electricity. What effect does each one have on our environment?

Fossil fuel problems

- When we burn coal, oil or gas, the chemical reaction makes 'greenhouse gases' such as carbon dioxide. These gases cause global warming. We only get a small percentage of our electricity from oil-fired power stations. We use much more oil to produce fuels for transport.
- Burning fossil fuels can also produce sulfur dioxide. This gas causes acid rain in the atmosphere. We can remove the sulfur from a fuel before burning it to stop acid rain. For example, natural gas has its sulfur impurities removed before we use it.
- Fossil fuels are not renewable. Sooner or later, we will have used up the Earth's reserves of fossil fuels. We will then have to find alternative sources of energy. But how soon? Oil and gas reserves could be used up within the next 50 years.

a) Which gas given off when we burn fossil fuels contributes towards:
   i) global warming?
   ii) acid rain?

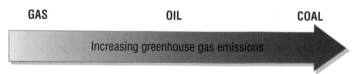

Figure 2 Greenhouse gases from fossil fuels

Nuclear *v* renewable

We need to cut back on fossil fuels to stop global warming. Should we rely on nuclear power or on renewable energy in future?

(1) Nuclear power
Advantages
- No greenhouse gases (unlike fossil fuel).
- Much more energy from each kilogram of uranium fuel than from fossil fuel.

Disadvantages
- Used fuel rods contain radioactive waste, which has to be stored safely for centuries.
- Nuclear reactors are safe in normal operation. However, an explosion at one could release radioactive material over a wide area. This would affect these areas for many years.

b) Why is nuclear fuel non-renewable?

(2) Renewable energy sources and the environment
Advantages
Renewable energy resources

- never run out,
- do not produce greenhouse gases or acid rain,
- do not create radioactive waste products.

Disadvantages
- Wind turbines are unsightly and create a whining noise that can upset people nearby.
- Tidal barrages affect river estuaries and the habitats of creatures and plants there.
- Hydroelectric schemes need large reservoirs of water, which can affect nearby plant and animal life. Habitats are often flooded to create dams.
- Solar cells would need to cover large areas to generate large amounts of power.

c) Do wind turbines affect plant and animal life?

DID YOU KNOW?

In 1986, some nuclear reactors at Chernobyl in Ukraine overheated and exploded. Radioactive substances were thrown high into the atmosphere. Chernobyl and the surrounding towns were evacuated. Many children from the area have since died from illnesses such as leukaemia. Radioactive material from Chernobyl was also deposited on parts of Britain.

Figure 3 Chernobyl, the site of the world's most serious accident at a nuclear power station

Figure 4 The effects of acid rain

KEY POINTS

1. Fossil fuels produce greenhouse gases.
2. Nuclear fuels produce radioactive waste.
3. Renewable energy resources can affect plant and animal life.

SUMMARY QUESTIONS

1. Choose words from the list to complete each of the sentences in a), b) and c).

   acid rain   fossil fuels   greenhouse gas   plant and animal life   radioactive waste

   a) Most of Britain's electricity is produced by power stations that burn ....
   b) A gas-fired power station does not produce ...... or much ....
   c) A tidal power station does not produce ...... like a nuclear power station does but it does affect ...... locally.

2. Match each energy source with a problem it causes.

   Problem: A Acid rain, B Noise, C Radioactive waste, D Takes up land
   Energy source: i) Coal, ii) Hydroelectricity, iii) Uranium, iv) Wind power

3. Make a leaflet for the general public explaining the issues involved in generating electricity using nuclear power.

P1a 4.5   Big energy issues

Nuclear or not?

Figure 1 A nuclear power station

FOR –

About a quarter of Britain's electricity comes from nuclear power stations. Many of these stations are due to close by 2020. A new nuclear power station takes several years to build. So the Uk government must build new nuclear power stations.

AGAINST –

We don't want new nuclear power stations. We can get our electricity from renewable devices like wind turbines. We believe that renewable energy devices can provide enough electricity. We don't need new nuclear power stations.

ACTIVITY

Who is right? Find out what your friends think. Then read on before forming your opinion.

Supply and demand

The demand for electricity varies during each day. It is also higher in winter than in summer. Our electricity generators need to match these changes in demand.

- Power stations can't just 'start up' instantly. The 'start up' time depends on the type of power station:

| **natural gas** | | **oil** | **coal** | | **nuclear** |

shortest start-up time———→———→———longest start-up time

- Renewable energy resources are unreliable. The amount of electricity they generate depends on the conditions:

Hydroelectric	Upland reservoir could run dry.
Wind, waves	Wind and waves too weak on very calm days.
Tidal	Height of tide varies both on a monthly and yearly cycle.
Solar	No solar energy at night and variable during the day.

The variable demand for electricity is met by:

- using nuclear, coal and oil-fired power stations to provide a constant amount of electricity (the *base load* demand),
- using gas-fired power stations and pumped-storage schemes to meet daily variations in demand and extra demand in winter,
- using renewable energy sources when demand is high and renewables are in operation (e.g. use of wind turbines in winter when wind speeds are suitable),
- using renewable energy sources when demand is low to store energy in pumped storage schemes.

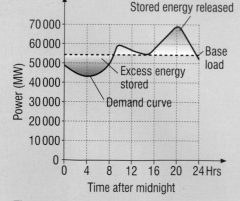

Figure 2 Electricity demand

QUESTION

1 We need to cut back on fossil fuels to reduce greenhouse gases. What would happen if we went over completely to:

a) renewable energy?

b) nuclear power?

Fusion power

Energy from the Sun is produced by a process called **fusion**. Deep inside the Sun, the enormous pressure and temperature force small nuclei to fuse together. These nuclei merge and form heavier nuclei. Energy is released in the process. Scientists have successfully built experimental fusion reactors that release energy – but not for long!

Research is continuing to find out how to turn small experimental reactors into a reactor that can supply large amounts of power. The benefits would be fantastic because we can get the fuel – hydrogen – from sea water. Even better, the reaction products, such as helium, are not radioactive!

Figure 3 Testing fusion

QUESTION

2 Explain why fusion reactors offer the promise of limitless fuel supplies and no pollution.

Nothing is free

Fossil fuels

Removing the sulfur from coal and oil is expensive.

Stopping greenhouse gases escaping, if it could be done, would be even more expensive.

Energy saving

Most home owners are unlikely to buy energy-saving improvements until energy bills go up even more.

Nuclear power

The cost of building and running a nuclear power station is very high.

So is the cost of decommissioning it (i.e. taking it out of use).

Also, radioactive waste products are expensive to store.

Renewables

There are no fuel costs for renewables but capital costs of setting up are high.

This is because lots of expensive equipment is needed to 'collect' large quantities of renewable energy.

ACTIVITY

Who pays? Should we pay through higher taxes or through higher energy bills?
Take a vote!

The big energy debate

Is it possible to generate enough electricity for everyone and to cut back on greenhouse gases? Here are some suggestions:

1. Develop renewable energy resources on a much larger scale.

2. Use energy more efficiently.

Build more nuclear power stations.

Continue to use fossil fuels but remove the greenhouse gases produced.

ACTIVITY

Add your own suggestions. Work in a group and narrow them down to the two most popular ones. Then use your scientific knowledge to debate which one is best!

SUMMARY QUESTIONS

1 Use the list of fuels below to answer a) to e).

coal   natural gas   oil   uranium   wood

a) Which fuels from the list below are fossil fuels?

b) Which fuels from the list cause acid rain?

c) Which fuels release chemical energy when they are used?

d) Which fuel releases the most energy per kilogram?

e) Which fuel produces radioactive waste?

2 a) Complete the following sentences using words from the list.

hydroelectric   tidal   wave   wind

i) …… power stations trap sea water.

ii) …… power stations trap rain water.

iii) …… generators must be located along the coast line.

iv) …… turbines can be located on hills or off-shore.

b) Which renewable energy resource transforms

i) the kinetic energy of moving air into electrical energy?

ii) the gravitational potential energy of water running downhill into electrical energy?

iii) the kinetic energy of water moving up and down into electrical energy?

3 a) Complete the sentences below using words from the list.

**coal-fired   geothermal
hydroelectric   nuclear**

i) A …… power station does not produce greenhouse gases and uses energy which is from inside the Earth.

ii) A …… power station uses running water and does not produce greenhouse gases.

iii) A …… power station releases greenhouse gases.

iv) A …… power station does not release greenhouse gases but does produce waste products that need to be stored for many years.

b) Wood can be used as a fuel. State whether it is

i) renewable or non-renewable,

ii) a fossil fuel or a non-fossil fuel.

EXAM-STYLE QUESTIONS

A hydroelectric power station uses two lakes.

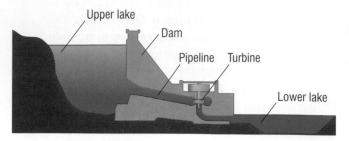

1 As water flows from the top to the bottom lake it turns a turbine coupled to a generator that produces electricity. What is the energy transformation that takes place as the water flows?

A Electrical energy to kinetic energy.

B Gravitational potential energy to kinetic energy.

C Kinetic energy to gravitational potential energy.

D Kinetic energy to heat energy.   (1)

2 Where does geothermal energy come from?

A Radioactive processes in nuclear power stations.

B Radioactive processes within the Earth.

C The decay of organic material.

D The movement of the tides.   (1)

3 Renewable energy sources can be used to generate electricity. However these sources are not always available.
Match words from the list with the numbers **1** to **4** in the table.

A hydroelectric scheme

B solar cells

C tidal barrage

D wind farm

Renewable energy source	Source is available to generate electricity . . .
1	only during the daylight
2	only when the weather is suitable
3	only during certain periods of the day and night
4	usually whenever it is needed

(4)

4 Wind energy, waves, tides, falling water and solar energy can all be used as energy sources to generate electricity.

(a) What do all these energy sources have in common?

A They are available at any time of the day or night.

B They are renewable energy sources.

C They do not affect wildlife.

D They do not cause any sort of pollution. (1)

(b) Which of these energy sources is most appropriate to generate electricity to run a well in a remote African village?

A falling water   **B** solar energy

C tides   **D** waves (1)

(c) Which of these energy sources is most likely to produce noise pollution when used to generate electricity?

A solar energy   **B** tides

C waves   **D** wind energy (1)

(d) Which of these energy sources is **least** likely to be associated with damaging wildlife or the habitat of wildlife when used to generate electricity?

A falling water   **B** tides

C waves   **D** wind energy (1)

5 In coal, gas and oil-fired power stations fuels are burnt to produce heat.

(a) How is heat produced in a nuclear power station? (1)

(b) How is the heat used to produce electricity? (4)

(c) Apart from the cost of the electricity what are the advantages and disadvantages of using a nuclear power station to produce electricity? (5)

6 In the UK there are three different fossil fuels burnt in power stations.

(a) Name the three fossil fuels. (3)

(b) During burning all fossil fuels release carbon dioxide into the atmosphere. Some also release sulfur dioxide.

(i) Why does the release of carbon dioxide into the atmosphere produce a problem for the environment? (3)

(ii) Why is the release of sulfur dioxide a problem for the environment? (2)

HOW SCIENCE WORKS QUESTIONS

Tamara was interested in solar cells. She had been given a solar cell panel in a physics kit and set out to find how surface area affected the voltage that the panel could produce. Her solar cell panel was 10 cm × 3 cm. She set up a circuit with a voltmeter and the solar cell panel.

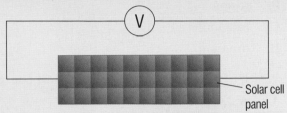

Solar cell panel

She took the circuit into the garden and covered different parts of the panel with black paper. Her preliminary work showed that it did not matter which part of the solar cell panel was covered, just how much was covered.

a) What do you think Tamara did in her preliminary work? (2)

Her final results are in this table.

Part of panel covered	Test 1 (V)	Test 2 (V)	Average (V)
None	0.3	0.4	0.35
A bit	0.3	0.3	0.3
A bit more	0.2	0.3	0.25
Most	0.1	0.1	0.1
All	0.1	0.1	0.1

b) Tamara was pleased that her results showed what she had expected.
What do you think these results show? (1)

c) Farzana said that Tamara's independent variable was not good enough.
What did Farzana mean by this? (1)

d) How could Tamara have improved her independent variable? (1)

e) Farzana looked at the voltage readings and suggested to Tamara that they were not very useful.
Why do you think Farzana thought that the readings were not very useful? (1)

f) Farzana suggested that she used a better voltmeter.
What type of voltmeter do you think Farzana suggested? (1)

g) Is there any evidence of a zero error in Tamara's results? (1)

h) What could Tamara do about this? (1)

EXAMINATION-STYLE QUESTIONS

Science A

1   In a nuclear power station the process that produces heat is called

   A   fission

   B   fusion

   C   radiation

   D   uranium   *(1 mark)*

See page 267

2   Which of these devices will transfer most energy?

   A   A 2 kW kettle used for 3 hours.

   B   A 3 kW heater used for 2 hours.

   C   A 4 kW motor used for 2 hours.

   D   A 7 kW shower used for 1 hour.   *(1 mark)*

See page 258

GET IT RIGHT!

The device with the biggest power rating may not transfer the most energy.

3   Which of these statements about solar cells is correct?

   A   In a solar cell, water is heated which produces steam and drives a turbine.

   B   Solar cells can produce electricity directly from the Sun's radiation.

   C   Solar cells produce electricity even in the dark.

   D   Solar cells transform geothermal energy into electrical energy   *(1 mark)*

See pages 270–1

4   Thermal energy can be transferred in different ways.
Match the words in the list with the numbers **1** to **4** in the sentences.

   A   electrons   **B**   liquids

   C   particles   **D**   solids

Conduction occurs mainly in ......**1**...... All metals are good conductors because they have a lot of free ......**2**....... Convection occurs in gases and ......**3**...... Radiation does not involve ......**4**......   *(4 marks)*

See pages 228–35

GET IT RIGHT!

Read through all of the sentences first and make sure that they all make sense with your choice of words before you select your answers in.

5   Electrical devices transform energy from electrical energy to other forms.
Match the words in the list with the numbers **1** to **4** in the sentences.

   A   kilowatt-hours   **B**   power

   C   joules   **D**   time

The ......**1**...... of a device is the rate at which it transforms energy. Energy is normally measured in ......**2**...... The amount of electrical energy a device transforms depends on the rate at which the device transforms energy and the ......**3**...... for which it used. The amount of electrical energy transferred from the mains is measured in ......**4**......   *(4 marks)*

See page 256–8

6   A student is doing an experiment on the rate of heat transfer from a beaker of hot water. Which of the following is true?

   A   The darker the colour of the beaker the slower the rate of heat transfer.

   B   The hotter the water the faster the rate of heat transfer.

   C   The shape of the beaker does not affect the rate of heat transfer.

   D   The temperature of the water does not affect the rate of heat transfer.

         (1 mark)

See pages 230, 236–7

Science B

1 The chart shows the energy resources used to produce electricity in Britain.

See pages 266–73

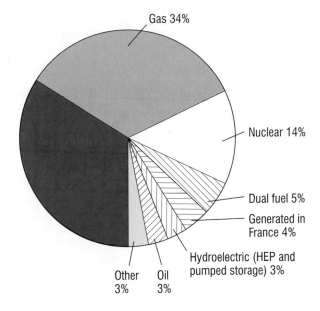

Gas 34%

Nuclear 14%

Dual fuel 5%

Generated in France 4%

Hydroelectric (HEP and pumped storage) 3%

Other 3% Oil 3%

(a) What percentage of the electricity is produced from coal? *(1 mark)*

(b) Give one advantage and one disadvantage, other than cost, of producing electricity from coal. *(2 marks)*

(c) In one power station, for every 1000 J of energy obtained from coal 300 J is wasted as heat. Use the following equation to work out the efficiency of the power station. Show clearly how you work out your answer.

$$\text{Efficiency} = \frac{\text{Useful energy transferred by the device}}{\text{Total energy transferred to the device}}$$

(3 marks)

See page 248

GET IT RIGHT!

When calculating efficiency make sure that you have correctly identified the useful energy transferred **by** the device and the total energy supplied **to** the device.

(d) In Britain 3% of the electricity is produced from *other* resources. One *other* resource is to use energy from the tides.
Discuss the advantages and disadvantages, other than cost, of producing electricity from tidal energy. *(3 marks)*

(e) Name one *other* resource, apart from the tides, not already given in the chart that is used to produce electricity. *(1 mark)*

2 A thermos flask is used to keep hot things hot and cold things cold. It does this by minimising heat transfer.
Explain how each of the following minimises heat transfer:

See page 236

(a) the tight fitting plastic stopper. *(2 marks)*

(b) the silver coating on the surfaces of the glass walls. *(2 marks)*

(c) the vacuum between the glass walls. *(2 marks)*

P1b | Radiation and the Universe

What you already know

Here is a quick reminder of previous work that you will find useful in this unit:

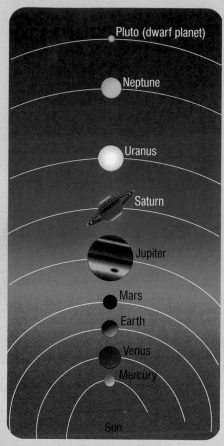

Figure 1 The Solar System showing the planets. The distances are not to scale. In 2006 Pluto was re-classified as a dwarf planet.

The Solar System
- The Moon orbits the Earth. The Earth and the other planets orbit the Sun.
- The Solar System consists of the Sun, the planets and their moons. The nearest star to the Sun is far beyond the Solar System.
- The Sun is a star. All stars give out their own light. They appear as pinpoints of light in the night sky because they are so far away.
- The Moon and the planets reflect sunlight. They do not give out their own light. We see them because they reflect sunlight to us.

Light
- We can use a prism to split sunlight into the colours of the spectrum (red, orange, yellow, green, blue, indigo and violet).
- Light that contains all the colours of the spectrum is called white light.
- We represent the path of light by light rays. Light travels in straight lines and is reflected by a mirror.

Gravity
- Gravity is a force of attraction between any two objects.
- The force of gravity between two objects depends on their masses and their distance apart.
- An orbiting object is kept in its orbit about a larger object by the force of gravity between it and the larger object. For example,
  - the Moon orbits the Earth,
  - satellites orbit the Earth,
  - the planets orbit the Sun.

RECAP QUESTIONS

1   a) Put the objects listed below in order of increasing size:

      the Earth   the Moon   the Sun

    b) Which is nearer to the Earth – the Sun or the Moon?

2   Complete the sentences below using words from the list.

      planet   satellite   star   Sun

    a) The Sun is a ....
    b) The Moon is a natural ...... of the Earth.
    c) The Earth is a ....
    d) Planets reflect light from the ....

3   The letters ROYGBIV represent the colours of the spectrum. What colour does each letter stand for?

4   a) What happens when a light ray strikes a mirror?
    b) An image of an object seen in a mirror always has the same colour as the object. What does this tell you about light and mirrors?
    c) How do we show the direction of a light ray?

5   a) What is the name for the force that keeps us on the Earth?
    b) What keeps the planets in their orbits round the Sun?
    c) What stops a rocket from leaving the Earth?

Making connections

Figure 2 A GPS receiver uses radio waves to tell you where you are

Radiation all around us

We use radiation in many different ways. Here are some different ways in which we use radiation.

On the move

If you want to know where you are or where you are heading, a GPS (Global Position Satellite) receiver will tell you. Satellites above you send out signals all the time. The signals are short-wave radio waves. You can pick them up with a hand-held receiver wherever you are.

Fixing a fracture

Have you or someone you know ever broken a limb? Doctors need to X-ray a broken bone before they reset it in plaster. X-rays are high-energy electromagnetic waves. The X-ray picture tells them exactly where it is broken and how badly it is broken. If it isn't reset correctly, the bones will re-grow crooked. Then they may need to be reset – and that can be very painful!

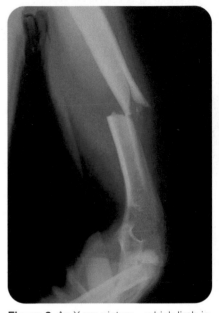

Figure 3 An X-ray picture – which limb is broken?

The Hubble Space Telescope

For almost twenty years, the Hubble Space Telescope has been orbiting the Earth. It's like a giant eye in space. Its pictures are beamed to the ground using radio waves.
It has sent us thousands of amazing images of objects in space. They range from nearby planets to distant galaxies at the edge of the Universe. In all this time, astronauts have only ever had to go once into space to repair it.

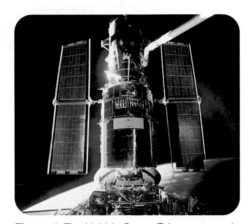

Figure 4 The Hubble Space Telescope sends us fantastic images using radio waves

ACTIVITY

The people in the village of Downdale want a mobile phone mast because they can't get a signal when they use their mobile phones. But where should the mast go?
Discuss with your friends the question of how they should decide where it should go.

Figure 5 No to mobile phone masts!

A mobile phone revolt!

Where would you be without your mobile phone? But what would you say if a mobile phone company wanted to put a mobile phone mast outside your house? Many people would object. They say that mobile phone radiation could be a risk to health. They don't want masts near their homes. But they won't give up their mobiles.

Chapters in this unit

Electromagnetic waves — Radioactivity — The origins of the Universe

P1b 5.1 The electromagnetic spectrum

LEARNING OBJECTIVES

1 What are the parts of the electromagnetic spectrum?
2 What do we mean by the frequency and wavelength of a wave and how can we calculate their values?
3 What do all electromagnetic waves have in common?

We all use waves from different parts of the electromagnetic spectrum. Figure 1 shows the spectrum and some of its uses.

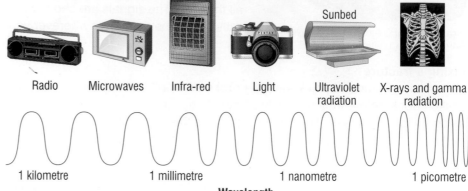

Radio   Microwaves   Infra-red   Light   Ultraviolet radiation   X-rays and gamma radiation

1 kilometre   1 millimetre   1 nanometre   1 picometre

Wavelength

(1 nanometre = 0.000 001 millimetres, 1 picometre = 0.001 nanometres)

Figure 1 Notice the spectrum is continuous. The frequencies/wavelengths at the boundaries are approximate as the different parts of the spectrum are not precisely defined.

Electromagnetic waves are electric and magnetic disturbances that transfer energy from one place to another.

Wavelength

Electromagnetic waves do not transfer matter. The energy they transfer depends on the **wavelength** of the waves. This is why waves of different wavelengths have different effects.

a) Which part of the electromagnetic spectrum causes sunburn?

The wavelength is the distance from one wave peak to the next wave peak along the waves.

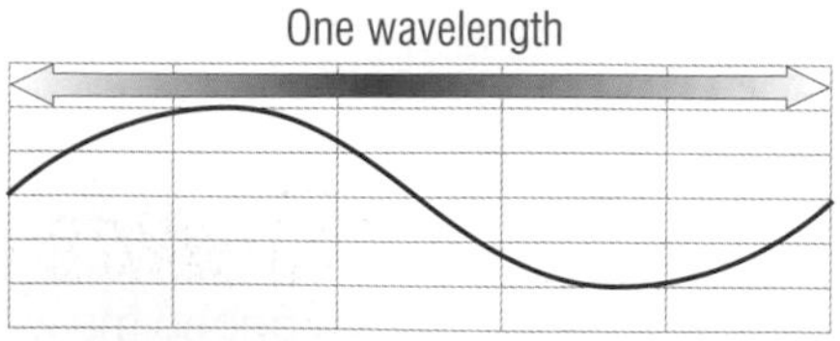

Waves from different parts of the electromagnetic spectrum have different wavelengths.

- Long-wave radio waves have wavelengths of more than 1000 m.
- X-rays and gamma rays have wavelengths of less than a millionth of a millionth of a metre (= 0.000 000 000 001 m).

One wavelength

Figure 2 Wavelength – take care **not** to measure the distance between a peak and a trough. Measure the distance between neighbouring peaks (or troughs).

b) Where in the electromagnetic spectrum would you find waves of wavelength 0.01 m?

Frequency

The **frequency** of electromagnetic waves of a certain wavelength is the number of complete waves passing a point each second.

The unit of frequency is the hertz (Hz), where:

- 1 hertz (Hz) = 1 complete wave per second,
- 1 kilohertz (kHz) = 1000 Hz (= 1 thousand hertz),
- 1 megahertz (MHz) = 1 000 000 Hz (= 1 million hertz).

DID YOU KNOW?

The spectrum of visible light covers just a very tiny part of the electromagnetic spectrum.

The speed of electromagnetic waves

All electromagnetic waves travel at a speed of 300 million m/s through space or in a vacuum. This is the distance the waves travel each second.

We can link the speed of the waves to their frequency and wavelength using this equation:

$$\text{wave speed} = \text{frequency} \times \text{wavelength}$$
$$\text{(metre/second, m/s)} \quad \text{(hertz, Hz)} \quad \text{(metre, m)}$$

1 We can work out the wavelength if we know the frequency and the wave speed. To do this, we rearrange the equation into:

$$\text{wavelength (in metres)} = \frac{\text{wave speed (in m/s)}}{\text{frequency (in Hz)}}$$

Worked example

A mobile phone gives out electromagnetic waves of frequency 900 million Hz. Calculate the wavelength of these waves.

The speed of electromagnetic waves in air = 300 million m/s.

Solution

$$\text{wavelength (in metres)} = \frac{\text{wave speed (in m/s)}}{\text{frequency (in Hz)}} = \frac{300\,000\,000 \text{ m/s}}{900\,000\,000 \text{ Hz}} = 0.33 \text{ m}$$

c) Work out the wavelength of electromagnetic waves of frequency 102 million Hz.

2 We can work out the frequency if we know the wavelength and the wave speed. To do this, we rearrange the equation into:

$$\text{frequency (in Hz)} = \frac{\text{wave speed (in m/s)}}{\text{wavelength (in metres)}}$$

d) Work out the frequency of electromagnetic waves of wavelength 1500 m.

SUMMARY QUESTIONS

1 Choose words from the list to complete each sentence in a), b) and c).

greater than   smaller than   the same as

a) The wavelength of light waves is ...... the wavelength of radio waves.
b) The speed of radio waves in a vacuum is ...... the speed of gamma rays.
c) The frequency of X-rays is ...... the frequency of infra-red radiation.

2 a) Fill in the missing parts of the electromagnetic spectrum in the list below.

radio   ......   infra-red   visible   ......   X-rays   ......

b) Work out:
  i) the wavelength of radio waves of frequency 600 million Hz,
  ii) the frequency of microwaves of wavelength 0.30 m.

KEY POINTS

1 The electromagnetic spectrum (in order of increasing wavelength) is:

gamma and X-rays
ultraviolet
visible
infra-red
microwaves
radio.

2 All electromagnetic waves travel through space at a speed of 300 million m/s.

3 Wave speed = wavelength × frequency.

5.2 Gamma rays and X-rays

1 What do we use X-rays and gamma rays for?
2 Why are X-rays and gamma rays dangerous?

X-rays

Have you ever broken an arm or a leg? If you have, you will have gone to your local hospital for an X-ray photograph.

X-rays pass through soft tissue but they are absorbed by bones and thick metal plates. To make a **radiograph** or X-ray picture, X-rays from an X-ray tube are directed at the patient. A light-proof cassette containing a photographic film is placed on the other side of the patient.

Figure 2 Spot the break

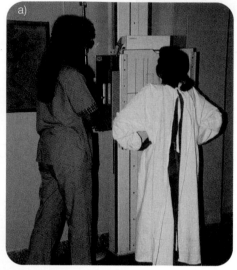

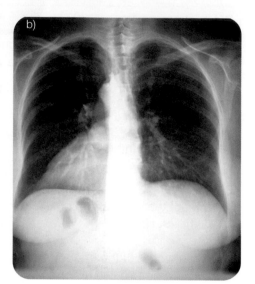

Figure 1 a) Taking a chest X-ray. b) A chest X-ray.

- When the X-ray tube is switched on, X-rays from the tube pass through the patient's body. They leave a 'shadow' image on the film showing the bones.
- When the film is developed, the parts exposed to X-rays are darker than the other parts. So the bones appear lighter than the surrounding tissue which appears dark. The developed film shows a 'negative image' of the bones.

a) Why is a crack in a bone visible on a radiograph (X-ray image)?

Gamma radiation

Gamma radiation is electromagnetic radiation from radioactive substances. Gamma rays and X-rays have similar wavelengths so they have similar properties. For example, a lead plate will stop gamma radiation or X-rays if it is several centimetres thick.

Gamma radiation is used:

- to kill harmful bacteria in food,
- to sterilise surgical instruments,
- to kill cancer cells.

b) Will gamma radiation pass through thin plastic wrappers?

We lose about 20% of the world's food through spoilage. One of the major causes is bacteria. The bacteria produce waste products that cause food poisoning. Exposing food to gamma radiation kills 99% of disease-carrying organisms, including *Salmonella* (found in poultry) and *Clostridium* (the cause of botulism).

Using gamma radiation

Doctors and medical physicists use gamma therapy to destroy cancerous tumours. A narrow beam of gamma radiation is directed at the tumour. The beam is aimed at it from different directions in order to kill the tumour but not the surrounding tissue. The cobalt-60 source, which produces the gamma radiation, is in a thick lead container. When it is not in use, it is rotated away from the exit channel.

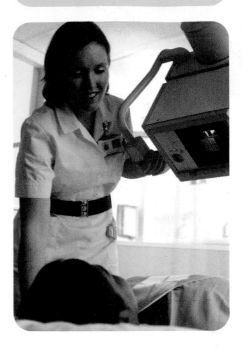

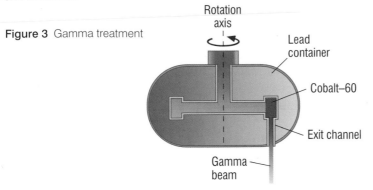

Figure 3 Gamma treatment

c) Why does the gamma beam need to be narrow?

Safety matters

Too much X-radiation or gamma radiation is dangerous and causes cancer. High doses kill living cells. Low doses cause cell mutation and cancerous growth. There is no evidence of a lower limit below which living cells would not be damaged.

Some people use equipment or substances that produce X-radiation or gamma radiation (or alpha or beta radiation) at work. (See page 301.) These workers must wear a film badge. If the badge is over-exposed to such radiation, its wearer is not allowed to continue working with the equipment.

d) Why does a film badge have a plastic case, and not a metal case?

Figure 4 A film badge tells you how much ionising radiation the wearer has received. Who might wear these?

SUMMARY QUESTIONS

1 Choose the correct words from the list to complete each sentence below.

absorb   damage   penetrate

a) X-rays and gamma rays …… thin metal sheets.
b) Thick lead plates will …… X-rays and gamma rays.
c) X-rays and gamma rays …… living tissue.

2 When an X-ray photograph is taken, why is it necessary:

a) to place the patient between the X-ray tube and the film cassette?
b) to have the film in a light-proof cassette?
c) to shield those parts of the patient not under investigation from X-rays? Explain what would happen to healthy cells.

KEY POINTS

1 X-rays and gamma radiation are absorbed by dense materials in bone and by 2metal.
2 X-rays and gamma radiation damage living tissue when they pass through it.
3 X-rays are used in hospitals to take radiographs.
4 Gamma rays are used to kill harmful bacteria in food, to sterilise surgical equipment and to kill cancer cells.

P1b 5.3 | Light and ultraviolet radiation

LEARNING OBJECTIVES

1 How does ultraviolet radiation differ from light?
2 Where do we use ultraviolet radiation?
3 Why is ultraviolet radiation dangerous?

Light from ordinary lamps and from the Sun is called **white light**. This is because it has all the colours of the visible spectrum in it. You see the colours of the spectrum when you look at a rainbow. You can also see them if you use a glass prism to split a beam of white light. But the human eye can't detect the ultraviolet radiation beyond the violet part of the spectrum or the infra-red radiation beyond the red part of the spectrum. (See page 228.)

- Coloured filters can be used to filter out colours from a beam of white light. For example, a red filter absorbs all colours except red from the light beam. So the beam is red when it comes out of the filter.
- Each colour of the visible spectrum has a different wavelength. The wavelength increases across the spectrum as you go from violet to red.
- Ultraviolet radiation is electromagnetic radiation between violet light and X-rays in wavelength.

a) Is the wavelength of ultraviolet radiation longer or shorter than the wavelength of light?

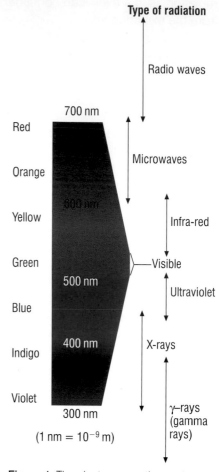

Figure 1 The electromagnetic spectrum with an expanded view of the visible range

Ultraviolet radiation

Ultraviolet radiation makes some chemicals emit light. Posters and ink that glow in ultraviolet light contain these chemicals. The chemicals absorb ultraviolet radiation and emit light as a result.

Ultraviolet radiation is harmful to human eyes and can cause blindness. Ultraviolet (UV) wavelengths are smaller than light wavelengths. UV rays carry more energy than light rays. Too much UV radiation causes sunburn and can cause skin cancer.

- If you stay outdoors in summer, use skin creams to block UV radiation and prevent it reaching the skin.
- If you use a sunbed to get a suntan, don't exceed the recommended time. You should also wear special 'goggles' to protect your eyes.

b) How does invisible ink work?

DID YOU KNOW?

The ozone layer in the atmosphere absorbs some of the Sun's ultraviolet radiation. Unfortunately, gases such as CFCs from old refrigerators react with ozone and make the ozone layer thinner. Many countries now have strict regulations about disposal of unwanted refrigerators.

Electromagnetic waves and substances

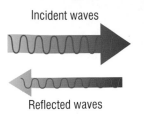
Incident waves

Reflected waves

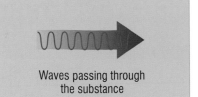
Waves passing through
the substance

Transmitted waves

Figure 2 Electromagnetic waves
and substances

When electromagnetic waves are directed at a substance:

- some or all of the waves may be **reflected** at the surface. The type of surface and the wavelength of the waves determine if the waves are totally or partly reflected or not reflected at all. For example,
  - light directed at ordinary glass is partly reflected,
  - X-rays directed at plastic are not reflected at all.
- some or all of the waves that go into a substance may be **absorbed** by it. The substance would become hotter due to the energy it gains from the absorbed radiation. Also, the radiation itself may create an alternating current at the same frequency as itself.
- The waves that are not absorbed by the substance are **transmitted** by it. The substance and the wavelength of the waves determine if the waves pass through it. For example:
  - ultraviolet radiation is mostly but not completely absorbed by glass,
  - X-rays mostly pass through human tissue and are mainly absorbed by bone.

c) Is the light from invisible ink due to ultraviolet light being reflected or absorbed or transmitted by the ink?

DEMONSTRATION

Ultraviolet radiation

Watch your teacher place different coloured clothes under an ultraviolet lamp. The lamp must point downwards so you can't look directly at the glow from it. Observe what happens.

- What do white clothes look like under a UV lamp?

d) How does a 'security' marker pen work?

Figure 3 Using an ultraviolet lamp to detect biological stains. The stain absorbs ultraviolet radiation from the lamp and gives out visible light as a result. The background red lighting makes the light from the stain more visible.

SUMMARY QUESTIONS

1. Choose words from the list for each of the spaces in the sentences below. Each option can only be used once.

   red light   **blue light**   **ultraviolet radiation**   **white light**

   a) …… has a longer wavelength than …… .
   b) …… from the Sun is absorbed by the ozone layer.
   c) …… includes all the colours of the spectrum.

2. a) Why is ultraviolet radiation harmful?
   b) i) How does the Earth's ozone layer help to protect us from ultraviolet radiation from the Sun?
   ii) Why do people outdoors in summer need suncream?

3. Explain what happens to the energy carried by electromagnetic waves when they are reflected, transmitted or absorbed.

KEY POINTS

1. Ultraviolet radiation is in the electromagnetic spectrum between violet light and X-radiation.
2. Ultraviolet radiation has a shorter wavelength than light.
3. Ultraviolet radiation harms the skin and the eyes.

ELECTROMAGNETIC WAVES

Infra-red, microwaves and radio waves

LEARNING OBJECTIVES

1 What do we use infra-red radiation for?
2 Why are microwaves used for heating?
3 When do we use infra-red radiation, microwaves and radio waves for communications?

Infra-red radiation

All objects emit infra-red radiation.

● The hotter an object is, the more infra-red radiation it emits. (See page 228 for more about the properties of infra-red radiation.)
● Infra-red radiation is absorbed by the skin. It damages or kills skin cells because it heats up the cells.

a) Can you remember where infra-red radiation lies in the electromagnetic spectrum? (Look back to page 282 if necessary.)

Infra-red devices

● **Heaters** in grills, toasters, and electric heaters all emit infra-red radiation to heat objects.
● **Infra-red scanners** are used in medicine to detect 'hot spots' on the body surface, which can mean the underlying tissue is unhealthy. You can use **infra-red cameras** to see people and animals in darkness.
● **Optical fibres** in communications systems use infra-red radiation instead of light. This is because infra-red radiation is absorbed less than light in the glass fibres.
● **Remote control handsets** for TV and video equipment transmit signals carried by infra-red radiation. When you press a button on the handset, it sends out a sequence of infra-red pulses.

Figure 1 Infra-red devices

PRACTICAL

Testing infra-red radiation

Can infra-red radiation pass through paper?

You can use a remote handset to find out.

● What happens?

Microwaves

Microwaves lie between radio waves and infra-red radiation in the electromagnetic spectrum. They are called '**micro**waves' because they are shorter in wavelength than radio waves.

We use microwaves for:

● **communications**, because they can pass through the atmosphere and reach satellites orbiting the Earth. We use them to 'beam' signals from one place to another, because they don't spread out as much as radio waves;
● **heating food in microwave ovens**. These heat the food from the inside as well as from the outside. Unlike infra-red radiation, microwaves penetrate substances like food.

Figure 2 A microwave oven heats food from the inside as well as from the outside

b) i) Can microwaves pass through plastic?
   ii) Why is it dangerous to put metal objects in a microwave oven?

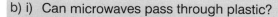

Radio waves

Radio wave frequencies range from about 300 000 Hz to 3000 million Hz (where microwave frequencies start). Radio waves are longer in wavelength and lower in frequency than microwaves.

We use radio waves to carry radio, TV and mobile phone signals.

- Radio waves are emitted from an aerial when we apply an alternating voltage to the aerial. The frequency of the radio waves produced is the same as the frequency of the alternating voltage.
- When the radio waves pass across a receiver aerial, they cause a tiny alternating voltage in the aerial. The frequency of the alternating voltage is the same as the frequency of the radio waves received.

Figure 3 A radio transmitter

PRACTICAL

Testing microwaves

Look at the demonstration shown.

- What does this show?

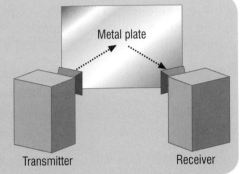

Figure 4 Testing microwaves

KEY POINTS

	Frequency	Wavelength	Applications
Infra-red	↑	↓	heaters, communications (remote handsets, optical fibres)
Microwaves			microwave oven, communications
Radio waves			communications

SUMMARY QUESTIONS

1 Use words from the list to complete the sentences below:

infra-red radiation   light   microwaves   radio waves

a) In a TV set, the aerial detects ...... and the screen emits ....
b) In a microwave oven, food absorbs ......, heats up and emits ....

2 Complete the table below showing the type of electromagnetic radiation produced by each device a) to d).

Device	Infra-red	Microwave	Radio
a) Electric toaster			
b) Microwave oven			
c) TV broadcast transmitter			
d) Remote handset			

3 Describe how a radio transmitter produces radio waves and what happens to the radio waves at a receiver aerial.

DID YOU KNOW?

The sinking of the *Titanic* in 1912 claimed over 1500 lives. There weren't enough lifeboats for everyone. A few years earlier, two ships collided and everyone was saved – because an SOS radio message was sent out. Nearby ships got there in good time. Ship owners decided they didn't need lifeboats for everyone. After the *Titanic* disaster, they were made to put all the lifeboats back!

P1b 5.5   Communications

LEARNING OBJECTIVES

1 Which waves do we use for mobile phones and satellite links?

2 Why can microwaves be used to carry satellite TV signals?

3 Why are optical fibres useful for communications?

Radio communications

The radio and microwave spectrum is divided into *frequency bands*. How we use each band depends on its frequency range. This is because the higher the frequency of the waves:

- the more information they can carry,
- the shorter their range (due to increasing absorption by the atmosphere),
- the less they spread out (because they diffract less).

Figure 1 Using a car radio

Wavebands

Waveband	Frequency range	Uses
Microwaves	greater than 3000 MHz	Satellite links (e.g. phone and TV) Mobile phones
UHF (ultra-high frequency)	300–3000 MHz	Terrestrial TV Mobile phones
VHF (very high frequency)	30–300 MHz	Local radio (FM), Emergency services Digital radio
HF (high frequency)	3–30 MHz	Amateur radio, CB
MF (medium frequency – also called 'medium wave' or MW)	300 kHz – 3 MHz	National radio (analogue)
LF (low frequency – also called 'long wave' or LW)	less than 300 kHz	International radio (analogue)
(M = mega = million)		

DID YOU KNOW?

You can 'tune in' to distant LF radio stations at any time if your radio receiver is powerful enough. Long wave radio waves stay near the ground and follow the Earth's curvature.

a) What is the difference between satellite TV and terrestrial TV?

Radio waves and the ionosphere

The ionosphere is a layer of gas in the upper atmosphere. It reflects radio waves that have frequencies less than about 30 MHz. So it reflects HF, MF and LF wavebands.

The ionosphere is stronger in summer than in winter. This is why you can listen to distant radio stations in summer but not in winter. Radio waves from these stations bounce back and forth between the ionosphere and the ground.

b) Why can't you listen to distant MF and HF radio stations in winter?

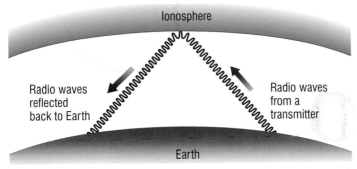

Figure 2 LF, MF and HF radio waves are reflected by the ionosphere. So they can be sent very long distances despite the curvature of the Earth.

c) Why can't you watch TV stations that are below the horizon?

Optical fibre communications

Optical fibres are very thin glass fibres. We use them to transmit signals carried by light or infra-red radiation. The light rays can't escape from the fibre. When they reach the surface of the fibre, they are reflected back into the fibre.

In comparison with radio waves and microwaves:

- optical fibres can carry much more information – this is because the frequency of light and infra-red radiation is much higher,
- optical fibres are more secure because the signals stay in the fibre.

d) Why can't we use optical fibres for satellite communications?

Satellite TV

Satellite TV signals are carried by microwaves. We can detect them on the ground because they pass through the ionosphere. But you can't watch distant TV stations because the signals go straight through the ionosphere into space – they are not reflected back.

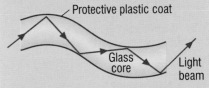
SUMMARY QUESTIONS

1 Choose the correct word from the list to fill in each of the spaces in the sentences below.

microwaves   **radio waves**

a) TV signals are carried by ....
b) Satellite TV signals are carried by ....
c) A beam of ...... can travel from the ground to a satellite but a beam of ...... waves cannot if its frequency is below 30 MHz.

2 a) Why is it not possible to tune your radio in to American local radio stations?
b) Why are signals in optical fibres more secure than radio signals?

3 Explain why we use microwaves for satellite communications.

KEY POINTS

1 The use we make of radio waves depends on the frequency of the waves.
2 Visible light and infra-red radiation are used to carry signals in optical fibres.

291

P1b 5.6   Analogue and digital signals

LEARNING OBJECTIVES

1 How do digital and analogue signals differ?
2 What are the advantages of digital signals?

In the next ten years, we will all have to change our TV sets or adapt them. At present, TV stations transmit *analogue* signals and *digital* signals. But after the 'Big Switchover', they will only send out digital signals.

- **A digital signal** is a sequence of pulses. The voltage level of each pulse is either high (a '1') or low (a '0') with no in-between levels. Each '0' or '1' is called a **bit**. Mobile phone signals and signals from computers are digital.
- **An analogue signal** is a wave that varies continuously in amplitude or frequency between zero and a maximum value. For example, a microphone generates electrical waves when it detects sound waves.

a) A 'digibox' changes an analogue signal to a digital signal. Why is a digibox needed to receive digital TV at present?

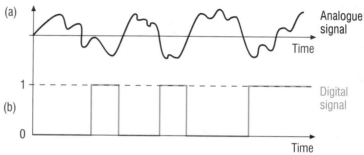

Figure 1 a) An analogue signal, b) a digital signal.

Sending signals

The waves we use to carry a signal are called **carrier waves**. They could be radio waves, microwaves, infra-red radiation or light.

- To send a digital signal, the pulses are used to switch the carrier waves on and off repeatedly. Digital radio transmitters, scanners and fax machines all convert analogue signals into digital signals which are then sent.

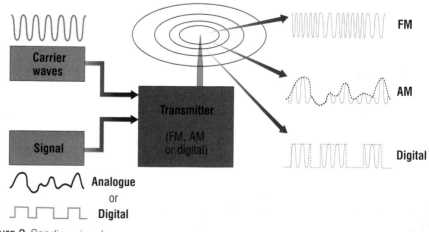

Figure 2 Sending signals

- To send an analogue signal, the signal waves are used to vary or **modulate** the carrier waves. The amplitude (amplitude modulation AM) or the frequen (frequency modulation FM) of the carrier waves is modulated in this proces

b) Is an e-mail message analogue or digital?

Why digital is better than analogue

- Digital signals suffer **less interference** than analogue signals. Interference causes a hissing noise when you listen to analogue radio. It doesn't happen with digital signals because regenerator circuits are used to clean 'noisy' pulses. So a digital signal has a higher quality than an analogue signal.

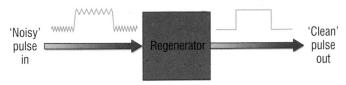

'Noisy' pulse in   Regenerator   'Clean' pulse out

Figure 3 Cleaning a noisy pulse

- **Much more information** can be sent using digital signals instead of analogue signals. Digital pulses can be made very short so more pulses can be carried each second.

c) Do video phones use analogue or digital signals?

DID YOU KNOW?

A single optical fibre can carry 0 000 phone calls or 30 TV nnels at the same time. In arison, a wire cable can rry about 2000 calls at an e time, and a owave beam only about 0 phone calls.

SUMMARY QUESTIONS

1 Choose a word from the list to complete the sentences below.

analogue   carrier   digital

a) AM and FM are forms of …… signals in which the signal modulates a …… wave.

b) …… signals have a higher quality than …… signals.

c) …… signals consist of pulses.

2 Look at the diagram below. It shows a communication system.

Match the boxes labelled **A–D** with the words from the list below.

carrier wave   receiver   signal   transmitter

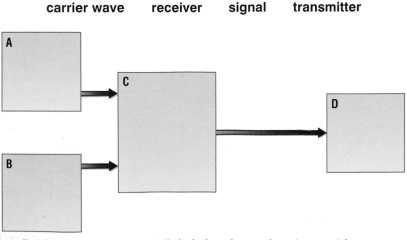

3 Describe how analogue and digital signals can be changed from one to the other.

KEY POINTS

1 Analogue signals vary continuously in amplitude.

2 Digital signals are either high ('1') or low ('0').

3 Digital transmission, when compared with analogue transmission, is free of noise and distortion. It can also carry much more information.

P1b 5.7   Microwave issues

A key scientific invention

The Battle of the Atlantic was one of the most important battles in the Second World War. Enemy U-boat submarines almost stopped supplies from North America reaching Britain by sinking so many British cargo ships. But the submarine menace was defeated as a result of an invention in 1940 by the British scientists, John Randall and Henry Boot.

They invented the high-power magnetron. It was the first high-power microwave transmitter. It replaced the low-power microwave transmitters used until then in radar sets. The new high-power radar sets were fitted into aircraft and warships to find submarines when they surfaced.

Randall and Boot's invention was invaluable. Now we use it in microwave ovens as well as in radar systems.

QUESTION

1 A high-power radar set has a longer range than a low-power set. Why was this important in the Battle of the Atlantic?

The Big Switchover

Digital TV is in a different waveband to analogue TV. Viewers have a greater choice of TV channels on digital. When analogue TV ends, its waveband will be sold off to mobile phone companies. They will get their money back from the charges their customers pay. But the Big Switchover could be costly for TV viewers because they will have to replace their TV set or buy a digibox to adapt it.

ACTIVITY

Either
Imagine you are a journalist on your local newspaper and the date for the Big Switchover has been set. Write a short article about it for your newspaper.
Or
Discuss this viewer's opinion on the 'Big Switchover':

I'm perfectly happy with my old TV. Why should I have to spend good money to switch from analogue to digital? It's just not right!

Using your mobile phone

When you use a mobile phone to talk to a friend, the handset sends out a radio signal. This is detected by the nearest receiver mast. The receiver is linked to a telephone exchange so your call is then routed to your friend's phone. The local mast sends out radio waves carrying the return signal from your friend.

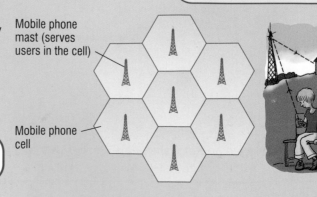

Mobile phone mast (serves users in the cell)

Mobile phone cell

ACTIVITY

Compare mobile phone costs from two different companies and say which is best for you.

Selling the radio spectrum

How can you sell radio waves? Mobile phone companies are only allowed to use a narrow band of frequencies in the UHF waveband. Mobile phone companies have to pay the Government to use these frequencies.

Who decides what each frequency band can be used for? Imagine the chaos if air traffic controllers and taxi drivers used the same frequency bands. Engineers advise the Government on the best use of each frequency band. Users are then required by law to keep to the band chosen for them.

ACTIVITY

Imagine a taxi firm uses an 'ambulance' radio channel by mistake. Write a short story about a 'mix-up' that happened when the taxi firm used the 'ambulance' radio channel.

Mobile phone hazards 12:00

Microwaves and short wave radio waves penetrate living tissues and heat the water inside living cells. This can damage or even kill the cells. Microwave ovens have safety switches that turn the microwaves off when the door is open. Also, the metal casing acts as a shield to stop microwaves from escaping.

Mobile phones send out radio waves when they are used. The waves are very low-power waves.

Is the radiation from a mobile phone hazardous? At the present time, there is no conclusive scientific evidence one way or the other. Some people believe that mobile phones can cause brain tumours. The government recommends caution, particularly for young people, until scientists and doctors can find out more.

Are mobile phones intrusive? People can get upset when mobile phone users share their views with everyone within hearing range. A videophone user who points the phone camera at someone else can be prosecuted. Mobile phone technology is here to stay but users should not upset other people or use mobile phones illegally.

Menu   Contacts

GREENFIELD SCHOOL

QUESTION

2 a) Why might young people be more affected by mobile phone radiation than older people?

b) How would scientists see if there were a link (correlation) between the use of mobile phones and brain tumours?

c) How might scientists show that mobile phones cause brain tumours?

d) Mobile phones can be dangerous in other ways as well. Why is driving and talking on your mobile phone dangerous? You might step off the pavement at the wrong moment if you are talking to someone on your mobile at the same time. Design a poster to warn people about the dangers of using a mobile phone.

ACTIVITY

A mobile phone company wants to put a mobile phone mast on the roof of your school. The teachers and the students don't want it. The school governors say the company will pay rent to the school and the money can be used for more computers.

Discuss your opinions on this proposal with your friends.

How would you advise your student council to respond to the proposal?

SUMMARY QUESTIONS

1 a) Place the four different types of electromagnetic waves listed below in order of increasing wavelength.

    A Infra-red waves   **B** Microwaves
    C Radio waves   **D** Gamma rays

b) The radio waves from a local radio station have a wavelength of 3.3 metres in air and a frequency of 91 million hertz.
  i) Write down the equation that links frequency, wavelength and wave speed.
  ii) Calculate the speed of the radio waves in air.

2 At the top of page 282 you will find the typical wavelengths of electromagnetic waves.

Match each of **A**, **B**, **C** and **D** below with **1** to **4** in the second list.

    A 0.0005 mm   **B** 1 millionth of 1 mm
    C 10 cm   **D** 1000 m

  1 X-rays   **2** light   **3** microwaves   **4** radio

3 a) Complete the following sentences using words from the list below.

    gamma radiation   **infra-red radiation**
    light   **ultraviolet radiation**

  i) The Earth's ozone layer absorbs …… .
  ii) An ordinary lamp gives out …… and …… .
  iii) …… passes through a metal object.

b) Which type of radiation listed above damages the following parts of the human body?
  i) the internal organs,
  ii) the eyes but not the internal organs,
  iii) the skin but not the eyes or the internal organs.

4 a) Complete each of the sentences below using words from the list.

    microwave   **mobile phone**
    radio waves   **TV**

  i) A …… beam can travel from a ground transmitter to a satellite, but a beam of …… cannot if its frequency is below 30 MHz.
  ii) …… signals and …… signals always come from a local transmitter.

b) i) Explain the difference between a digital signal and an analogue signal.
  ii) State and explain two advantages of digital transmission compared with analogue transmission.

EXAM-STYLE QUESTIONS

1 Electromagnetic waves can be grouped according to their wavelength and frequency.
Match the words in the list with the spaces **1** to **4** in the diagram.

    A gamma rays   **B** microwaves
    C ultraviolet rays   **D** visible light

Increasing wavelength, decreasing frequency

⟶

  1   X-rays   **2**   **3**   Infra red rays   **4**   Radio waves

(4)

2 The number of waves passing a point each second is the …
  A amplitude   **B** frequency
  C speed   **D** wavelength   (1)

3 Which of the following statements about the waves of the electromagnetic spectrum is true?
  A They all have the same frequency.
  B They all have the same wavelength.
  C They all travel at the same speed through space.
  D They cannot travel through a vacuum.   (1)

4 The uses of the radiations in different parts of the electromagnetic spectrum depend on their wavelength and frequency.

(a) Shadow pictures of the bones can be produced using …
  A microwaves.   **B** ultraviolet rays.
  C visible light.   **D** X-rays.   (1)

(b) Which type of electromagnetic radiation is used to send signals from a TV remote control?
  A infra-red rays.   **B** microwaves.
  C radio waves.   **D** ultraviolet rays.   (1)

(c) Which type of electromagnetic radiation is used to sterilise surgical instruments?
  A gamma rays   **B** microwaves
  C ultraviolet rays   **D** visible light   (1)

(d) What is the equation that relates the speed, wavelength and frequency of the waves of the electromagnetic spectrum?
  A Speed = frequency $\times$ wavelength
  B Speed = frequency $\div$ wavelength
  C Speed = wavelength $\div$ frequency
  D Speed = wavelength + frequency   (1)

5 (a) Information can be transmitted through optical fibres.
Name two types of electromagnetic wave used to carry information through an optical fibre. (2)

(b) Information can be sent as a digital signal or an analogue signal. What is the difference between a digital and an analogue signal? (2)

(c) A signal gets weaker as it travels and needs to be amplified. Explain why an amplified analogue signal will have deteriorated compared with the original signal. (3)

6 Astronauts in space wear special suits designed to prevent dangerous radiation from the Sun reaching their bodies.

Explain how each of these types of electromagnetic radiation can harm the body:

(a) gamma rays (2)

(b) ultraviolet rays (3)

(c) microwaves. (2)

HOW SCIENCE WORKS QUESTIONS

Your teacher has set up a demonstration of light radiation. She used a slide projector to shine light onto a prism. The prism split the light into the colours of the rainbow – a spectrum. Then a thermistor was placed into the spectrum of light.

Thermistors can be used to measure heat. When put into a circuit, the hotter the thermistor the greater the voltage, measured by a voltmeter.

You can get a better idea of what she did from this diagram.

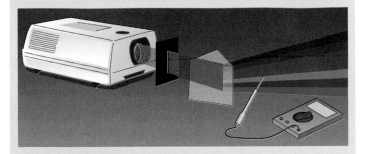

The thermistor was gradually moved through the spectrum from the violet end to the red end and beyond. The voltage was taken every 10 seconds and the colour of light was also recorded.

Here are the results.

Time (seconds)	Voltage (mV)	Colour
0	745	none
10	750	violet
20	760	indigo
30	770	blue
40	780	green
50	790	yellow
60	800	orange
70	810	red
80	990	none

a) What is the pattern in these results? Complete the sentence:
The longer the wavelength of light ...... (1)

b) Which result could be an anomaly? (1)

c) What should be done with this anomaly? (2)

d) If you wanted to draw a graph showing how the voltage varied with time, explain what type of graph you would use. (2)

P1b 6.1

Observing nuclear radiation

LEARNING OBJECTIVES

1 How can we observe radioactivity?
2 When does a radioactive source give out radiation (radioactivity)?
3 Why does a radioactive source give out radiation (radioactivity)?

A key discovery

Figure 1 Becquerel's key

If your photos showed a mysterious image, what would you think? In 1896, the French physicist, **Henri Becquerel**, discovered the image of a key on a film he developed. He remembered the film had been in a drawer under a key. On top of that there had been a packet of uranium salts. The uranium salts must have sent out some form of radiation that passed through paper (e.g. the film wrapper) but not through metal (e.g. the key).

Becquerel asked a young research worker, **Marie Curie**, to investigate. She found that the salts gave out radiation all the time. It happened no matter what was done to them. She used the word **radioactivity** to describe this strange new property of uranium.

She and her husband, Pierre, did more research into this new branch of science. They discovered new radioactive elements. They named one of the elements **polonium**, after Marie's native country, Poland.

a) You can stop a lamp giving out light by switching it off. Is it possible to stop uranium giving out radiation?

Marie Curie 1867-1934

Becquerel and the Curies were awarded the Nobel prize for the discovery of radioactivity. Pierre died in a road accident. Marie went on with their work. She was awarded a second Nobel prize in 1911 for the discovery of polonium and radium. She died in 1934 from leukaemia. This is a disease of the blood cells and was caused by the radioactive materials she worked with.

Figure 2 Marie Curie 1867–1934

PRACTICAL

Investigating radioactivity

We can use a **Geiger counter** to detect radioactivity. Look at Figure 3. The counter clicks each time a particle of radiation from a radioactive substance enters the Geiger tube.

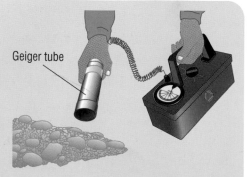

Figure 3 Using a Geiger counter

What stops the radiation? Ernest Rutherford carried out tests to answer this question about a century ago. He put different materials between the radioactive substance and a 'detector'.

He discovered two types of radiation:

- One type (**alpha radiation**, symbol α) was stopped by paper.
- The other type (**beta radiation**, symbol β) went through it.

Scientists later discovered a third type, **gamma radiation** (symbol γ), even more penetrating than beta radiation.

b) Can gamma radiation go through paper?

A radioactive puzzle

Why are some substances radioactive? Every atom has a nucleus made up of protons and neutrons. Electrons move about in the space surrounding the nucleus.

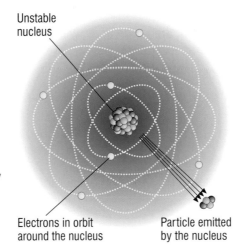

Unstable nucleus

Figure 4 Radioactive decay

Electrons in orbit around the nucleus

Particle emitted by the nucleus

Most atoms each have a stable nucleus that doesn't change. But the atoms of a radioactive substance each have a nucleus that is unstable. An unstable nucleus becomes stable by emitting alpha, beta or gamma radiation. We say an unstable nucleus *decays* when it emits radiation.

We can't tell when an unstable nucleus will decay. It is a **random** event that happens without anything being done to the nucleus.

c) Why is the radiation from a radioactive substance sometimes called nuclear radiation?

SUMMARY QUESTIONS

1 Complete the sentences below, using words from the list.

   proton   neutron   nucleus   radiation

   a) The …… of an atom is made up of …… and …… .
   b) When an unstable …… decays, it emits …… .

2 a) The radiation from a radioactive source is stopped by paper. What type of radiation does the source emit?
   b) The radiation from a different source goes through paper. What can you say about this radiation?

3 Explain why some substances are radioactive.

KEY POINTS

1 A radioactive substance contains unstable nuclei.
2 An unstable nucleus becomes stable by emitting radiation.
3 There are 3 main types of radiation from radioactive substances – alpha, beta and gamma radiation.
4 Radioactive decay is a random event – we cannot predict or influence when it will happen.

299

P1b 6.2

Alpha, beta and gamma radiation

LEARNING OBJECTIVES

1 How far can each type of radiation travel in air and what stops it?
2 What is alpha, beta and gamma radiation?
3 Why is alpha, beta and gamma radiation dangerous?

FOUL FACTS

Radium is an element. It is so radioactive that it glows with light on its own. Many years ago, the workers in a US factory used radium paint on clocks and aircraft dials. They often licked the brushes to make a fine point to paint very accurately on the dials. Many of them later died from cancer.

Penetrating power

Alpha radiation can't penetrate paper.

But what stops beta and gamma radiation? And how far can each type of radiation travel through air? We can use a Geiger counter to find out.

● To test different materials, we need to place each material between the tube and the radioactive source. Then we can add more layers of material until the radiation is stopped.

● To test the range in air, we need to move the tube away from the source. When the tube is beyond the range of the radiation, it can't detect it.

Look at the table below:

It shows the results of the two tests.

Type of radiation	Absorber materials	Range in air
alpha (α)	paper	about 10 cm
beta (β)	aluminium sheet (1 cm thick) lead sheet (2–3 mm thick)	about 1 m
gamma (γ)	thick lead sheet (several cm thick) concrete (more than 1 m thick)	unlimited

Gamma radiation spreads out in air without being absorbed. It does get weaker as it spreads out.

a) Why is a radioactive source stored in a lead-lined box?

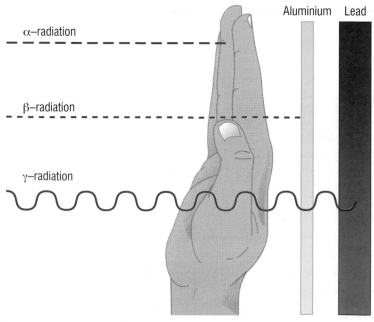

Figure 1 The penetrating powers of α, β and γ radiation

The nature of alpha, beta and gamma radiation

What are these mysterious radiations? They can be separated using a magnetic field. Look at Figure 2.

We use magnetic fields to deflect electron beams in a TV tube. The beams create the picture as they scan across the inside of the tube.

- β-radiation is easily deflected, in the same way as electrons. So it has a negative charge. In fact, a β-particle is a fast-moving electron. It is emitted by an unstable nucleus containing too many neutrons when it decays.
- α-radiation is deflected in the opposite direction to β-radiation because an α-particle has a positive charge. α-radiation is harder to deflect than β-radiation. This is because an α-particle is a lot heavier than a β-particle. It has a much greater mass. In fact, an alpha particle is two protons and two neutrons stuck together, the same as a helium nucleus.
- γ-radiation is not deflected by a magnetic field or an electric field. This is because gamma radiation is electromagnetic radiation so is uncharged.

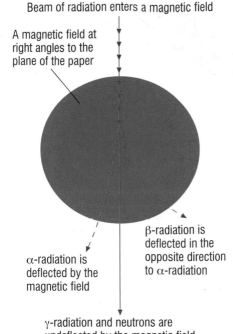

Beam of radiation enters a magnetic field

A magnetic field at right angles to the plane of the paper

β-radiation is deflected in the opposite direction to α-radiation

α-radiation is deflected by the magnetic field

γ-radiation and neutrons are undeflected by the magnetic field

Figure 2 Radiation in a magnetic field

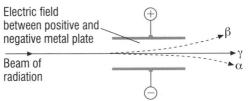

Electric field between positive and negative metal plate

Beam of radiation

Note that α and β-particles passing through an electric field are deflected in opposite directions.

Figure 3 Radiation passing through an electric field

b) How do we know that gamma radiation is not made up of charged particles?

Radioactivity dangers

The radiation from a radioactive substance can knock electrons out of atoms. The atoms become charged because they lose electrons. The process is called **ionisation**.

X-rays also cause ionisation. Ionisation in a living cell can damage or kill the cell. Damage to the genes in a cell can be passed on if the cell generates more cells. Strict rules must always be followed when radioactive substances are used.

Alpha radiation has a greater ionising effect than beta or gamma radiation. This makes it is more dangerous in the body than beta or gamma radiation. See page 307 for more information.

Figure 4 Radioactive warning

c) Why should long-handled tongs be used to move a radioactive source?

SUMMARY QUESTIONS

1 Choose words from the list to complete the sentences below:

alpha   beta   gamma

a) Electromagnetic radiation from a radioactive substance is called ...... radiation.

b) A thick metal plate will stop ...... and ...... radiation but not ...... radiation.

2 Which type of radiation is: a) uncharged? b) positively charged? c) negatively charged?

3 Explain why ionising radiation is dangerous.

KEY POINTS

1 **α-radiation** is stopped by paper or a few centimetres of air.
2 **β-radiation** is stopped by thin metal or about a metre of air.
3 **γ-radiation** is stopped by thick lead and has an unlimited range in air.

P1b 6.3   Half-life

LEARNING OBJECTIVES

1 What happens to the activity of a radioactive isotope as it decays?
2 What do we mean by the 'half-life' of a radioactive source?

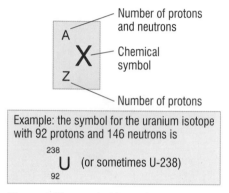

Example: the symbol for the uranium isotope with 92 protons and 146 neutrons is

$$^{238}_{92}U \text{ (or sometimes U-238)}$$

Figure 1 The symbol for an isotope

Every atom of an element always has the same number of protons in its nucleus. However, the number of neutrons in the nucleus can differ. Each type of atom is called an **isotope**. (So isotopes of an element contain the same number of protons but different numbers of neutrons.)

The **activity** of a radioactive substance is the number of atoms that decay per second. Each unstable atom (the 'parent' atom) forms an atom of a different isotope (the 'daughter' atom) when its nucleus decays. Because the number of parent atoms goes down, the activity of the sample decreases.

We can use a Geiger counter to monitor the activity of a radioactive sample. We need to measure the **count rate** due to the sample. This is the number of counts per second (or per minute). The graph below shows how the count rate of a sample decreases.

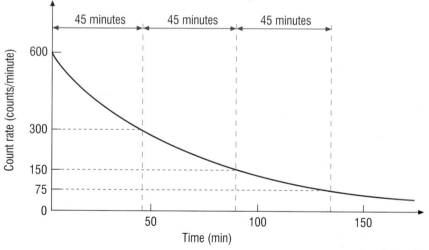

Figure 2 Radioactive decay: a graph of count rate against time

The graph shows that the count rate decreases with time. The count rate falls from:

● 600 counts per minute (c.p.m.) to 300 c.p.m. in the first 45 minutes,
● 300 counts per minute (c.p.m.) to 150 c.p.m. in the next 45 minutes.

The time taken for the count rate (and therefore the number of parent atoms) to fall by half is always the same. This time is called the **half-life**. The half-life shown on the graph is 45 minutes.

a) What will the count rate be after 135 minutes from the start?

The half-life of a radioactive isotope is the time it takes:

● **for the number of nuclei of the isotope in a sample (and therefore the mass of parent atoms) to halve,**
● **for the count rate due to the isotope in a sample to fall to half its initial value.**

The random nature of radioactive decay

We can't predict **when** an individual atom will suddenly decay. But we **can** predict how many atoms will decay in a certain time – because there are so many of them. This is a bit like throwing dice. You can't predict what number you will get with a single throw. But if you threw 1000 dice, you would expect one-sixth to come up with a particular number.

Suppose we start with 1000 unstable atoms. Look at the graph below:

NEXT TIME YOU...

... help someone choose numbers for the lottery, think about whether this is something you can predict. The balls come out of the machine at random; is there any way of predicting what they will be?

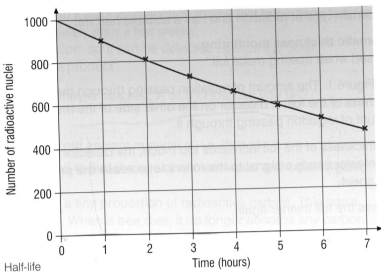

Figure 3 Half-life

If 10% disintegrate every hour,

100 atoms will decay in the first hour, leaving 900.
90 atoms (= 10% of 900) will decay in the second hour, leaving 810.

The table below shows what you get if you continue the calculations.
The results are plotted as a graph in Figure 3.

Time from start (hours)	0	1	2	3	4	5	6	7
No. of unstable atoms present	1000	900	810	729	656	590	530	477
No. of unstable atoms that decay in the next hour	100	90	81	73	66	59	53	48

b) Use the graph in Figure 3 to work out the half-life of this radioactive isotope.

SUMMARY QUESTIONS

1 Complete the following sentences using words from the list below.

       half-life   **stable**   **unstable**

a) In a radioactive substance, ... atoms decay and become ....
b) The ...... of a radioactive isotope is the time taken for the number of ...... atoms to decrease to half.

2 A radioactive isotope has a half-life of 15 hours. A sealed tube contains 8 milligrams of the substance.

What mass of the substance is in the tube:

a) 15 hours later?
b) 45 hours later?

KEY POINTS

1 The **half-life** of a radioactive isotope is the time it takes for the number of nuclei of the isotope in a sample to halve.
2 The number of unstable atoms and the activity decreases to half in one half-life.

P1b 6.5   Radioactivity issues

Nuclear waste

The fuel rods in nuclear power stations are radioactive. Used fuel rods are very hot and are still very radioactive when they are removed from a nuclear reactor. They contain many radioactive isotopes that are formed when the uranium nuclei split.

Figure 1 Storage of nuclear waste

- After removal from a reactor, used fuel rods are stored in large tanks of water for up to a year. The water cools down the rods.
- Remote-control machines are then used to open the fuel rods. The machines remove unused uranium. This is stored in sealed containers so it can be used again.
- The remaining material contains many radioactive substances with long half lives. This radioactive waste must be stored in secure conditions for many years.

ACTIVITY

a) Why does radioactive waste need to be stored:
   i) securely?
   ii) for many years?
b) Some people say that nuclear power stations are better for the environment than power stations that burn fossil fuels. Discuss this issue.

Chernobyl

When the nuclear reactors in Ukraine exploded in 1986, emergency workers and scientists struggled for days to contain the fire. More than 100 000 people were evacuated from Chernobyl and the surrounding area. Over thirty people died in the accident. Many more have developed leukaemia or cancer. It was, and remains (up to now), the world's worst nuclear accident.

Could it happen again?

- Most nuclear reactors are of a different design.
- The Chernobyl accident did not have a high-speed shutdown system like most reactors have.
- The operators at Chernobyl ignored safety instructions.

Figure 2 Chernobyl

There are thousands of nuclear reactors in the world. They have been working for many years. Countries such as Sweden wanted to 'phase' them out after Chernobyl. Now they are planning new ones because they need electricity.

ACTIVITY

Should the UK government replace our existing nuclear reactors with new ones? Debate this question with your friends and take a vote on it.

Radioactivity all around us

When we use a Geiger counter, it clicks even without a radioactive source near it. This is due to **background radioactivity**. Radioactive substances are found naturally all around us.

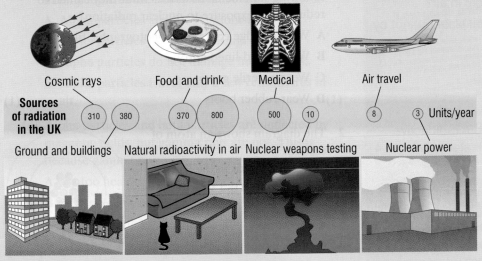

Figure 3 Sources of background radiation in the UK

Figure 3 shows the sources of background radioactivity. The numbers tell you how much radiation each person gets on average in a year from each source.

Radioactive risks

The effect on living cells of radiation from radioactive substances depends on:

- the type and the amount of radiation received (the dose), and
- whether the source of the radiation is inside or outside the body.

	Alpha radiation	Beta radiation	Gamma radiation
Source inside the body	**very dangerous!!!** – affects all the surrounding tissue	**dangerous!!** – reaches cells throughout the body	
Source outside the body	some **danger!** – absorbed by the skin; damages skin cells		

- The larger the dose of radiation someone gets, the greater the risk of cancer. High doses kill living cells.
- The smaller the dose, the less the risk – but the dose is never zero. So there is a very low level of risk to every one of us because of background radioactivity.

Radioactivity on the move

A nuclear power company needs to move radioactive waste from its nuclear power stations around the country to a specially designed storage site.

They intend to move the waste in strong metal containers which can withstand high-speed crashes. They plan to move the containers by train on main lines passing through towns and cities.

Lots of people are protesting about these plans. They want the waste moved by sea on ships. The company thinks that would be unsafe, as a ship might sink.

ACTIVITY

a) i) What is the biggest source of background radioactivity?
   ii) Which source contributes least to background radioactivity?
   iii) List the sources and say which ones could be avoided.

b) For a large part of the population, the biggest radiation hazard comes from radon gas, which gets into their homes. The dangers of radon gas can be minimised by building new houses that are slightly raised on brick pillars and modifying existing houses.
   Who should pay for alterations to houses, the government or the householder?
   Discuss this question.

QUESTIONS

1 Why is a source of alpha radiation very dangerous inside the body but not outside it?

2 Find out the hazard warning sign used on radioactive sources.

ACTIVITY

Imagine you and your friends are at a public inquiry into the company's plans. One of your group is to put forward the case for the company and someone else is to oppose it on behalf of other rail users. Other interested parties can also have their say. You need someone to 'chair' the meeting.
After you have heard all the evidence for and against the company's plans, take a vote on the plans.

P1b 7.2

The Big Bang

LEARNING OBJECTIVES

1 Why is the Universe expanding?
2 What is the Big Bang theory of the Universe?

The Universe is expanding, but what is making it expand? The **Big Bang theory** was put forward to explain the expansion. This states that:

- the Universe is expanding after exploding suddenly in a Big Bang from a very small initial point,
- space, time and matter were created in the Big Bang

Many scientists disagreed with the Big Bang theory. They put forward an alternative theory, the Steady State theory. The scientists said that the galaxies are being pushed apart. They thought that this is caused by matter entering the Universe through 'white holes' (the opposite of black holes).

Figure 1 The Big Bang

Which theory is weirder – everything starting from a Big Bang or matter leaking into the Universe from outside? Until 1965, most people backed the Steady State theory.

It was in 1965 that scientists first detected microwaves coming from every direction in space. The existence of this **background microwave radiation** can only be explained by the Big Bang theory.

a) Scientists think the Big Bang happened about 13 000 million years ago. What was before the Big Bang?

Background microwave radiation

- It was created as high-energy gamma radiation just after the Big Bang.
- It has been travelling through space since then.
- As the Universe has expanded, it stretched out to longer and longer wavelengths and is now microwave radiation.
- It has been mapped out using microwave detectors on the ground and on satellites.

DID YOU KNOW?

You can detect background microwave radiation very easily – just disconnect your TV aerial. The radiation causes lots of fuzzy spots on the screen.

b) What will happen to background microwave radiation as the Universe expands?

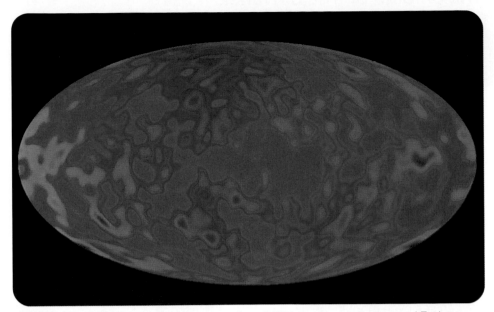

Figure 2 A microwave image of the Universe from COBE, the Cosmic Background Explorer satellite

The future of the Universe

Will the Universe expand forever? Or will the force of gravity between the distant galaxies stop them from moving away from each other? The answer to this question depends on their total mass and how much space they take up – in other words, the density of the Universe.

- If the density of the Universe is less than a certain amount, it will expand forever. The stars will die out. So will everything else as the Universe heads for a Big Yawn!
- If the density of the Universe is more than a certain amount, it will stop expanding and go into reverse. Everything will head for a Big Crunch!

Recent observations by astronomers suggest that the distant galaxies are accelerating away from each other. It looks like we're in for a Big Ride followed by a Big Yawn.

c) What could you say about the future of the Universe if the galaxies were slowing down?

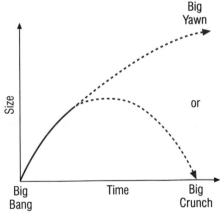

Figure 3 The future of the Universe

SUMMARY QUESTIONS

1 Complete the sentences below using words from the list.

    created   detected   expanded   stretched

  a) The Universe was …… in an explosion called the Big Bang.
  b) The Universe …… suddenly in and after the Big Bang.
  c) Microwave radiation from space can be …… in all directions.
  d) Radiation created just after the Big Bang has been …… by the expansion of the Universe and is now microwave radiation.

2 What will happen to the Universe:

  a) if its density is less than a certain value?
  b) if its density is greater than a certain value?

KEY POINTS

1 The Universe started with the Big Bang, a massive explosion from a very small point.
2 Background microwave radiation is radiation created just after the Big Bang.

313

P1b 7.3

Looking into space

LEARNING OBJECTIVES

1 What can we see in space?
2 What electromagnetic radiations, other than light, can we detect from space?
3 Why are observations from satellites better than those we make from the ground?

When we look at the night sky, we sometimes see unexpected objects in the sky, as well as planets and stars. Such objects include:

- shooting stars which are small objects from space that burn up when they enter the Earth's atmosphere,
- comets which are frozen rocks that orbit the Sun – we only see them when they get near the Sun because then they get so hot that they emit light,
- stars that explode (supernova) or flare up then fade (nova).

a) Which is nearer to us, a comet bright enough to see, or a shooting star?

We can see even more in the night sky with a telescope.

- A telescope makes stars appear much brighter. Because it is much wider than your eye, it collects much more light than your eye can. All the light it collects is channelled into your eye. So you can see stars too faint to see without a telescope.
- A telescope makes the Moon and the planets appear bigger. A telescope with magnification ×20 would make Venus appear 20 times wider. As well as that, you can see more detail. For example, you can see the Great Red Spot on the surface of Jupiter. This was first observed by Galileo almost 400 years ago.

b) Why can you see more stars by using a telescope?

The Earth's atmosphere affects telescopes on the ground. It scatters the light from space objects and makes their images fuzzy.

c) Why doesn't the Earth's atmosphere affect the Hubble Space Telescope?

Figure 1 A comet

Figure 2 Jupiter's Great Red Spot

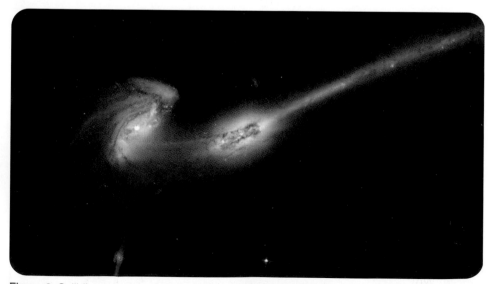

Figure 3 Colliding galaxies – an image from Hubble Space Telescope. The Hubble Space Telescope (HST) is in orbit around the Earth. It gives us amazing images of objects in space. Compared with telescopes on the ground, HST enables us to see objects in much more detail. We can also see things that are much further away.

PRACTICAL

Telescopes

You can make a simple astronomical telescope with two lenses.

Look at Figure 4. The objective lens forms an image of a distant object in front of the eyepiece. You see a magnified picture of this image when you look through the eyepiece.

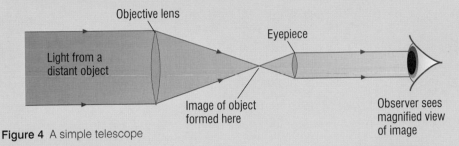

Figure 4 A simple telescope

Make and test a simple telescope to look at a distant object. What difference would it make if you looked at an object through the wrong end of a telescope?

Never use binoculars or a telescope to look at the Sun. You would be blinded permanently because far too much light would be channelled into your eyes.

Beyond the visible spectrum

(1) *Radio telescopes* are used to map out sources of radio waves, such as distant galaxies. Radio waves, as well as light (and some ultraviolet radiation), can reach the ground. The bigger a radio telescope is, the more detail it can map out and the further away it can detect radio sources.

(2) *Satellites* are used to carry detectors of electromagnetic waves that can't penetrate the Earth's atmosphere. Specially designed detectors are used for each type of radiation. Using these detectors, astronomers have discovered many unusual space objects such as:

- massive stars that suddenly explode and emit bursts of gamma rays,
- planets beyond the Solar System that give off infra-red radiation as they orbit nearby stars,
- black holes that destroy stars at the centre of galaxies.

d) Why can't we detect infra-red radiation from space using ground-based detectors?

SUMMARY QUESTIONS

1 Complete the following sentences using words from the list below.

   gamma rays   infra-red radiation   light   radio waves

   a) ...... and ...... can reach the ground from space.
   b) ...... and ...... from space are absorbed by the Earth's atmosphere.

2 a) Why do we get much better images from the Hubble Space Telescope than from telescopes on the ground?
   b) Why do gamma ray detectors need to be on satellites to detect gamma rays from space whereas radio telescopes are based on Earth?
   c) Which objects in space produce i) gamma rays ii) infra-red radiation.

Figure 5 Jodrell Bank Radio Telescope

KEY POINTS

1 The Earth's atmosphere absorbs all electromagnetic waves (except visible light, radio waves and some ultraviolet radiation).

2 Satellite detectors are used to make observations outside the visible and the radio spectrum.

3 We also get clearer images from telescopes on satellites detecting visible light.

P1b 7.4

Looking into the unknown

A short history of the Universe

Light we can see from the most distant galaxies has been travelling through space for thousands of millions of years. Big telescopes looking at these galaxies are looking back in time.

- In the first moments after the Big Bang, matter formed from the radiation given out in the Big Bang.
- After the first few seconds, protons and neutrons started to clump together to form nuclei. At this stage, the Universe was opaque, very hot and about the size of the Solar System.
- After about 100 000 years, as it expanded further, it cooled and atoms formed. This was when it became clear. The background microwave radiation we now detect was released.
- After a thousand million years or so, the Universe turned lumpy as galaxies formed. Stars formed in the galaxies and lit them up.

ACTIVITY

Make a time line on a poster to illustrate the history of the Universe.

Galileo (1564–1642)

Galileo was already a well-known scientist in 1609 when he found out about the newly-invented telescope. Within a short time, he had made his own. He used it to observe the Moon, the stars and the planets.

He discovered:

- the Moon's surface is covered with craters,
- moons orbiting the planet Jupiter,
- many more stars too faint to be seen with the unaided eye.

He published his observations in support of the Copernican model of the Solar System. This was the theory that the Earth and the planets orbit the Sun. The theory went against the teaching of the Church which said the Earth is at the centre of the Universe.

ACTIVITY

Imagine what Galileo's trial would have been like.
Write a short play about it and act it out with your friends.

Galileo was reprimanded by the Church in 1613 but he continued to teach the Copernican model. He published his work in 1630. In 1632, he was summoned to the Inquisition in Rome and forced to say his work was wrong. He was placed under house arrest for the rest of his life. But his discoveries and his teaching were taken up by other scientists in Britain and Northern Europe.

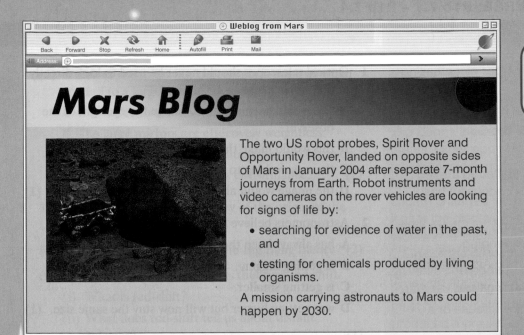

Mars Blog

The two US robot probes, Spirit Rover and Opportunity Rover, landed on opposite sides of Mars in January 2004 after separate 7-month journeys from Earth. Robot instruments and video cameras on the rover vehicles are looking for signs of life by:

- searching for evidence of water in the past, and
- testing for chemicals produced by living organisms.

A mission carrying astronauts to Mars could happen by 2030.

ACTIVITY

Imagine you're an astronaut on Mars. Write a weblog about a day in your life. Include some background music and photos.

Space invaders!

Asteroids are massive chunks of rock orbiting the Sun. They are found mostly between Mars and Jupiter. Sometimes, an asteroid gets pulled toward the Sun and crosses the Earth's orbit. In 1992, an asteroid narrowly missed the Earth. On a scale where the Sun is 1 metre away, it came within just 3 cm of the Earth! If it had hit the Earth, the impact could have killed millions of people.

ACTIVITY

Make a poster to convince people that the Government should pay for an asteroid 'early warning' system.

The search for extra-terrestrial intelligence (SETI)

The search for extra-terrestrial intelligence has gone on for more than 40 years. We can use radio telescopes to detect signals in a band of wavelengths. Such signals would indicate the existence of living beings. They would be at least as advanced as we are, perhaps on a planet in a distant solar system. Radio astronomers have searched without success for such radio signals. However, just one SETI signal would change our outlook for ever!

ACTIVITY

Imagine SETI signals have been detected from Andromeda. It is in all the newspapers, on radio and on TV. You do a survey of 100 people to see how people have reacted to this news. The results are in the table below.

Question	Yes	No
1. Have you heard about the discovery of signals from space aliens?	55	45
2. Should we send peaceful signals back?	70	30
3. Do you think the signals are from a more advanced civilisation than us?	30	70

a) Make a bar chart to display these findings.
b) Discuss with your friends what you would put into a 5-minute return signal. It could include video and audio, as well as text.

Sense organ Collection of special cells known as receptors which respond to changes in the surroundings (e.g. eye, ear).

Sensitivity The smallest change that an instrument can measure, e.g. 0.1 mm.

Slag The waste produced when iron is made in a blast furnace.

Slaked lime Calcium hydroxide.

Smart alloy An alloy which returns to its original shape when it is heated.

Social How science influences and is influenced by its effects on our friends and neighbours. E.g. building a wind farm next to a village.

Solar cell Electrical cell that produces a voltage when in sunlight; solar cells are usually connected together in solar cell panels.

Solar heating panel Sealed panel designed to use sunlight to heat water running through it.

Speed of a wave Distance travelled per second by a wave.

Stabiliser A substance with molecules that produce large 'cages' full of air when they are mixed with water.

Statins Drugs which lower the blood cholesterol levels and improve the balance of HDLs to LDL.

Steels Alloys of iron containing controlled amounts of carbon and/or other metals.

Stimulus A change which causes a response in the body.

Stomata Small holes (pores) in the leaves of a plant that can be opened or closed.

Sustainable development Using natural resources in a way which also conserves them for future use.

Synapses The gaps between neurones where the transmission of information is chemical rather than electrical.

Systematic error If the data is inaccurate in a constant way e.g. all results are 10 mm more than they should be. This is often due to the method being routinely wrong.

T

Tar Thick, black chemical found in tobacco smoke which can cause cancer.

Tectonic plates Huge sections of the Earth's crust and upper mantle.

Technology Scientific knowledge can be used to develop equipment and processes that can in turn be used for scientific work.

Telescope, optical Instrument consisting of lenses (and/or a mirror) used to make distant objects appear larger or brighter.

Telescope, radio Large concave metal dish and aerial used to detect radio waves from space.

Territory An area where an animal lives and feeds which it may mark out or defend against other animals.

Theory A theory is not a guess or a fact. It is the best way to explain why something is happening. E.g. Sea levels are rising, and the global warming theory is the best way to describe why they are. Theories can be changed when better evidence is available.

Thermal decomposition Splitting a substance using heat.

Thermal radiation Energy transfer by electromagnetic waves emitted by objects due to their temperature.

Thermosetting A polymer that hardens or sets permanently when it is formed by heating the monomers of which it is made.

Thermosoftening A polymer that softens when it is heated.

Transformer Electrical device used to change an (alternating) voltage. A **step-up transformer** is used to step the voltage up from a power station to the grid voltage.

A **step-down transformer** is used to step the voltage down from the grid voltage to the mains voltage used in homes and offices.

Transition metals The large block of metallic elements in the middle of the periodic table.

Transpiration stream The constant movement of water through a plant from the roots to the leaves where it is lost by evaporation from the leaf surface.

Trial run Carried out before you start your full investigation to find out the range and the interval measurements for your independent variable.

Tsunami A large wave caused by an underwater earthquake or volcanic eruption.

U

Unsaturated oils Oils in which the molecules contain carbon atoms joined together by carbon-carbon double bonds.

Unsaturated A hydrocarbon which contains a carbon–carbon double bond.

Useful energy Energy transferred to where it is wanted in the form it is wanted.

V

Valid data Evidence that can be reproduced by others and answers the original question.

W

Wasted energy Energy that is not usefully transferred or transformed.

Wavelength The distance from one wave peak to the next wave peak along the waves.

Z

Zero error A systematic error, often due to the measuring instrument having an incorrect zero. E.g. forgetting that the end of the ruler is not at zero.

Index

Acknowledgements

Action +/Glyn Kirk 51.3; AEA Technology 306.1; Alamy fotolincs 129.3; Alamy/Adrian Sherratt 145.3, /Steve Bloom 228.1; Ann Fullick 116tl; Bananastock T (NT) 22tl; Clare Marsh/John Birdsall 33r; Corbis V84 (NT) 57; Corbis V98 (NT) 48.1; Corbis V257 (NT) 188.2; Corbis/Jose Luis Pelaez 190.2, /Pete Saloutos/zefa 24.1, /Tom Brakefield 117mr; Corel 11 (NT) 39.3; Corel 21 (NT) 5.3; Corel 41 (NT) 140.4, 151ml; Corel 46 (NT) 84.2; Corel 50 (NT) 81tl; Corel 60 (NT) 110.2; Corel 148 (NT) 98.1; Corel 178 (NT) 122.2; Corel 219 (NT) 52.1; Corel 243 (NT) 234.4; Corel 250 (NT) 201.3; Corel 284 (NT) 3.1; Corel 318 (NT) 158.1; Corel 320 (NT) 152.2; Corel 337 (NT) 32.1, 109.3; Corel 342 (NT) 273.4; Corel 437 (NT) 140.3, 147.4; Corel 444 (NT) 270.1; Corel 456 (NT) 217.4; Corel 465 (NT) 50.1; Corel 467 (NT) 126.1; Corel 511 (NT) 39.2; Corel 584 (NT) 242.1; Corel 587 (NT) 140.1; Corel 588 (NT) 41.2; Corel 599 (NT) 94.1; Corel 624 (NT) 229b; Corel 632 (NT) 155.3; Corel 638 (NT) 100bl; Corel 640 (NT) 244.1; Corel 657 (NT) 128.2; Corel 706 (NT) 108.1; Corel 753 (NT) 91tr; CSIRO 161.4; David Buffington/Photodisc 67 (NT) 32.2; Digital Vision 1 (NT) 154.1, 156.1, 226.1, 260.1, 261.4; Digital Vision 3 (NT) 234.1; Digital Vision 4 (NT) 80tl; Digital Vision 5 (NT) 87.2; Digital Vision 6 (NT) 257.2, 281.4; Digital Vision 7 (NT) 120.1, 130, 211.2; Digital Vision 9 (NT) 313.2, 314.1; Digital Vision 14 (NT) 182tl, 214.1; Digital Vision 15 (NT) 122.1, 173.3; Empics/Jerome Delay/AP Photos 147.3; Empics/Malcolm Croft/PA News 30.2; Gerry Ellis/Digital Vision JA (NT) 82.1, 86.1b; Image 100 22 (NT) 284.1b, 285tr; Imagin/London (NT) 140.2; Ingram ILP V1CD5 (NT) 91br; Ingram IL V1CD2 (NT) 42.1; Ingram PL V1 CD2 (NT) 7.2; James Cook 81br, 81tr; James Lauritz/Digital Vision C (NT) 162.1; Jim Breithaupt 258.1, 259.2; Karl Ammann/Digital Vision AA (NT) 83.4; Kilmer McCully 45m; Martyn F. Chillmaid 7.3, 121.4; Mary Evans Picture Library 118l; NASA 314.2; NASA Goddard Space Flight Centre 314.3; Patrick Fullick 206tl; Peter Adams/Digital Vision BP (NT) 152.1; Photodisc 4 (NT) 214.2; Photodisc 10 (NT) 245.3; Photodisc 17 (NT) 268.1; Photodisc 17B (NT) 149m; Photodisc 18 (NT) 163.2; Photodisc 19 (NT) 8.1, 196.1, 205.4; Photodisc 21 (NT) 39.4, 115.3; Photodisc 22 (NT) 157.3, 176.2; Photodisc 29 (NT) 112.1, 128.1; Photodisc 31 (NT) 125.3, 127.2, 177.3; Photodisc 38A (NT) 28.1; Photodisc 40 (NT) 35; Photodisc 44 (NT) 112.2, 172.1; Photodisc 50 (NT) 24.3; Photodisc 54 (NT) 19.3, 34l, 70.1, 269.3; Photodisc 59 (NT) 34r, 230.1; Photodisc 60 (NT) 145.2; Photodisc 66 (NT) 295ml; Photodisc 67 (NT) 38.1, 190.1, 200.2a, 257.3; Photodisc 70 (NT) 232.1; Photodisc 79 (NT) 149b; Photolink/Photodisc 18 (NT) 305mr; PowerStudies 256.1; Rex Features/Sipa Press 41.3; Roslin Institute 100.1; Ryan McVay/Photodisc 67 (NT) 246.1; Science Photo Library 63.3, 238.2, 251.3, 316br, /A. Barrington Brown 95.4, 319, /A. Crump, TDR, WHO 284.1a, /Adam Hart-Davis 72.1, 85.3, /Adrian Thomas 146.1, /Alan Sirulnikoff 161.3, /Alex Bartel 274.1, /Alfred Pasieka 62.1, /Andrew Lambert 20tr, 142.1, 186.2, 203.2, /Andrew Syred 64.2, /Annabella Bluesky 203.3, /Anthony Mercieca 83.3, /Athenais, ISM 284.2, /BSIP VEM 42.2, /Charles D Winters 138.1, /Chris Priest & Mark Clarke 238.3, /CNRI 12.1, 53.2a, 53.2b, 200.2b, /Cordelia Molloy 45t, 67.3, 90m, 104t, 146.2, 186.1, 193tl, 193tr, 196.2, 198.1, 198.2, 200.1, 202.1, 236.1, 255.4, 281.2, 288.1, 288.2, 290.1, 294tm, /Cristina Pedrazzini 96.1, /David Nunuk 89.2, /David Scharf 215.4, /Dirk Wiersma 216.1, /Div. of Computer Research and Technology, National Institute of Health 95.3, /Dr Jeremy Burgess 121.3, /Dr Kari Lounatmaa 62.2, /Eamonn McNulty 27, /EFDA-JET 275.3, /Francoise Sauze 113.3, /G. Brad Lewis 270.3, /Gary Parker 97.3, /Georgette Douwma 215.3, /Gregory Dimijian 81bl, /J.C. Revy 5.2, /John Kaprielian 6.1, /John Mead 289.3, /Josh Sher 30.1, /Kaj R Svensson 184.1, /Kenneth W. Fink 114.2, /Kent Wood 64.1, /Lawrence Lawry 188.1, /Martin Bond 176.1, 269.4, /Martyn F. Chillmaid 4.1, 199.4, 254.1, 285.4, /Matt Meadows/Peter Arnold Inc. 281.3, /Mauro Fermariello 104b, 287.3, /Maximillian Stock Ltd 219.2, 246.2, /NASA 50.2a, 50.2b, 317tl, /NASA/ESA/STScl 310.1, /National Library of Medicine 298.2, /Novosti 273.3, /Pascal Goetgheluck 86.1a, 159.2, /Paul Whitehill 66.1, /Phillipe Plailly 103.2, 109.2, /Robert Brook 103.3, 177.4, 266.2, /Rod Planck 90br, /Sheila Terry 239tr, 240l, /Simon Fraser 129.4, /St Mary's Hospital School 67.2, /Steve Allen 271.5, /Steve Gschmeissner 94.2, /Sue Baker 20ml, /TH Foto-Werbung 18.1, /Tony Craddock 54.1, 229.3, 315.5, /TRL LTD 168.1, /USA Library of Medicine 69.3; Steve Mason/Photodisc 46 (NT) 206br; Stockbyte 9 (NT) 43.3; Stockbyte 28 (NT) 127.3; Stockbyte 29 (NT) 189.4; Stockhaus UKBS (NT) 149br, 149ml; Stuart Sweatmore 33.3; The Charcoal Burners Camp at the Weald & Downland Open Air Museum, Singleton, Chichester, West Sussex 148.1; The Syndics of Cambridge University Library 111.3; Topfoto.co.uk/WDS 136tl, /AP Photos 306.2, /National Pictures 49.3, /The ArenaPAL Library 71.3, /The Image Works 40.1,; Trevor Baylis 254.2; Trustees, Royal Botanic Garden, Kew 117bl, 117ml; USDA 205.2

Picture research by Stuart Silvermore, Science Photo Library and johnbailey@ntlworld.com.

Every effort has been made to trace all the copyright holders, but if any have been overlooked the publisher will be pleased to make the necessary arrangements at the first opportunity.